The Logica Yearbook
2016

The Logica Yearbook 2016

Edited by

Pavel Arazim
and
Tomáš Lávička

© Individual authors and College Publications 2017
All rights reserved.

ISBN 978-1-84890-243-5

College Publications
Scientific Director: Dov Gabbay
Managing Director: Jane Spurr

www.collegepublications.co.uk

Original cover design by Laraine Welch
Printed by Lightning Source, Milton Keynes, UK

All rights reserved. No part of this publication may be reproduced, stored in a retrieval system or transmitted in any form, or by any means, electronic, mechanical, photocopying, recording or otherwise without prior permission, in writing, from the publisher.

Preface

The book that you are holding in your hands is a further entry in a series of volumes that aspires to make some of the ideas presented at the annual international symposium Logica permanently accessible to both the conference participants and the wider public.

The symposium, which took place at Hejnice Monastery in the Czech Republic from June 20 to June 24, 2016, brought together logicians from many different countries. This volume contains a representative sample of the contributions made at the conference. The Logica symposium is an event with a long tradition. Last year it could celebrate not only its thirtieth anniversary but also the fact that it has evolved into a respected conference which possesses a firm place in the annual schedule of the international community of logicians. The symposium is open to researchers of both a mathematical and a philosophical bent, which is reflected by the diversity of its audience. The informal atmosphere provides a space for a stimulating exchange of ideas among logicians of all generations, including students. As the editors of this volume we are proud that we can contribute to the successful completion of the annual symposium cycle by presenting this collection to you.

Last year's Logica was – as were all previous Logica symposia – organized by the Department of Logic of the Institute of Philosophy of the Czech Academy of Sciences. More than thirty lectures were presented during the conference, including those given by a distinguished list of invited speakers: Sara Negri, Kit Fine, Nicholas Smith, and Neil Tennant. As happens every year, the conference was enriched by a social programme that provided room for friendly debates concerning professional topics as well as for starting and developing personal friendships. The proceedings, which are traditionally published within one year of the conference, unfortunately offer only a limited record of the topics discussed and cannot hope to even partially convey its atmosphere. In spite of that, we hope that you will find this book worthy of your attention.

Both the Logica symposium and The Logica Yearbook are the result of a joint effort by many people to whom we would like to express our gratitude. We are, of course, very grateful to the Institute of Philosophy for all the important support that made the event possible. We express our thanks to the staff of Hejnice Monastery for their hospitality and friendly assistance. Special thanks from the organizers and also, we believe, from the guests go to the Bernard Family Brewery of Humpolec, which has traditionally sponsored the social programme of the symposium by providing three barrels of its excellent beer. We owe thanks to the Czech Science Foundation, which provided significant support for the meeting and for the publication of this volume with the funding of the grant project no. 13-21076S. We would like to express our gratitude to Olga Bažantová, who is a key member of the organizing crew. Our thanks also go to College Publications and its managing director, Jane Spurr, for their very pleasant cooperation during the preparation of this book. Last, but not least, we would like to thank all of the authors for their exemplary collaboration during the editorial process.

Prague, May 2017

Pavel Arazim and Tomáš Lávička

Contents

Determinate Truth in Fuzzy Plurivaluationism 1
   Libor Běhounek

Validity and Applicability of Leibniz's Law of Substitution of
Identicals 17
   Marie Duží and Pavel Materna

Making Sense of Higher-Level Plurals 37
   Berta Grimau

When Structural Principles Hold Merely Locally 53
   Ulf Hlobil

Four Remarks on Relations and Predication 69
   Haythem O. Ismail

Truthmaker Semantics: Fine Versus Martin-Löf 87
   Ansten Klev

Ignorance Without K(nowledge) 109
   Ekaterina Kubyshkina

Non-Normal Modal Logics: A Challenge to Proof Theory 125
   Sara Negri

Revisiting Dummett's Proof-Theoretic Justification Procedures.. 141
   Harmógenes Oliveira

Double-Line Harmony in a Sequent Setting 157
   Norbert Gratzl and Eugenio Orlandelli

A Normal Paradox 173
   Mattia Petrolo and Paolo Pistone

The Definitional View of Atomic Systems in Proof-Theoretic
Semantics 185
   Thomas Piecha and Peter Schroeder-Heister

Two Accounts of Pairs 201
   Martin Pleitz

A General Framework for Logics of Questions 223
   Vít Punčochář

Non-Classical PDL on the Cheap 239
   Igor Sedlár

LP, K3, and FDE as Substructural Logics 257
   Lionel Shapiro

Truth via Satisfaction? 273
   Nicholas J.J. Smith

Free Choice Permission in STIT 289
   Frederik Van De Putte

A Hierarchy of Logical Constants 305
   Alexandra Zinke

Determinate Truth in Fuzzy Plurivaluationism

LIBOR BĚHOUNEK[1]

Abstract: Degree-theoretical accounts of vagueness usually equate determinate truth with truth to degree 1. However, by rendering determinate truth as a sharp notion, this identification disregards the phenomenon of higher-order vagueness. In this paper I propose a more adequate degree-theoretical model of determinate truth as a vague notion, based on the logical indistinguishability of truth degrees and measured by the length of the shortest formal argument separating a given degree from 1.

Keywords: Vagueness, Sorites paradox, Truth degree, Fuzzy logic, Higher-order vagueness, Determinate truth, Fuzzy plurivaluationism

1 Vagueness and fuzzy plurivaluationism

The phenomenon of vagueness is usually characterized by several interrelated attributes: susceptibility to the sorites paradox, the existence of borderline cases, and higher-order vagueness (cf., e.g., Keefe, 2000, pp. 6–9; Smith, 2009, pp. 1–2). Vague predicates can thus be loosely defined as such predicates P that:

1. A *sorites series* for P can be constructed: i.e., a series of objects $x_1, \ldots, x_N$ such that each two adjacent objects x_i, x_{i+1} differ only negligibly in respects relevant to P (making it implausible that they would differ in being P), and yet x_0 is clearly P and x_N is clearly not P.

2. *Borderline cases* of P exist: i.e., there are objects x that are neither determinately P nor determinately not P.

3. P manifests *higher-order vagueness:* i.e., the property of being borderline P is itself vague, and so has borderline cases.

---

[1] The work was supported by MŠMT ČR, programme NPU II, project LQ1602 'IT4I XS'. The author would like to thank participants in the Logica conference (Hejnice, June 2016) and the Pukeko Logic Group workshop (Wellington, January 2016) for fruitful discussions.

Vague predicates abound in natural language; prototypical examples include such properties as *tall, red, warm, large,* etc.

There are several approaches to modelling vagueness: for an overview see, e.g., Williamson (1994) or Keefe (2000). This paper deals with a particular problem related to the *fuzzy plurivaluationistic* approach to vagueness proposed by Smith (2009, ch. 6.1). Before introducing the problem, let us first illustrate fuzzy plurivaluationism and its closest relatives by describing how they model the vague predicate *tall*. For simplicity let us assume that (i) in a given context (e.g., that of Central European men), the only feature relevant for the predicate *tall* is the person's height; we also understand that (ii) it is a part of the meaning of *tall* that if a person x is tall and the height of a person y is no less than that of x, then y is tall as well; and that in the given context, (iii) people of height 150 cm are not tall, while (iv) people of height 200 cm are tall. We shall call conditions (i)–(iv) the *meaning postulates* of the predicate *tall*.

1.1 The classical semantics of *tall*

Since the standard semantics of classical logic construes all predicates as bivalent, there must be a sharp breaking point between the heights of people who are *tall* and those who are *not tall* (as depicted in Figure 1).

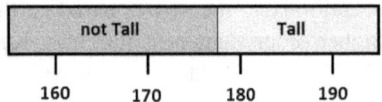

Figure 1: The classical semantics of *tall*

While employing classical bivalent semantics for modelling vague predicates such as *tall* is often satisfactory and has the undeniable advantage of simplicity, there are some drawbacks to it as well:

1. It is not clear where the breaking point should occur: in fact, no particular position of a sharp breaking point is supported by either the linguistic practice or any reasonable explication of the meaning of *tall*.

2. The breaking point creates an implausible discontinuity: it does not conform with the meaning of *tall* that an imperceptible change in height (e.g., of 0.1 cm) should make a difference.

Determinate Truth in Fuzzy Plurivaluationism

3. On the other hand, denying the discontinuity generates the sorites paradox.

4. There are no borderline cases of *tall* in the bivalent account, while in fact there clearly are some.

Even though these disadvantages may in many situations be harmless, they indicate that the classical modelling of vague predicates is rather crude, and sometimes (e.g., in soritical settings) inadequate. One approach to improving the model is supervaluationism.

1.2 The supervaluationistic account of *tall*

Placing a particular breaking point between *tall* and *not tall* in the classical semantics can be viewed as a *precisification* of the meaning of *tall*. The basic idea of *supervaluationism* (Fine, 1975) is to consider not just one, but *all* possible precisifications of the meaning of *tall*.[2] The supervaluationistic semantics of *tall* thus consists of all classical models of *tall* that satisfy its meaning postulates:[3]

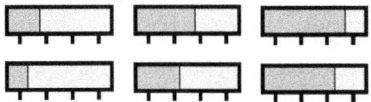

Figure 2: The supervaluationistic semantics of *tall*

This clearly addresses objection 1 of Section 1.1: the breaking point is no longer arbitrarily fixed, as we are considering all of its possible positions. Supervaluationism furthermore distinguishes between truth under a particular precisification and *supertruth,* or truth under all admissible precisifications. Since truth can be an artefact of precisification, it is, naturally, just supertruth that matters. This solves the sorites paradox: its inductive premise (if Px_i then Px_{i+1}) is not supertrue, as there is an admissible precisification that puts the breaking point between x_i and x_{i+1}. The solution avoids introducing a point of discontinuity (objection 2), as no statement

---

[2]The version of supervaluationistic semantics presented here is called *plurivaluationism* by Smith (2009); however, Smith's distinction between supervaluationism and plurivaluationism is immaterial for our purposes.

[3]In the context of supervaluationism, meaning postulates are called *penumbral connections* (as coined by Fine, 1975).

"The change occurs at height h" is supertrue. Another virtue of supervaluationism is that it retains classical logic, as it is concerned with truth under all *classical* precisifications of vague predicates.

One issue which remains unresolved by supervaluationism is the *jolt problem* (as termed by Smith, 2009): even though none of the statements "The change occurs at height h" is supertrue, the *existence* of a sharp breaking point (a 'jolt') is still supertrue, as each admissible classical precisification (see Figure 2) does indeed contain a jolt. This establishes the supertruth of such implausible statements as "There are no borderline cases of *tall*" and "There is no gradual transition between *tall* and *not tall*" in supervaluationistic models.

A different approach to vagueness that aims specifically at modelling the gradual transition between P and $\neg P$ is the degree-theoretic (or *fuzzy*) semantics (e.g., Williamson, 1994, ch. 4; Keefe, 2000, ch. 4).

1.3 The fuzzy semantics of *tall*

In order to accommodate the gradual transition between P and $\neg P$, fuzzy models of vagueness admit intermediary *degrees of truth*. Most often, the degrees of truth are represented by real numbers from the unit interval $[0, 1]$, where the degree 1 represents full truth and the degree 0 full falsity. A vague predicate P is then modelled by a *function*[4] $\|P\| \colon X \to [0, 1]$ which assigns to each object x from a domain X a number $\|Px\| \in [0, 1]$ representing the degree to which x has the property P.

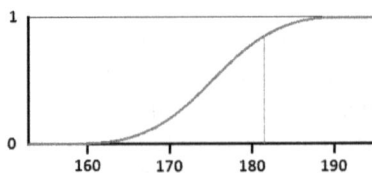

Figure 3: A fuzzy semantics of *tall*

A fuzzy model for our paradigmatic predicate *tall* is depicted in Figure 3. In this model, the membership function $\|tall\|$ assigns degrees from $[0, 1]$

---

[4] In fuzzy set theory, $\|P\|$ is called the *membership function* of P or the *fuzzy set* delimited by P. Thus in fuzzy models, vague predicates are represented by $[0, 1]$-valued 'fuzzy' sets instead of ordinary two-valued 'crisp' sets. Crisp sets can be identified with special membership functions that only assign the degrees 0 and 1.

to people according to their height (in cm on the horizontal axis). In this model, if John is, for instance, 182 cm tall, then $\|tall\|$(John) $= 0.84$; i.e., the statement "John is tall" is assigned the truth degree 0.84. Observe that the model honours the meaning postulates (i)–(iv) for *tall* (see page 2) by (i) making $\|tall\|$ functional in height, (ii) assigning larger truth degrees to taller people, (iii) assigning the degree 0 to people of height 150 cm, and (iv) assigning the degree 1 to people of height 200 cm.

By admitting the degrees of truth, fuzzy models capture the graduality of vague predicates and remove the jolt between P and $\neg P$. The sorites paradox can now be solved by letting the truth degree of Px_i gradually decrease along the sorites series, which makes the inductive premise almost (though not quite) true. Nevertheless, fuzzy semantics still suffers from several problems. Let us mention just two of them:

1. A fuzzy model specifies a particular membership function $\|P\|$, which assigns exact truth degrees to objects; e.g., $\|tall\|$(John) $= 0.84$. However, such precision seems unwarranted: there are many functions compatible with the meaning postulates of *tall*, and nothing in language appears to determine whether John's tallness is 0.84 or 0.83. In general it seems rather incongruous to model a vague predicate with unclear and imprecise boundaries by means of a completely precise real-valued function. This objection to fuzzy semantics is known as the problem of *artificial precision* (e.g., Williamson, 1994, p. 127; Smith, 2009, p. 277).

2. Relatedly, assigning precise degrees on a linearly ordered scale such as the real unit interval makes the degrees of all properties comparable. However, many properties are qualitatively so different that there is hardly any way to compare their intensities; for instance, it is largely meaningless to say that John is more *tall* than *young*. Let us call this the *linearity objection* to fuzzy semantics (e.g., Smith, 2009, p. 293; Williamson, 1994, p. 128).

A remedy to these problems for fuzzy semantics is to do the supervaluationist trick in the fuzzy setting (as proposed by Smith, 2009, ch. 6). The rationale for this move is the very reason behind the problem of artificial precision: namely that neither the meaning postulates nor any other linguistic facts determine membership functions uniquely. Rather, the meaning postulates (or more generally, the meaning-determining facts) only establish *constraints*[5] on membership functions in fuzzy models. This approach, which combines the merits of supervaluationism and fuzzy semantics, has been called *fuzzy plurivaluationism* by Smith (2009, p. 9).

1.4 The fuzzy plurivaluationistic semantics of *tall*

Following the above motivation, fuzzy plurivaluationistic semantics assigns to vague predicates, in general, not just a single fuzzy model, but *all* fuzzy models satisfying the constraints given by the meaning-determining facts. In the particular case of the predicate *tall,* the fuzzy plurivaluationistic semantics consists of all membership functions that satisfy its meaning postulates:

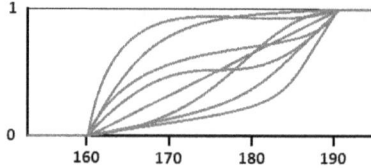

Figure 4: The fuzzy plurivaluationistic semantics of *tall*

Let us call the class of admissible fuzzy models of a vague predicate its *fuzzy plurivaluation.* Just like supervaluationism, fuzzy plurivaluationism understands particular models in the plurivaluation as *admissible precisifications* of the meaning of the vague predicate; only this time, the precisifications are gradual. Consequently, it is again just supertruth, or truth in all admissible fuzzy models, that matters, as truth in a particular fuzzy model may be an artefact of the precisification (i.e., of the particular choice of the membership function in that model).

By combining the merits of supervaluationism and fuzzy semantics, fuzzy plurivaluationism addresses several problems of either account:

- Like fuzzy models, it avoids the jolt by letting the truth degree gradually decrease from 1 to 0 along the sorites series.[6]

---

[5] or "fuzzy penumbral connections", to use the supervaluationistic terminology

[6] In fact, fuzzy models and fuzzy plurivaluationism make it possible to require, as another meaning postulate for *tall,* that (v) very small changes in height should only result in very small changes in the truth (degree) of *tall.* This requirement would only admit (Lipschitz) continuous membership functions for *tall,* making the absence of a jolt and the presence of borderline cases supertrue. Cf. Smith's principle of Closeness for vague predicates (2009, ch. 3.3–5).

- Like supervaluationism, it removes the arbitrariness of a particular model by considering all admissible models. This addresses the problem of arbitrary precision in fuzzy semantics: since John's tallness has different degrees in different admissible models, no statement assigning to it a precise degree (such as "John is tall to degree 0.84") is supertrue. The linearity objection is answered in a similar manner: since the degree of "John is tall" exceeds that of "John is young" only in *some* models, neither the statement "John is more *tall* than *young*" nor its converse is supertrue.

From the model-theoretic point of view, the fuzzy plurivaluation for a set of vague predicates can be explained as the class of models of a theory that formalizes the meaning postulates of the vague predicates involved (Běhounek, 2014).[7] Supertruths in the plurivaluation then coincide with *logical consequences* of this theory; i.e., (super)true statements about vague predicates are exactly the logical consequences of their meaning postulates.

An apparent problem is that taken *prima facie,* the meaning postulates of most vague predicates are inconsistent, as they typically entail the sorites paradox. One option, then, is to translate them into the language of membership functions, as we did with the meaning postulate (ii) for *tall,* interpreting it as the monotonicity condition for the membership function.[8] Another option, which additionally offers a recipe for the translation, is to use a non-classical logic designed specifically for fuzzy models—a fuzzy logic.

2 A logic for fuzzy plurivaluationism

Fuzzy logics are logics tailored to fuzzy models of Section 1.3. There is a whole family of fuzzy logics, corresponding to different possible meaning postulates for fuzzy connectives. Only some of them are suitable for modelling the logical aspects of the sorites paradox, and so for modelling vagueness; of those, perhaps the most prominent is the infinite-valued logic of Łukasiewicz (e.g., Hájek, 1998, ch. 3). For simplicity, we shall restrict our attention to this logic and particularly to its standard $[0, 1]$-valued semantics.

---

[7]The explanation is based on understanding the meaning postulates as (regularized) characterizations of meanings, abstracted from (irregular) linguistic meaning-determining facts (Běhounek, 2011). It is assumed that they can be formalized in a suitable logic.

[8]Similarly, the continuity (or Closeness) postulate (v) of footnote 6 can be viewed as a degree-theoretical reinterpretation of Wright's principle of Tolerance, making it consistent with the postulates (i)–(iv) for *tall.* Cf. Smith's discussion of Closeness vs. Tolerance (2009, ch. 3.5).

In the standard $[0,1]$-valued semantics, the propositional connectives of Łukasiewicz logic combine the degrees from $[0,1]$ in the following manner:

$$\|\varphi \wedge \psi\| = \min(\|\varphi\|, \|\psi\|) \tag{1}$$
$$\|\varphi \vee \psi\| = \max(\|\varphi\|, \|\psi\|) \tag{2}$$
$$\|\neg \varphi\| = 1 - \|\varphi\| \tag{3}$$
$$\|\varphi \mathbin{\&} \psi\| = \max(\|\varphi\| + \|\psi\| - 1, 0) \tag{4}$$
$$\|\varphi \to \psi\| = \min(1 - \|\varphi\| + \|\psi\|, 1). \tag{5}$$

Note that like other contraction-free logics, Łukasiewicz logic possesses *two* conjunctive connectives, $\wedge$ and $\&$.[9] It is the non-idempotent $\&$ that represents iterative cumulation of premises (as in the sorites paradox), since

$$\|\varphi_1 \to (\varphi_2 \to \cdots (\varphi_{n-1} \to \varphi_n) \cdots)\| =$$
$$= \|(\varphi_1 \mathbin{\&} \varphi_2 \mathbin{\&} \ldots \mathbin{\&} \varphi_{n-1}) \to \varphi_n\|. \tag{6}$$

The following connective, often added to Łukasiewicz logic, indicates the full truth of its argument:

$$\|\triangle \varphi\| = \begin{cases} 1 & \text{if } \|\varphi\| = 1 \\ 0 & \text{otherwise.} \end{cases}$$

First-order models for Łukasiewicz logic are defined analogously to classical first-order models, the only difference being the interpretation of predicate symbols: just like in fuzzy models of Section 1.3, each n-ary predicate P is interpreted by a $[0,1]$-valued membership function $\|P\| \colon M^n \to [0,1]$, where M is the universe of the model, instead of a two-valued characteristic function as in classical models. The quantifiers are interpreted as follows:

$$\|(\forall x)\varphi\| = \inf_{a \in M} \|\varphi(a)\|, \qquad \|(\exists x)\varphi\| = \sup_{a \in M} \|\varphi(a)\|.$$

A sentence is considered true in a model if it is evaluated to degree 1. As usual, tautologicity is defined as truth in all models; and the logical consequences of a theory T are defined as the sentences true in all models of T (i.e., all models in which all formulae from T are true).[10]

---

[9]Distinguishing between them actually answers most common objections raised against truth-functional fuzzy logic, including those offered by Williamson (1994, pp. 118, 136–8) and Keefe (2000, pp. 96–98).

[10]The notions of tautologicity and finitary entailment in Łukasiewicz logic turn out to be finitely axiomatizable (e.g., Hájek, 1998), while infinitary entailment requires an infinitary rule of inference or, alternatively, relaxing somewhat the standard semantics of Łukasiewicz logic to fit its finitary approximation.

Determinate Truth in Fuzzy Plurivaluationism

Notice that the notion of *logical consequence* in Łukasiewicz logic exactly matches the (fuzzy plurivaluationistic) notion of *supertruth* in the class of models of a theory. Thus, if a theory T formalizes the meaning postulates of some vague predicates, then its logical consequences in Łukasiewicz logic are precisely the supertruths about these predicates. In this sense, (Łukasiewicz) fuzzy logic can be viewed as the logical background of fuzzy plurivaluationism. From this perspective, fuzzy plurivaluationistic modelling of vague predicates amounts to expressing their meaning postulates as a theory in fuzzy logic;[11] the class of its models then constitutes the fuzzy plurivaluation, and the logic's consequence relation captures the notion of supertruth.

This approach to modelling vague predicates has been taken, e.g., by Hájek and Novák (2003). We shall illustrate it by a (simplified) formalization of the predicate *large natural number*.

A minimalistic set of meaning postulates for the predicate *large* on natural numbers can be listed as follows: (i) zero is not large; (ii) numbers larger than large numbers are large, too; (iii) there are large numbers. These postulates can be straightforwardly formalized in Łukasiewicz logic as follows:[12]

$$\neg \operatorname{Large}(0) \tag{7}$$

$$(\forall m, n)(m \geq n \ \&\ \operatorname{Large}(n) \to \operatorname{Large}(m)) \tag{8}$$

$$(\exists n)\operatorname{Large}(n) \tag{9}$$

Notice that in accordance with fuzzy plurivaluationism, these axioms do not specify a unique membership function for Large, but only constrain the class of admissible fuzzy models of Large. In particular, by the standard semantics of Łukasiewicz logic, the class of fuzzy models admitted by these axioms consists of all membership functions $\|\operatorname{Large}\| = \ell \colon \mathbb{N} \to [0,1]$ (where $\mathbb{N}$ is a model of crisp Peano Arithmetic) such that (i) $\ell(0) = 0$, (ii) ℓ is non-decreasing, and (iii) $\sup_n \ell(n) = 1$. It can be observed that already this minimal set of meaning postulates ensures that many intuitively plausible statements (e.g., that there are large prime numbers) are supertrue,

---

[11] The formalization of meaning postulates in fuzzy logic can moreover be done in a straightforward manner, since the vague predicates will automatically get interpreted by gradual $[0,1]$-valued membership functions as intended.

[12] Since Large is a predicate on natural numbers, the meaning postulates for (non-vague) natural numbers need be formalized in Łukasiewicz logic, too. This can be done, e.g., by means of Crisp Peano Arithmetic, which consists of the classical axioms of Peano Arithmetic together with the axioms enforcing the bivalence of its primitive predicates, e.g., $(\forall m, n)((m \leq n) \vee \neg(m \leq n))$.

while many intuitively implausible statements (e.g., that there is the least large number) are not. Various additional meaning postulates can be added in order to model the vague predicate *large* more accurately;[13] the axioms (7)–(9), nevertheless, suffice for our illustrative purposes.

3 The jolt problem for determinate truth

Recall the jolt problem for supervaluationism (Section 1.2): in the supervaluationistic model of a sorites series for a vague predicate P, the proposition

$$(\exists n)(Px_n \wedge \neg Px_{n+1}) \qquad (10)$$

comes out supertrue. Fuzzy models of Section 1.3 solve the problem by letting the truth degree of Px_n decrease gradually with increasing n. Consequently, the jolt sentence (10) is no longer valid in the fuzzy plurivaluation for P, and so not (super)true of P.

However, the problem reappears for determinate truth, or the truth of P to the full degree 1: in fuzzy models of the sorites series for P, the jolt sentence for the predicate *determinately P*, which can be formalized in Łukasiewicz logic as

$$(\exists n)(\triangle Px_n \wedge \neg \triangle Px_{n+1}),$$

is still supertrue, despite the fact that the boundary between borderline and determinate cases of vague predicates is typically vague as well (that is, vague predicates typically manifest higher-order vagueness; see page 1).

One option is to admit that fuzzy plurivaluationism cannot model higher-order vagueness any better (without jolts) than supervaluationism; after all, for bivalent predicates such as *determinately P*, fuzzy models reduce to bivalent ones, so for these predicates fuzzy plurivaluationism reduces to classical supervaluationism. Another route, which we intend to pursue here, is to acknowledge that due to higher-order vagueness, the predicate *determinately P* is vague as well, and therefore should not be represented in fuzzy models by a bivalent membership function. This, however, means to contest the common identification of determinate truth with truth to degree 1 in

---

[13]E.g., postulates ensuring the existence of borderline cases or enforcing Lipschitz-style conditions on the membership function (cf. footnote 6). These postulates are expressible in Łukasiewicz logic, too (e.g., the former by the axiom $\neg(\forall n)(\text{Large}(n) \vee \neg \text{Large}(n))$); they would make the theory classically inconsistent, but still consistent in Łukasiewicz logic.

degree-theoretical accounts of vagueness, and search for a more adequate representation of determinate truth in fuzzy models. We will follow this route by considering the inferential rôle of determinate truth in arguments involving vagueness.

4 Indistinguishability of close truth degrees

Notice that by (1)–(5), the basic connectives of Łukasiewicz logic (without the connective $\triangle$, whose legitimacy we contest) are Lipschitz continuous, with the Lipschitz constant 1. Therefore, if the difference between the truth degrees $\|p\|$ and $\|q\|$ is at most ε, then for any formula φ of Łukasiewicz logic (without $\triangle$) containing at most n occurrences of basic propositional connectives,[14] the difference between the degrees $\|\varphi(p)\|$ and $\|\varphi(q)\|$ is at most $n \cdot \varepsilon$. Consequently, truth degrees differing by at most ε cannot be distinguished (i.e., one made true and the other false) by any formula shorter than $1/\varepsilon$ connectives; or by any argument that can be formalized in Łukasiewicz logic by a formula shorter than $1/\varepsilon$. In other words, very close truth degrees can only be distinguished by very long arguments.

In particular, truth degrees that are very close to 1 can be distinguished from 1 only by very long arguments, such as a sufficiently long sorites argument. For instance, consider the degree 0.9: it can indeed be distinguished from 1 by a 10-step sorites argument;[15] but due to the Lipschitz property of Łukasiewicz connectives, 0.9 and 1 cannot be distinguished by any shorter argument. Similarly, the truth degree 0.99 can only be distinguished from 1 by a (sorites)[16] argument of length 100; and the degree 0.999 999 999 requires a sorites argument over a series of one thousand million elements to distinguish it from 1.

Thus, degrees too close to 1 cannot be distinguished from 1 by arguments of ordinary lengths. Moreover, in arguments of ordinary lengths, such degrees make no observable difference from 1: for instance, even if used 100 times in any argument that can be formalized by basic connectives of Łukasiewicz logic, using the degree 0.999 999 999 instead of 1 changes

---

[14] Since negation preserves the distance of degrees, counting just binary basic connectives or atomic subformulae would yield a tighter estimate; however, the accuracy of this upper estimate will not be essential for our considerations.

[15] Each application of the 0.9-true inductive premise decreases the truth degree of the conclusion by 0.1, therefore 10 applications are needed to reach 0; cf. (4) and (6) in Section 2.

[16] Since the degree of iterated conjunction decreases most rapidly of all basic Łukasiewicz connectives, the sorites argument is actually optimal for distinguishing between 1 and $1 - \varepsilon$.

the resulting truth value only by $0.000\,000\,1$, or one ten-millionth. Only in very long arguments such as the sorites of enough length, their difference from 1 may become apparent. Degrees extremely close to 1, such as $1 - 10^{-100}$, thus behave just like 1 in any argument of feasible length.

This suggests that determinate truth should not be be represented just by the degree 1, but should include also the degrees observationally indistinguishable from 1. Even though such degrees can in principle be distinguished from 1 by very (often, unfeasibly) long arguments, in practice they behave just like determinate truth, and should therefore be regarded as representing determinate truth (almost) as well as the degree 1.

Obviously, the borderline between the degrees that can be regarded as indistinguishable from 1 and those that cannot, is not sharp: the indistinguishability of degrees depends on the length of arguments that can distinguish them. For instance, the degree 0.999 is much better distinguishable from 1 than 0.999 999 (as the latter requires a sorites argument of length one million, while the former just one thousand, in order to be distinguished from 1), even though in ordinary arguments of lengths up to a few hundred connectives they both behave like fully true. This in fact conforms very well with the observation that motivated these considerations—namely that because of higher-order vagueness, determinate truth must be a gradual notion, or else the jolt problem and the sorites paradox reappear for *determinately P*. The minimal length of a distinguishing argument then provides a natural measure of practical indistinguishability between a given degree and 1, and so the suitability of that degree for representing determinate truth.

5 Determinate truth redefined

The above considerations lead us to a revised definition of determinate truth in fuzzy models. The commonly accepted definition that *determinately φ* (henceforth denoted by $\mathrm{Det}\,\varphi$) is true in fuzzy models if and only if $\|\varphi\| = 1$ turns out to be unsuitable, since it does not accommodate higher-order vagueness (specifically, being bivalent, it is subject to the jolt problem just like classical supervaluationism). Our analysis corroborates that determinate truth is a gradual notion, and suggests that the truth degree of *determinately φ* is suitably measured by the length of the shortest argument that can distinguish the truth of φ from 1: the *larger* the length, the less $\|\varphi\|$ is distinguishable from 1 and so, the more φ is determinately true. Since, moreover, the quickest way to distinguish $\|\varphi\|$ from 1 in standard Łukasiewicz logic

Determinate Truth in Fuzzy Plurivaluationism

is by means of iterated conjunction (cf. footnote 16), our analysis identifies the degree of *determinately* φ with the degree of largeness of the minimal number n of conjuncts making the iterated conjunction $\varphi^n \equiv_{\text{def}} \varphi \& \ldots \& \varphi$ (n times) false. This characterization is expressed by the following definition:

$$\text{Det } \varphi \equiv_{\text{def}} \text{Large}(\min\{n \mid \|\varphi^n\| = 0\}). \tag{11}$$

In the setting of standard Łukasiewicz logic, the minimal distinguishing n can be explicitly calculated. By (4) we obtain:

$$\|\varphi^n\| = 1 - \min(n \cdot (1 - \|\varphi\|), 1),$$

whereby the definition (11) is equivalent to:

$$\text{Det } \varphi \equiv \text{Large}\lceil 1/(1 - \|\varphi\|) \rceil. \tag{12}$$

Thus, if $\|\varphi\| = 1 - \varepsilon$, then $\|\text{Det } \varphi\| = \|\text{Large}\lceil 1/\varepsilon \rceil\|$. (For $\|\varphi\| = 1$ we set $\|\text{Det } \varphi\| = 1$ as the limit case.)

It can be observed that definition (11) involves the vague predicate Large on natural numbers. Consequently, our notion of determinate truth is vague as well. Recall that in fuzzy plurivaluationism, we do not specify a particular membership function for Large; rather, its semantics is delimited by its (formalized) meaning postulates. A set of such meaning postulates for Large has been presented in Section 2; the corresponding axioms (7)–(9) are, thus, part of the meaning of the vague notion of determinate truth. In other words, there is a *penumbral connection* between the vague notions of *determinate* truth and *large* natural number, because determinacy depends on how large a sorites argument is needed to ascertain that a determinate-looking case is in fact borderline.

It can be readily seen that modelling determinate truth by $\text{Det } \varphi$ (rather than $\triangle \varphi$) succeeds in removing the jolt problem for *determinately P,* as the jolt sentence

$$(\exists n)(\text{Det } Px_n \wedge \neg \text{Det } Px_{n+1})$$

is no longer supertrue. This follows easily from the observation that the jolt in Large, i.e.,

$$(\exists m)(\text{Large}(m+1) \wedge \neg \text{Large}(m))$$

is not supertrue in the fuzzy plurivaluation of Large, i.e., the class of fuzzy models satisfying the meaning postulates (7)–(9).

6 Conclusions

We have seen that our redefinition of determinate truth solves several problems faced by its traditional identification with truth to degree 1. Unlike the connective $\triangle$, the operator Det accounts for higher-order vagueness by admitting borderline determinate cases; being gradual, it removes the jolt problem for determinate cases; and it is based on a fairly natural concept of (graded) logical indistinguishability of truth degrees in standard Łukasiewicz logic. Consonant with fuzzy plurivaluationism, the membership function of Det is not uniquely determined, but depends on the fuzzy plurivaluation for the vague predicate Large on natural numbers.

These facts indicate that determinate truth is better modelled by Det than $\triangle$, and that truth to degree 1 is an artefact of a similar kind as the precise truth degrees contested by the objection of artificial precision. In fuzzy plurivaluationism, this artefact is analogously eliminated by restriction to supertrue statements on Det, or (cf. Section 2) the logical consequences of its meaning postulates embodied in (11). The connective $\triangle$, on the other hand, directly represents the very artefact, and so should be avoided in the logical analysis of vague predicates.

Admittedly, our treatment of determinate truth in this paper has been somewhat simplified, and several details still need to be elaborated. For instance, the following consideration may require a modification of definition (11): in our exposition, we have not considered the possibility of introducing defined propositional symbols in the course of the distinguishing argument. If this is allowed, then the descent towards falsity can be much faster; for instance, if $\|p_0\| = 0.999$, then we can distinguish it from 1 in just 10 steps by letting $p_1 \equiv_{\text{def}} p_0 \mathbin{\&} p_0$, $p_2 \equiv_{\text{def}} p_1 \mathbin{\&} p_1$, etc.: then already $\|p_{10}\| = 0$. While this accelerates the distinction between $\|p_0\|$ and 1 exponentially,[17] the main point still stands: for each truth degree close to 1 there is a minimal number of steps required for distinguishing it from 1, and for degrees very close to 1 (such as $1 - 10^{-100\,000}$), the number of steps can be infeasibly large, making the degree practically indistinguishable from 1, and thereby allowing it to represent determinate truth in ordinary arguments (almost) as well as does the degree 1. Essentially, the only modification to (11) required by this acceleration is replacing $\|\varphi^n\|$ by $\|\varphi^{2^n}\|$ and adjusting the explicit calculation (12) accordingly.

---

[17]The corresponding 'accelerated sorites argument' could be formulated as follows: one grain makes no difference (premise, apply twice); therefore two grains make no difference; analogously, if two grains make no difference, then four grains make no difference; etc.

Determinate Truth in Fuzzy Plurivaluationism

A further detail that requires elaboration is a more complete account of the meaning postulates for Large, as hinted in Section 2 (esp. footnote 13). Another missing item is a formalization of Det in Łukasiewicz logic: note that definition (11) refers to the semantic value of φ, which cannot be straightforwardly formalized by first-order means. Nevertheless, it can be shown, but is omitted here for reasons of space, that Det can be formalized in Łukasiewicz logic of a higher order (which has enough expressive power to internalize truth values of formulae). Yet another topic for future study is a generalization of Det to some other fuzzy logics besides standard Łukasiewicz.[18] All of these topics are left for future work.

References

Běhounek, L. (2011). Comments on "Fuzzy logic and higher-order vagueness" by Nicholas J.J. Smith. In P. Cintula, C. Fermüller, L. Godo, & P. Hájek (Eds.), *Understanding Vagueness: Logical, Philosophical, and Lingustic Perspectives* (pp. 21–28). College Publications.

Běhounek, L. (2014). In which sense is fuzzy logic a logic for vagueness? In T. Łukasiewicz, R. Peñaloza, & A.-Y. Turhan (Eds.), *Logics for Reasoning about Preferences, Uncertainty, and Vagueness (PRUV 2014)*. CEUR Workshop Proceedings (Vol. 1205, pp. 26–38).

Fine, K. (1975). Vagueness, truth and logic. *Synthese, 30*, 265–300. (Reprinted in Keefe and Smith (1999), pp. 119–150)

Hájek, P. (1998). *Metamathematics of Fuzzy Logic*. Dordrecht: Kluwer.

Hájek, P., & Novák, V. (2003). The sorites paradox and fuzzy logic. *International Journal of General Systems, 32*, 373–383.

Keefe, R. (2000). *Theories of Vagueness*. Cambridge University Press.

Keefe, R., & Smith, P. (Eds.). (1999). *Vagueness: A Reader*. MIT Press.

Smith, N. J. (2009). *Vagueness and Degrees of Truth*. Oxford University Press.

Williamson, T. (1994). *Vagueness*. London: Routledge.

Libor Běhounek
CE IT4Innovations/IRAFM, University of Ostrava
The Czech Republic
E-mail: `libor.behounek@osu.cz`

---

[18] Not all fuzzy logics are suitable for modelling vagueness, though: for instance, idempotent elements of & between 0 and 1 prevent solving the sorites paradox within such fuzzy semantics.

Validity and Applicability of Leibniz's Law of Substitution of Identicals

MARIE DUŽÍ AND PAVEL MATERNA

Abstract: In this paper we examine Leibniz's principle of the indiscernibility of identicals. Our goal is to show that this principle is universally valid in all kinds of context, provided it is properly applied. This principle is subject to restrictions, because invalid applications are rooted in confounding different levels of abstraction. We distinguish three kinds of context, to wit extensional, intensional and hyperintensional ones, and the distinction makes it possible to define valid rules of substitution depending on the relevant kind of context.

Keywords: Leibniz, Substitution of identicals, Three kinds of context, Transparent Intensional Logic

1 Introduction

Leibniz's Law is a principle that has a mirror image. One principle is the strong metaphysical principle of the (i) *identity of indiscernibles*; the other is the principle of the (ii) *indiscernibility of identicals*. They are both formally defined in second-order predicate logic as follows:

(i) $\forall x \forall y (\forall P(Px \leftrightarrow Py) \rightarrow x = y)$

(ii) $\forall x \forall y (x = y \rightarrow \forall P(Px \leftrightarrow Py))$

In this paper we deal with principle (ii). It is commonly taken as a logical truth in direct (extensional) contexts, yet alleged to fail in indirect (opaque) contexts. Quine thought that the failure of substitutivity in opaque contexts shows that non-extensional logic is an impossible project.[1] Kripke (1979) holds that this failure may be the result of an implicit use of the disquotational principle. Many other logicians and philosophers (e.g. Frege and Montague) explain substitutivity failure by way of 'reference shift'. We

---

[1] See, for instance, Quine (1953). A summary of the debate on Quine's view on substitutivity in opaque contexts can be found in Sayward (2007).

are going to show that the principle is *universally valid* independently of which sort of context it is applied to, *provided* it is applied correctly. Hence, we want to show that it is valid also when applied to intensional and hyperintensional contexts, and that this can be accomplished in a principled, non-*ad-hoc* manner.

The application of the principle, as we just said, is subject to restrictions. Invalid application is rooted in confounding different levels of abstraction, according to our diagnosis. An assumption of ours is that variables are abstract mini-procedures, not terms. Let x, y be procedures whose respective products are their respective values. The values are assigned to them in a Tarski-style sequence, with the difference that variables are not terms, but logical procedures. What does "$x = y$" express? Taken one way, it expresses the necessary falsehood that two different procedures are the same procedure. Taken another way, it expresses the possible truth that two different procedures share the same product (value). Ours is the latter reading. What does then

$$"\forall x \forall y [x = y \to \forall P(Px \leftrightarrow Py)]"$$

mean? It means that if x and y are procedures co-producing the same object O then this object O has all the properties Pi if and only if it has the properties Pi. Is it valid? Sure it is, *provided* Pi are *applicable* to O; that is, provided O is of the right type to serve as an argument for Pi.

Our background theory is Tichý's Transparent Intensional Logic (TIL)[2] that comes with a *procedural* rather than model-theoretic (i.e. set-theoretic) semantics. Roughly, the meaning of a term or expression is a procedure that details how to operate on input objects in order to produce an output object, if any. These procedures are rigorously defined as TIL constructions that are assigned to expressions as their context-invariant meanings. TIL operates with a fundamental dichotomy between procedures and their products, i.e. functions. This dichotomy corresponds to two fundamental ways in which a construction (meaning) can occur, to wit, *displayed* or *executed*. If the construction occurs in the execution mode, then it is a constituent of another procedure, and an additional distinction can be found at this level. The *constituent* presenting a function may occur within another procedure either *intensionally* (de dicto) or *extensionally* (de re). If intensionally, then the whole function is the object of predication; if extensionally, then a functional value is the object of predication. These distinctions enable us to

---

[2] See Tichý (1988, Chapters 4 and 5).

distinguish between three kinds of context, namely *hyperintensional, intensional* and *extensional*. Distinguishing the three kinds of context makes it possible to define context-dependent rules of substitution for all the three kinds of context, which is the main goal of this paper.

The rest of the paper is organised as follows. In Section 2 we briefly introduce the technical apparatus of TIL that we need to define valid rules of substitution for the three kinds of context. These rules are introduced in Section 3. Concluding remarks are in Section 4.

2 Basic principles of TIL

The terms of the formal TIL symbolism denote abstract procedures that produce set-theoretical mappings (functions-in-extension) or lower-order procedures. These procedures are rigorously defined as TIL *constructions*.

Definition 1 (Construction)

(i) *Variables* $x, y, \ldots$ are constructions *that construct objects (elements of their respective ranges) dependently on a valuation v; they v-construct.*

(ii) *Where X is an object whatsoever (even a* construction*), ^{0}X is the* construction Trivialization *that* constructs X *without any change of X.*

(iii) *Let* $X, Y_1, \ldots, Y_n$ *be arbitrary* constructions. *Then* Composition $[X\ Y_1 \ldots Y_n]$ *is the following* construction. *For any v, the Composition $[X\ Y_1 \ldots Y_n]$ is v-improper if at least one of the constructions $X, Y_1, \ldots, Y_n$ is v-improper, or if X does not v-construct a function that is defined at the n-tuple of objects v-constructed by $Y_1, \ldots, Y_n$. If X does v-construct such a function, then $[X\ Y_1 \ldots Y_n]$ v-constructs the value of this function at the n-tuple.*

(iv) *(λ-) Closure* $[\lambda x_1 \ldots x_m\ Y]$ *is the following* construction. *Let $x_1, x_2, \ldots, x_m$ be pair-wise distinct variables and Y a* construction. *Then $[\lambda x_1 \ldots x_m\ Y]$ v-constructs the function f that takes any members $B_1, \ldots, B_m$ of the respective ranges of the variables $x_1, \ldots, x_m$ into the object (if any) that is $v(B_1/x_1, \ldots, B_m/x_m)$-constructed by Y, where $v(B_1/x_1, \ldots, B_m/x_m)$ is like v except for assigning B_1 to $x_1, \ldots, B_m$ to x_m.*

(v) *Where X is an object whatsoever, ^{1}X is the* construction Single Execution *that v-constructs what X v-constructs. Thus if X is a v-improper* construction *or not a* construction *as all, ^{1}X is v-improper.*

(vi) Where X is an object whatsoever, ^2X is the construction Double Execution. *If X is not itself a construction, or if X does not v-construct a construction, or if X v-constructs a v-improper construction, then ^2X is v-improper. Otherwise ^2X v-constructs what is v-constructed by the construction v-constructed by X.*

(vii) Nothing is a construction, unless it so follows from (i) through (vi).

Note that for any valuation v the $(\lambda-)$ Closure $[\lambda x_1 \ldots x_m\ Y]$ is not v-improper, as it always v-constructs a function. Even if the constituent Y is v-improper for every valuation v, the Closure is not v-improper. Yet in such a case the so-constructed function is a bizarre object; it is a degenerate function that lacks a value at any argument.

With constructions of constructions, constructions of functions, functions, and functional values in our stratified ontology, we need to keep track of the traffic between multiple logical strata. The ramified type hierarchy does just that. The type of first-order objects includes all non-procedural objects. Therefore, it includes not only the standard objects of individuals, truth-values, sets, etc., but also functions defined on possible worlds (i.e., the intensions germane to possible-world semantics). The type of second-order objects includes constructions of first-order objects and functions with such constructions in their domain or range. The type of third-order objects includes constructions of first- and second-order objects and functions with such constructions in their domain or range. And so on, ad infinitum.

Definition 2 *Let B be a base, where a base is a collection of pair-wise disjoint, non-empty sets. Then:*

$\mathbf{T_1}$ *(types of order 1)*

  (i) Every member of B is an elementary type of order 1 over B.
  (ii) Let $\alpha, \beta_1, \ldots, \beta_m$ $(m>0)$ be types of order 1 over B. Then the collection $(\alpha\ \beta_1, \ldots, \beta_n)$ of all m-ary partial mappings from $\beta_1 \times \ldots \times \beta_m$ into α is a functional type of order 1 over B.
  (iii) Nothing is a type of order 1 over B unless it so follows from (i) and (ii).

$\mathbf{C_n}$ *(constructions of order n)*

  (i) Let x be a variable ranging over a type of order n. Then x is a construction of order n over B.
  (ii) Let X be a member of a type of order n. Then $^0X, ^1X, ^2X$ are constructions of order n over B.

(iii) *Let $X_1, \ldots, X_m$ ($m > 0$) be constructions of order n over B. Then $[X\ X_1, \ldots, X_m]$ is a* construction of order n over B.
(iv) *Let $x_1, \ldots, x_m, X$ ($m > 0$) be constructions of order n over B. Then $[\lambda x_1 \ldots x_m\ X]$ is a* construction of order n over B.
(v) *Nothing is a* construction of order n over B *unless it so follows from C_n (i)-(iv).*

T$_{n+1}$ *(types of order $n+1$)*
Let $*_n$ be the collection of all constructions of order n over B. Then
 (i) *$*_n$ and every type of order n are* types of order $n+1$.
 (ii) *If $m > 0$ and $\alpha, \beta_1, \ldots, \beta_m$ are types of order $n+1$ over B, then $(\alpha\ \beta_1 \ldots \beta_m)$ (see* **T$_1$** *(ii)) is a* type of order $n+1$ over B.
 (iii) *Nothing is a* type of order $n+1$ over B *unless it so follows from (i) and (ii).*

For the purposes of natural-language analysis, we are assuming the following base of ground types:

- o the set of truth-values $\{\mathbf{T}, \mathbf{F}\}$
- ι the set of individuals (the universe of discourse);
- τ the set of real numbers (doubling as discrete times);
- ω the set of logically possible worlds (the logical space).

We model sets and relations by their characteristic functions. Thus, for instance, $(o\iota)$ is the type of a set of individuals, while $(o\iota\iota)$ is the type of a relation-in-extension between individuals. Empirical expressions denote *empirical conditions* that may or may not be satisfied at the particular world/time pair of evaluation. We model these empirical conditions as possible-world-semantic *(PWS-) intensions*. PWS intensions are entities of type $(\beta\omega)$: mappings from possible worlds to an arbitrary type β. The type β is frequently the type of the *chronology* of α-objects, i.e., a mapping of type $(\alpha\tau)$. Thus α-intensions are frequently functions of type $((\alpha\tau)\omega)$, abbreviated as '$\alpha_{\tau\omega}$'. *Extensional entities* are entities of a type α where $\alpha \neq (\beta\omega)$ for any type α. Where w ranges over ω and t over τ, the following logical form essentially characterizes the logical syntax of empirical language: $\lambda w \lambda t [\ldots w \ldots t \ldots]$.

Examples of frequently used PWS intensions are: propositions of type $o_{\tau\omega}$, properties of individuals of type $(o\iota)_{\tau\omega}$, binary relations-in-intension between individuals of type $(o\iota\iota)_{\tau\omega}$, individual offices (or roles) of type $\iota_{\tau\omega}$, attitudes to constructions of type $(o\iota *_n)_{\tau\omega}$.

Logical objects like *truth-functions* and quantifiers are extensional:

∧ (conjunction), ∨ (disjunction) and ⊃ (implication) are of type (ooo), and ¬ (negation) of type (oo). *Quantifiers* $\forall^\alpha, \exists^\alpha$ are type-theoretically polymorphic total functions of type $(o(o\alpha))$, for an arbitrary type α, defined as follows. The *universal quantifier* $\forall^\alpha$ is a function that associates a class A of α-elements with **T** if A contains all elements of the type α, otherwise with **F**. The *existential quantifier* $\exists^\alpha$ is a function that associates a class A of α-elements with **T** if A is a non-empty class, otherwise with **F**. Below all type indications will be provided outside the formulae in order not to clutter the notation. Moreover, the outermost brackets of the Closure will be omitted whenever no confusion can arise. Furthermore, 'X/α' means that an object X is (a member) of type α. '$X \to_v \alpha$' means that X is typed to v-construct an object of type α, if any. We write '$X \to \alpha$' if what is v-constructed does not depend on a valuation v. Throughout, it holds that variables w, t range over ω and τ, respectively; $w \to_v \omega$, $t \to_v \tau$. If $C \to_v \alpha_{\tau\omega}$ then the frequently used Composition $[[Cw]t]$, which is the intensional descent (a.k.a. extensionalization) of the α-intension v-constructed by C, will be encoded as 'C_{wt}'.

To operate in a hyperintensional context where a *construction* figures as an argument to operate on, we need two special functions, Sub and Tr.

Definition 3 (functions *Sub* and *Tr*) *The polymorphic function Sub of type* $(*_n *_n *_n *_n)$ *operates on constructions as follows. When applied to constructions* C_1, C_2, C_3, *Sub returns as its value the construction D that is the result of substituting* C_1 *for* C_2 *in* C_3.

The likewise polymorphic function Tr returns as its value the Trivialization of its argument.

For instance, the Composition $[^0Wife\_of_{wt}\ ^0John]$ is the result of the Composition $[^0Sub\ ^{00}John\ ^0him\ ^0[^0Wife\_of_{wt}\ him]]$. The result of $[^0Tr\ ^0John]$ is ^0John. If the variable x ranges over ι, the Composition $[^0Tr\ x]$ $v(John/x)$-constructs ^0John. Note one essential difference between the function Tr and the construction Trivialization. Whereas the variable x is *free* in $[^0Tr\ x]$, the Trivialization ^0x binds the variable x by constructing just x independently of valuation.

3 Substitution of identicals

The rigorous definitions of the three kinds of contexts can be found in Duží, Jespersen, and Materna (2010, § 2.6). The details are rather complicated, though the basic ideas are fairly simple. Thus, here we only explain the main ideas. There is a fundamental distinction between an occurrence of a construction within another construction in a *displayed* mode and *executed* mode. If the former, then the *construction* itself becomes the object of predication; we say that it occurs *hyperintensionally*. If the latter, then the *product* of the construction is an object of predication. In this case the executed construction is a *constituent* of its super-construction, and an additional distinction can be found at this level. The constituent presenting a function may occur either *intensionally* (de dicto) or *extensionally* (de re). If intensionally, then the produced function is the object of predication; if extensionally, then the value of the produced function is the object of predication. The two distinctions, between displayed/executed and intensional/extensional occurrence, enable us to distinguish between three kinds of *context*:

- *hyperintensional* context: a construction occurs in *displayed* mode (though another construction at least one order higher needs to be executed in order to produce the displayed construction)

- *intensional context*: a construction occurs in *executed* mode in order to produce a function rather than its value (moreover, the executed construction does not occur within another hyperintensional context)

- *extensional context*: a construction occurs in an *executed* mode in order to produce a particular value of a function at a given argument (moreover, the executed construction does not occur within another intensional or hyperintensional context).

The basic idea underlying the above trifurcation is that the same set of logical rules applies to all three kinds of context, but these rules operate on different complements: procedures, produced functions, and functional values, respectively. Substitution is, of course, invalid if something coarser-grained is substituted for something finer-grained.

3.1 Rules for substitution in an extensional context

In an extensional context, the *value* of a constructed function is an object of predication. Hence *v-congruent* constructions, which *v*-construct the same object for a given valuation v, are substitutable. Thus, we can formulate the extensional rules of substitution.

Marie Duží and Pavel Materna

Let $C \to_v (\alpha\beta)$, $D \to_v (\alpha\gamma)$, $A_1 \to_v \beta$, $A_2 \to_v \gamma$, $[C\ A_1]$, $[D\ A_2] \to_v \alpha$, $a \to_v \alpha$: an atomic construction of an α-object, $p \to_v (o\alpha)$: a *total characteristic function*.[3] Then the following rules are valid.

$$(\text{R}_{1e}) \quad \frac{[C\ A_1] = [D\ A_2] \qquad [p\ [C\ A_1]]}{[p\ [D\ A_2]]}$$

$$(\text{R}_{2e}) \quad \frac{[C\ A_1] = a \qquad [p\ [C\ A_1]]}{[p\ a]}$$

$$(\text{R}_{3e}) \quad \frac{[D\ A_2] = a \qquad [p\ [D\ A_2]]}{[p\ a]}$$

Proof of (R_{1e}).
1. $[C\ A_1] = [D\ A_2]$ $\hspace{6cm}\emptyset$
2. $[p\ [C\ A_1]]$ $\hspace{7.5cm}\emptyset$
3. $[p\ [D\ A_2]]$ $\hspace{4cm}$ substitution of identicals, 1, 2

Proofs of (R_{2e}) and (R_{3e}) are like the proof of (R_{1e}), hence obvious.

Examples.

$$\frac{\text{The number of degrees Celsius in Ostrava is equal to the temperature in Brno}}{\text{The number of degrees Celsius in Ostrava is less than zero}}$$
$$\overline{\text{The temperature in Brno is less than zero}}$$

1. $\lambda w \lambda t\ [[^0Degrees\_in_{wt}\ ^0Ostrava] = [^0Temp\_in_{wt}\ ^0Brno]]$ $\hspace{1cm}\emptyset$
2. $\lambda w \lambda t\ [^0<\ [^0Degrees\_in_{wt}\ ^0Ostrava]\ ^00]$ $\hspace{2.5cm}\emptyset$
3. $[^0Degrees\_in_{wt}\ ^0Ostrava] = [^0Temp\_in_{wt}\ ^0Brno]$ $\hspace{1cm}$ λ-elim., 1
4. $[^0<\ [^0Degrees\_in_{wt}\ ^0Ostrava]\ ^00]$ $\hspace{3cm}$ λ-elim., 2
5. $[^0<\ [^0Temp\_in_{wt}\ ^0Brno]\ ^00]$ $\hspace{3cm}$ (R_{1e}), 3, 4
6. $\lambda w \lambda t\ [^0<\ [^0Temp\_in_{wt}\ ^0Brno]\ ^00]]$ $\hspace{2.5cm}$ λ-abst., 5

Types. $Degrees\_in, Temp\_in/(\tau\iota)_{\tau\omega}$: attributes; $Ostrava, Brno/\iota$; $=$, $</(o\tau\tau)$; $[^0Degrees\_in_{wt}\ ^0Ostrava], [^0Temp\_in_{wt}\ ^0Brno] \to_v \tau$

---

[3] For the sake of simplicity, here we deal only with *total* characteristic functions and unary functions. Generalization to n-ary functions is straightforward. Partiality can be dealt with by means of the special propositional properties True, False and Undef. For details, see, e.g., Duží (2007) or Duží et al. (2010, § 1.4.3).

Remark. In step 5, we substituted the constructions [$^0Degrees\_in_{wt}$ $^0Ostrava$], [$^0Temp\_in_{wt}$ ^0Brno] $\rightarrow_v \tau$ for the schemes [C A_1], [D A_2], respectively, $p = \lambda x[^0< x\ ^00] \rightarrow_v (o\tau)$, $x \rightarrow_v \tau$.

The Mayor of Ostrava is Ivo Vondrák

The Mayor of Ostrava is wise

Ivo Vondrák is wise

$\lambda w \lambda t\ [[^0Mayor\_of_{wt}\ ^0Ostrava] = {}^0Vondrak]$

$\lambda w \lambda t\ [^0Wise_{wt}\ [^0Mayor\_of_{wt}\ ^0Ostrava]]$

$\lambda w \lambda t\ [^0Wise_{wt}\ ^0Vondrak]$

Types. $Mayor\_of/(\iota\iota)_{\tau\omega}$, $Ostrava$, $Vondrak/\iota$, $[^0Mayor\_of_{wt}\ ^0Ostrava]$ $\rightarrow_v \iota$, $Wise/(o\iota)_{\tau\omega}$.

For a simple mathematical example, consider this argument. We start with the well-known formula "$Sin(x) = Cos(\pi/2 - x)$". Now we are going to apply the above rules in order to compute the value of the cosine function at the argument 0 on the assumption $Sin(\pi/2) = 1$.[4]

$[\lambda x\ Sin(x)\ \pi/2] = [\lambda x\ Cos(\pi/2 - x)\ \pi/2]$ $[\lambda x\ Sin(x)\ \pi/2] = 1$

$[\lambda x\ Cos(\pi/2 - x)\ \pi/2] = 1$

$[\lambda x\ Cos(\pi/2 - x)\ \pi/2] = Cos(\pi/2 - \pi/2) = Cos(0) = 1$

3.2 Rules for substitution in an intensional context

In an intensional context, the whole *function* is the object of predication. Hence equivalent constructions, i.e. constructions that v-construct the same object for any valuation v, are substitutable. Thus, we can formulate the extensional rules of substitution in an intensional context.

Let $C \rightarrow_v (\alpha\beta_1 \ldots \beta_n)$, $D \rightarrow_v (\alpha\beta_1 \ldots \beta_n)$, $q \rightarrow_v (o(\alpha\beta_1 \ldots \beta_n))$: the *total* characteristic function, and let C v-construct a function $f/(\alpha\beta_1 \ldots \beta_n)$, D v-construct a function $g/(\alpha\beta_1 \ldots \beta_n)$. Then the following rules are valid.

---

[4] Here we stick to common mathematical notation without transcription into the symbolism of TIL constructions to make the argument easier to read.

$$(\text{R}_{1i}) \quad \frac{[C = D] \quad [q\ C]}{[q\ D]}$$

$$(\text{R}_{2i}) \quad \frac{[C = {}^0f] \quad [q\ C]}{[q\ {}^0f]}$$

$$(\text{R}_{3i}) \quad \frac{[D = {}^0g] \quad [q\ D]}{[q\ {}^0g]}$$

The proofs are obvious.

Example.

The existence of an abominable snowman has not been demonstrated

Every abominable snowman is a yeti
___
The existence of a yeti has not been demonstrated

To get the above example off the ground, we are stipulating that 'is a yeti' and 'is an abominable snowman' are a pair *not* of synonymous but merely equivalent predicates. Their respective meanings are *co-intensional* by denoting one and the same property, but not co-hyperintensional (*procedurally isomorphic*).[5] The rationale for this stipulation is that the latter predicate has a molecular structure thanks to the application of the property modifier denoted by 'abominable' to the property denoted by 'snowman', whereas the former predicate is atomic.[6] One could object that it seems reasonable to assume that there is a meaning postulate in place to the effect that 'is a yeti' is shorthand for, or a notational variant of, 'is an abominable snowman', the same way 'lasts a fortnight' is arguably short for 'lasts two weeks'. What speaks against this assumption, at least through the lens of TIL, is that the Trivialization ^0Yeti and the Composition [$^0Abominable\ ^0Snowman$] are not procedurally isomorphic, but only equivalent constructions. Furthermore, from a formal point of view at least, it is questionable what semantic

---

[5] We will deal with the relation of procedural isomorphism on the set of constructions below. Here just a brief explanation. Synonymous expressions have the same meaning; hence they encode the same procedure. But TIL constructions assigned to expressions as their context-invariant meaning are a bit too fine-grained from the procedural point of view. To wipe out minor unimportant differences between those constructions that are procedurally almost identical, we introduce the relation of procedural isomorphism.

[6] For details on property modifiers, see Jespersen (2015b, § 4).

and inferential gain might be accrued from introducing a redundant predicate like 'is a yeti', on its construal as a mere notational variant of 'is an abominable snowman'.[7] Hence the analysis of this argument comes down to these constructions:

$\lambda w \lambda t \ \neg [^0 Demonstrated_{wt} \ \lambda w \lambda t [^0 Exist_{wt} \ [^0 Abominable \ ^0 Snowman]]]$

$[^0 = [^0 Abominable \ ^0 Snowman] \ ^0 Yeti]$
───
$\lambda w \lambda t \ \neg [^0 Demonstrated_{wt} \ \lambda w \lambda t [^0 Exist_{wt} \ ^0 Yeti]]$

Types: $Yeti, Snowman/(o\iota)_{\tau\omega}$; $Demonstrated/(oo_{\tau\omega})_{\tau\omega}$: the propositional property of being demonstrated as true; $Abominable/((o\iota)_{\tau\omega}(o\iota)_{\tau\omega})$: a property modifier; $= /(o(o\iota)_{\tau\omega}(o\iota)_{\tau\omega})$: identity of properties; $[^0 Abominable \ ^0 Snowman] \rightarrow (o\iota)_{\tau\omega}$, $Exist_{wt}/(o(o\iota)_{\tau\omega})_{\tau\omega}$: the property of an individual property of being instantiated.

For a mathematical example, consider this simple argument.

$Sin(x)$ is a periodic function

$Sin(x) = Cos(\pi/2 - x)$
───────────────────────────────
$Cos(\pi/2 - x)$ is a periodic function

$[^0 Periodic \ \lambda x \ [^0 Sin \ x]] \qquad \lambda x [^0 Sin \ x] = \lambda x [^0 Cos \ [^0 - [^0 / \ ^0\pi \ ^0 2] \ x]]$
──
$[^0 Periodic \ \lambda x [^0 Cos \ [^0 - [^0 / \ ^0\pi \ ^0 2] \ x]]]$

Types. $Periodic/(o(\tau\tau))$: the class of unary periodic functions; the other types are obvious.

3.3 Substitution in a hyperintensional context

3.3.1 Synonymy of semantically simple terms

As stated above, in a hyperintensional context the object of predication is a *construction* C that occurs in the displayed mode within a given superconstruction. Hence, the substitution of a merely equivalent construction C' is *not* valid.

───────────────────────
[7] See Jespersen (2015b, § 5) for the parallel example of 'is a bachelor', 'is an unmarried man'.

For instance, consider the following argument.

Tilman is seeking an abominable snowman
―――――――――――――――――――――――――――――――
Tilman is seeking a yeti

Is this argument valid? In our opinion, it is not, because Tilman can be seeking an abominable snowman without seeking a yeti, though both 'abominable snowman' and 'yeti' denote one and the same property. As explained above, the terms 'abominable snowman' and 'yeti' are not synonymous.

Hence this is the case where the mode of presentation of one and the same property matters. In this case an intensional analysis will yield a contradiction, because Tilman would be related, and at the same time not related, to one and the same property by the *seeking* relation. Here is the proof.

1. $\lambda w \lambda t\ [^0 Seek_{wt}\ ^0 Tilman\ [^0 Abominable\ ^0 Snowman]]$ $\emptyset$
2. $\lambda w \lambda t \neg [^0 Seek_{wt}\ ^0 Tilman\ ^0 Yeti]$ $\emptyset$
3. $[^0 Seek_{wt}\ ^0 Tilman\ [^0 Abominable\ ^0 Snowman]]$ λ-elim., 1
4. $\neg [^0 Seek_{wt}\ ^0 Tilman\ ^0 Yeti]$ λ-elim., 2
5. $[^0 = [^0 Abominable\ ^0 Snowman]\ ^0 Yeti]$ $\emptyset$
6. $[^0 Seek_{wt}\ ^0 Tilman\ ^0 Yeti]$ (R_{1e}), 3, 5
7. contradiction   4, 6

Additional types. $Seek/(o\iota(o\iota)_{\tau\omega})_{\tau\omega}$: the relation-in-intension of an individual to a property an instance of which the seeker wants to find; $Tilman/\iota$; $[^0 Seek_{wt}\ ^0 Tilman\ ^0 Yeti] \to o$, $[^0 Seek_{wt}\ ^0 Tilman\ [^0 Abominable\ ^0 Snowman]] \to o$; $=/(o(o\iota)_{\tau\omega}(o\iota)_{\tau\omega})$: the identity of individual properties.

Thus a truthful report of such a situation must be *hyperintensional*. To block the above invalid inference, a hyperintensional analysis has to be applied. Here is how.

1. $\lambda w \lambda t\ [^0 Seek^*{}_{wt}\ ^0 Tilman\ ^0[^0 Abominable\ ^0 Snowman]]$ $\emptyset$
2. $\lambda w \lambda t \neg [^0 Seek^*{}_{wt}\ ^0 Tilman\ ^{00} Yeti]$ $\emptyset$
3. $[^0 Seek^*{}_{wt}\ ^0 Tilman\ ^0[^0 Abominable\ ^0 Snowman]]$ λ-elim., 1
4. $\neg [^0 Seek^*{}_{wt}\ ^0 Tilman\ ^{00} Yeti]$ λ-elim., 2
5. $\neg [^0 =^*\ ^0[^0 Abominable\ ^0 Snowman]\ ^{00} Yeti]$ $\emptyset$

Leibniz's Law of Substitution of Identicals

Additional types. $Seek^*/(o\iota *_n)_{\tau\omega}$: the relation-in-intension of an individual to a *construction* of a property an instance of which the seeker wants to find; $=^*/(o *_n *_n)$: identity of constructions.

No contradiction arises. When construed hyperintensionally, Tilman's search is only ostensibly inconsistent. By this example, we illustrated how invalid inferences can be blocked in hyperintensional contexts. But there is the other side of the coin, which is the *positive* topic of which inferences *should be validated*. That is, how hyper are hyperintensional meanings? If there is one central question permeating hyperintensional logic and semantics, then it is this one.

There is a simple but overly simple answer. Sure, if a construction C occurs displayed in a hyperintensional context, then, trivially, an *identical* (not merely equivalent) construction is substitutable.

Example. Suppose that 'cerulean' and 'azure' are synonymous terms. Since synonymous terms have the same meaning, they express the same construction; thus, the Trivializations $^0Cerulean/*_1 \to (o\iota)_{\tau\omega}$ and $^0Azure/*_1 \to (o\iota)_{\tau\omega}$ are identical. One and the same property has been Trivialized, regardless of the name we use for that property. Let now $Believe^*/(o\iota *_n)_{\tau\omega}$ be a hyperintensional relation-in-intension of an individual to a hyperproposition (i.e., to a construction of the proposition). Then the following argument is valid:

Tilman believes* that the Italian national football team wear azure shirts
_____
Cerulean is azure
_____
Tilman believes* that the Italian national football team wear cerulean shirts

One might object that the argument is not valid, because it can be true that Tilman believes that the shirts are azure without him believing that they are cerulean. Yes, it is possible, but then it is due to Tilman's linguistic incompetence. On this issue, Richard says:

> It is impossible for a (normal, rational) person to understand expressions which have identical senses but not be aware that they have identical senses. (Richard, 2001, pp. 546–547)

Hence Mates's (1952) puzzle is not a problem of hyperintensionality; it is a problem of linguistic incompetence rather than a problem of logical incompetence.

On the other hand, this example is from the logical point of view too trivial. It is obvious that *identical* constructions are always mutually substitutable, because there is nothing to substitute; there is just one construction. Yet, at the linguistic level, strictly *synonymous* expressions (expressions encoding one and the same *procedure*) are substitutable in a hyperintensional context. The problem is this. TIL constructions are a bit too fine-grained tool to explicate the sense of an expression. Hence, those constructions which are considered isomorphic from the procedural point of view (though possibly slightly different) should be substitutable in a hyperintensional context.

3.3.2 Identity of procedures and synonymy

The problem how fine-grained hyperintensional entities, hence meanings should be was important already for Carnap (1947, §§13ff) who introduced the relation of intensional isomorphism. However, Church (1954) found a counterexample of two terms that are obviously not synonymous, yet intensionally isomorphic. Church himself considered several so-called Alternatives of how to constrain these entities.[8] Senses are identical if the respective expressions are (A0) 'synonymously isomorphic', (A1) mutually λ-convertible, (A2) logically equivalent. (A2), the weakest criterion, was refuted already by Carnap, and was not acceptable to Church, either. (A1) was considered to be the right criterion of synonymy. Yet it was subjected to a fair amount of criticism, in particular due to the involvement of unrestricted β-reduction ('by name'). For instance, Salmon (2010) adduces examples of expressions that should intuitively not be taken to be synonymous, yet their meanings are mutually β-convertible.[9] Moreover, partiality throws a spanner in the works; β-conversion by name is not guaranteed to be an equivalent transformation as soon as partial functions are involved.[10] Church also considered Alternative (A1'), which is (A1) plus η-convertibility. Yet η-convertibility is plagued by similar defects as those of β-convertibility by name. The alternative (A0) arose from Church's criticism of intensional isomorphism, and it is synonymy resting on α-equivalence and meaning postulates for semantically simple terms. Of course, we need meaning postulates to fix synonymy for pairs of semantically simple terms (possibly even of different languages). Now we are, however, interested in the synonymy of

---

[8] For details see Anderson (1998); Church (1993).
[9] For a discussion of Salmon's arguments, see Jespersen (2015a).
[10] For details see Duží and Jespersen (2013).

molecular terms, which depends on structural isomorphism.[11]

Similar work in TIL has been done in (Materna, 1998, § 5.3) and in (Materna, 2004, § 1.4.2.2) where the relation of *quasi-identity* of closed constructions is defined. It includes α- and η- conversion. As a nod to Carnap and Church, this criterion has been coined *procedural isomorphism* and incorporated into Duží et al. (2010) as Alternative (A$^{1}/_{2}$). Duží and Jespersen (2013) and Duží (2014) put forward a new definition of the criterion of structured synonymy called (A1″). Close to Church's (A1), it includes an adjusted version of α-conversion and β-conversion by value, while η-conversion is excluded. Using the functions Sub and Tr (Def. 3), β-conversion by value is defined as follows. [12]

Definition 4 (β-**conversion by value,** β-**equivalence**)
Let $Y \to_v \alpha$; $x_1, D_1 \to_v \beta_1, \ldots, x_n, D_n \to_v \beta_n$, $[\lambda x_1 \ldots x_n\ Y] \to_v (\alpha \beta_1 \ldots \beta_n)$. Let C, D be constructions $[[\lambda x_1 \ldots x_n\ Y]\ D_1 \ldots D_n]$, and $^2[^0Sub\ [^0Tr\ D_1]\ ^0x_1 \ldots [^0Sub\ [^0Tr\ D_n]\ ^0x_n\ ^0Y]]$, *respectively. Then the conversion* $C \Rightarrow_\beta D$ *is* β*-reduction by value. The reverse conversion is* β*-expansion by value. We will say that constructions* C *and* D *are* β*-equivalent.*

It should be obvious now that the problem is not simple, and a question arises whether there is a unique universal solution. We now formulate several conditions that should be met by constructions C and D, in order that these constructions could be taken as procedurally isomorphic, hence substitutable in a hyperintensional context.

(a) *Strict equivalence*; for any valuation v constructions C and D v-construct the same object or are both v-improper.
(b) Constructions C and D have the same number of constituents.

Both conditions are met only by α-equivalence and by meaning postulates. Condition (a) is met by β_r-conversion (i.e. the restricted β-conversion by name that only substitutes variables for λ-bound variables of the same type) and by β_v-conversion by value. However, both β_r-conversion and β_v-conversion by value do not meet condition (b). Finally, η-conversion does not meet either of them. We might formulate weaker requirements like this.

---

[11] See Duží (2014).
[12] Duží and Jespersen (2015) proves that this conversion is strictly equivalent in the sense of C, D v-constructing the same object or both being v-improper for any valuation v.

(c) *Weak equivalence*; for any valuation v constructions C and D v-construct the same object, *provided* none of them is v-improper
(d) Constructions C and D have the same number of *closed proper constituents*.

Both these requirements are met by η-conversion, β-conversion by name meets only (c). β-conversion by value meets, of course, (c) because it satisfies (a), but it does not meet (d). However, if we postulated that the term '$[[\lambda x_1 \ldots x_n \ Y] \ D_1 \ldots D_n]$' is just a notational shorthand for '$^2[^0Sub \ [^0Tr \ D_1] \ ^0x_1 \ldots [^0Sub \ [^0Tr \ D_n] \ ^0x_n \ ^0Y]]$', then β-conversion by value would trivially meet also the condition (d). Such a notational convention is well justified, because the conversion by value specifies the correct and proper way of executing the procedure of applying a function to its argument.

Based on the above considerations, we are convinced that it is philosophically wise to adopt several notions of procedural isomorphism, and we agree with Faroldi (2016) that there is no *universal criterion* for synonymy that would be applicable in any kind of language. The definition of Alternative (A0) is presumably the strongest criterion of synonymy, yet in a strongly logical language such as, for instance, a programming language, even this criterion would not be acceptable. In such a language λ-bound variables play the role of formal parameters for which actual arguments should be substituted when calling a procedure for execution. If these variables are public, then α-conversion is not acceptable. On the other hand, in an ordinary vernacular we usually do not explicitly use λ-bound variables. Thus, for an ordinary vernacular the Alternative (A1″) + β_r-reduction might be the right one.

Here is an example. Consider the iterated hyperintensional attitude

(Att)   Tilman (explicitly) knows that Tom believes *of* the Pope that *he* is wise.

Here explicit knowing is meant as a hyperintensional relation of an individual to a hyperproposition, $Know^*/(o\iota*_n)_{\tau\omega}$. Tilman's knowing is sensitive to the way in which a given proposition is conceptualized. The analysis of the complement clause "Tom believes (implicitly) *of* the Pope that *he* is wise" in its *de re* sense (the *occupant*, if any, of the papal office is believed by Tom to be wise) comes down to these three variants.

Leibniz's Law of Substitution of Identicals

(1) $\lambda w \lambda t\ [^0Believe\ ^0Tom\ [^0Sub\ [^0Tr\ ^0Pope_{wt}]\ ^0he\ ^0[\lambda w^* \lambda t^*[^0Wise_{w^*t^*}\ he]]]]$
(2) $\lambda w \lambda t\ [\lambda w' \lambda t' \lambda x\ [^0Believe_{w't'}\ ^0Tom\ [\lambda w^* \lambda t^*\ [^0Wise_{w^*t^*}\ x]]]_{wt}\ ^0Pope_{wt}]$
(3) $\lambda w \lambda t\ [\lambda x\ [^0Believe_{wt}\ ^0Tom\ [\lambda w^* \lambda t^*\ [^0Wise_{w^*t^*}\ x]]]_{wt}\ ^0Pope_{wt}]$

Additional types. $Believe/(o\iota o_{\tau\omega})_{\tau\omega}$: implicit belief relating an individual to a given proposition regardless of the way in which this proposition is conceptualized; $Pope/\iota_{\tau\omega}$; $Wise/(o\iota)_{\tau\omega}$; $x \to_v \iota$.

The first construction is the result of applying α-conversion and β-conversion by value to (3); this third construction is the contractum of (2) by β_r-reduction by name. The differences between these three variants are hard to formulate in an ordinary non-technical vernacular. Hence, these three constructions can be considered to be procedurally isomorphic analyses of the sentence "Tom believes *of* the Pope that *he* is wise" and thus substitutable in the analysis of the whole iterated attitude (Att).

$$\lambda w \lambda t [^0Know^*_{wt}\ ^0Tilman\ ^0(1)] = \lambda w \lambda t [^0Know^*_{wt}\ ^0Tilman\ ^0(2)]$$
$$= \lambda w \lambda t [^0Know^*_{wt}\ ^0Tilman\ ^0(3)]$$

Thus, we are convinced that several degrees of hyperintensional individuation are called for, depending on which sort of discourse happens to be analysed. What appears to be synonymous in an ordinary vernacular might not be synonymous in a professional language like the language of, for instance, logic, mathematics, the theory of algorithms, or physics. The weakest criterion is the Alternative (A1″) + β_r-reduction. On the other hand, the strongest and most restrictive criterion is synonymy resting only on meaning postulates for semantically simple terms understood as a shorthand for another (possibly complex) term.

Hence substitution in a hyperintensional context can be formulated only as a conditional rule:

(**R**$_{\text{hyperint}}$) If constructions A, B are procedurally isomorphic per one of the existing criteria of procedural isomorphism then A, B can be validly substituted within a hyperintensional context, provided the discourse has been defined to obey that particular calibration of procedural isomorphism.

It is a philosophical and linguistic decision which sorts of discourse obey which calibrations of procedural isomorphism.

4 Conclusion

In this paper, we examined the validity of Leibniz's law of substitution of identicals. We have shown that the law is universally valid, provided it is correctly applied to an object of a proper type. In an extensional or intensional context the rules of substitution are smoothly applicable. However, a hyperintensional context is rather restrictive, because the very meaning of an expression becomes the object of predication. Thus, we encounter the problem of synonymy, which in TIL corresponds to the problem of procedural isomorphism among TIL constructions. As a result, the rule of substitution in a hyperintensional context has been defined only conditionally.

Acknowledgements This research has been supported by the Grant Agency of the Czech Republic, project No. GA15-13277S, „Hyperintensional logic for natural language analysis", and also by the internal grant agency of VSB_TU Ostrava, project SGS No. SP2016/100. We are grateful to B. Jespersen for valuable comments that improved the quality of this paper.

References

Anderson, C. A. (1998). Alonzo Church's contributions to philosophy and intensional logic. *The Bulletin of Symbolic Logic*, *4*, 129–171.

Carnap, R. (1947). *Meaning and Necessity*. Chicago University Press.

Church, A. (1954). Intensional isomorphism and identity of belief. *Philosophical Studies*, *5*, 65–73.

Church, A. (1993). A revised formulation of the logic of sense and denotation. Alternative (1). *Noûs*, *27*, 141–157.

Duží, M. (2007). Presuppositions and two kinds of negation. In D. Chiffi (Ed.), *Logique et Analyse*. Amsterdam: Elsevier.

Duží, M. (2014). Structural isomorphism of meaning and synonymy. *Computacion y Sistemas*, *18 (3)*, 439–453.

Duží, M., & Jespersen, B. (2013). Procedural isomorphism, analytic information, and β-conversion by value. *Logic Journal of the IGPL*, *21*, 291–308.

Duží, M., & Jespersen, B. (2015). Transparent quantification into hyperintensional objectual attitudes. *Synthese*, *192 (3)*, 635–677.

Duží, M., Jespersen, B., & Materna, P. (2010). *Procedural Semantics for Hyperintensional Logic. Foundations and Applications of Transparent Intensional Logic.* Berlin: Springer.

Faroldi, F. (2016). Co-hyperintensionality. *Ratio.* (Published online, DOI: 10.1111/rati.12143)

Jespersen, B. (2015a). Should propositions proliferate? *Thought, 4,* 243–251.

Jespersen, B. (2015b). Structured lexical concepts, property modifiers, and transparent intensional logic. *Philosophical Studies, 172,* 321–345.

Kripke, S. (1979). A puzzle about belief. In A. Margalit (Ed.), *Meaning and Use* (pp. 239–283). Dordrecht and Boston: Reidel, Synthese Language Library. (Reprinted in Kripke, S., Philosophical Troubles, Collected Papers, vol. 1, Oxford University Press, pp. 125–161.)

Materna, P. (1998). *Concepts and Objects.* Helsinki: Acta Philosophica Fennica, Vol. 63.

Materna, P. (2004). *Conceptual Systems.* Berlin: Logos Verlag.

Mates, B. (1952). Synonymity. In L. Linsky (Ed.), *Semantics and the Philosophy of Language.* Urbana: University of Illinois Press.

Quine, W. (1953). Reference and modality. In *From a Logical Point of View* (pp. 139–159). Cambridge: Harvard University Press.

Richard, M. (2001). Analysis, synonymy and sense. In C. A. Anderson & M. Zelëny (Eds.), *Logic, Meaning and Computation* (pp. 545–572). Dordrecht: Kluver.

Salmon, N. (2010). Lambda in sentences with designators: an ode to complex predication. *Journal of Philosophy, 107,* 445–468.

Sayward, C. (2007). Quine and his critics on truth-functionality and extensionality. *Logic and Logical Philosophy, 16,* 45–63.

Tichý, P. (1988). *The Foundations of Frege's Logic.* Berlin, New York: de Gruyter.

Marie Duží
VSB-Technical University Ostrava, Department of Computer Science FEI
The Czech Republic
E-mail: marie.duzi@vsb.cz

Pavel Materna
The Czech Academy of Sciences, Institute of Philosophy
The Czech Republic
E-mail: materna@flu.cas.cz

Making Sense of Higher-Level Plurals

BERTA GRIMAU[1]

Abstract: Higher-level plural terms are said to stand to plural terms as plural terms stand to singular terms. If plural endings were iterable in English, we could say that 'catses' is the higher-level plural of 'cats'. But what could 'catses' mean? And, given that plural endings are not iterable in English, do any English terms at all count as higher-level plurals? Higher-level plurals have received some attention in the literature on plurals. However, all of these issues remain open. In this article, I will try to elucidate the existing debate around higher-level plurals as well as give my own proposal as to how to characterize them.

Keywords: Plurals, Higher-level plurals, Ontological commitment, Natural language semantics, Reference, Intensionality

1 Why higher-level plural logic?

Plural terms refer to the same objects singular terms refer to, but do so plurally, that is, refer to more than one of them at once. Today the notions of plural term and plural quantification enjoy a widespread acceptance within the philosophical community. Accordingly, formal languages incorporating these elements have been developed (for example, in Rayo, 2006). These formal languages have been put to work in a variety of contexts: plural interpretations of second-order set theory (Boolos, 1984), nominalist reconstructions of set theory (Lewis, 1991), neo-Fregean programmes (Boccuni, 2013) or atomistic programmes (Hossack, 2000).

Higher-level plural logic has been proposed as an extension of plural logic. This formal language incorporates not only plural terms and quantifiers, but also higher-level plural ones. Higher-level plural[2] terms are often informally introduced as terms that stand to plural terms like the latter stand

---

[1] I would like to thank my supervisors, Fraser MacBride, Adam Rieger and Stewart Shapiro, as well as audiences at Leeds and Hejnice for helpful comments.

[2] In this article, I will use 'HLP' interchangeably with 'higher-level plural'.

to singular terms. However, unlike its plural counterparts, the notion of HLP reference and the formal languages incorporating it have been received with hostility from the philosophical community. In the next section, we will examine the criticisms they have received, but first let us briefly explain why one may want to develop a HLP formal language in the first place.

There are at least three different motivations for developing a HLP logic. The first one is the development of a nominalist mathematics. A logic containing all finite levels of plural variables and quantifiers would be isomorphic with a simple type theory, but, arguably, its ontological commitments would consist only of the values of the singular first-order variables.

Secondly, one might want to use HLP logic in order to extend Boolos' interpretation of monadic second-order logic to its non-monadic fragment. Boolos[3] argues that second-order set theory ought to be interpreted plurally, if we are to be able to quantify over all sets. Note that plural logic cannot account for the non-monadic fragment of second-order logic, since plural terms cannot express order. This was not a problem for his purposes. However, if one believes that second-order logic should be interpreted plurally in contexts in which the non-monadic fragment is also required, then she may want to use higher-level plurals to that end. One possible strategy would be to use one of the set-theoretic definitions of ordered pair and with that convention at hand, consider relational predicates as higher-level plural terms. One of the applications falling within this approach can be found in neo-logicism. Some attempts at reviving Fregeanism replace the second-order notation with that of plural logic. In that context, HLP terms come into play in order to interpret equinumerosity.

Finally, one might want to develop HLP logic as a tool for natural language analysis. From the point of view of linguistics, one might find that a theory that considers higher-level plurals as a semantic category in its own right makes the right predictions. As a matter of fact, at least one voice in linguistics defends this view (see Moltmann, 2016). On the other hand, one might want to add some contraints to the aim of the linguist on a philosophical basis. For example, a nominalist not only about mathematics, but about other types of discourse, might want to provide an analysis of the ordinary fragment of language which does not imply the existence of abstract entities. She could use HLP as a means to that end.

Another way in which HLP terms may be used for natural language analysis is to provide semantics for collective predication. The idea would be

---

[3] In Boolos (1984) and Boolos (1985).

that in order to account for the meaning of a statements of the form P(aa), where P stands for a collective predicate, one could use HLP terms in the metalanguage and claim that the statement is true iff the aa are some of the ppp, where 'ppp' is a HLP term.

In this paper, I will do three things. First, I will examine and attempt to shed some light on the debate around higher-level plurals. My aim will be to facilitate a more fruitful discussion of the issues involved. Secondly, a proposal as to how to interpret HLP terms will be put forward. Finally, I will argue that some ordinary HLP statements are ineliminable, that is, cannot be paraphrased away as singular or plural statements.

2 The debate on higher-level plurals

Higher-level plural logic has received three main objections, which we discuss in some detail in the next two sections. Put in a slogan form, the objections are:

(i) **The natural language objection:** There aren't higher-level plurals in natural language or, at least, it's unclear whether there are any, given that no precise syntactic and semantic descriptions of them have been provided.

(ii) **The notational disguise objection against ontological innocence:** Higher-level plural terms are singular terms denoting sets, groups or sums[4] in disguise. Thus, HLP logic is set theory, mereology or some other theory of collective entities in disguise.

(iii) **The metaphysical objection:** Higher-level plurals denote higher-level pluralities, but it is not possible to make sense of higher-level pluralities. Therefore, there cannot be higher-level plurals.

2.1 Some remarks on natural language and the notational disguise objection

The objection from natural language carries little force for two reasons. Firstly, the objection is usually raised in contexts where no precise characterisation of the HLP is being considered. Sometimes, a very vague approximation to the notion is given, but such an imprecise picture does not

---

[4]Or some other sort of complex entity.

provide necessary and sufficient conditions for the identification of higher-level plurals in ordinary language.

Moreover, as a matter of fact, nowadays there are a few precise characterizations of the HLP phenomenon available. In the philosophical context, there is a recent proposal, in Linnebo and Nicolas (2008). On the linguistics side, the concept has been around for longer (see Link, 1984; Landman, 1989a). A more recent approach can be found in Moltmann (2016). The fact that not only logicians, but also linguists have deemed the HLP a genuine semantic category provides independent support to the hypothesis that we do find such expressions in natural language.

But why is it important to the philosopher whether there are HLP terms in natural language?[5] One reason why it is important has to do with the notational disguise objection. Whether there are such elements in natural language is relevant if one endorses semantics à la Boolos, where the plural idiom is re-used in the metalanguage in order to express the truth-conditions for plurally quantified statements. The same applies to HLP logic: were one to re-use the HLP in the semantics for higher-level plurals, she would need to show, at least, that there are HLP terms in natural language.

Typically one endorses this sort of semantics in order to ensure ontological innocence, i.e. in order to avoid appealing to set-like entities in the model theory. However, this strategy is open to the objection that natural language occurrences of plurals and higher-level plurals belong merely to a certain manner of speaking and that ordinary English is *really* singular and thus committed to set-like entities. This is indeed one way of putting the notational disguise objection.

In response, the advocate of the ontological innocence of plural and HLP language may argue in one of the following ways:

(a) We should take ordinary language to mean mean whatever ordinary speakers believe it means.

(b) We should take ordinary language to mean whatever our best linguistic semantics deems correct.

(c) We should take ordinary language to mean whatever its ineliminable paraphrases indicate.

Answer (a) seems wrongheaded, because speakers may very well be wrong about the semantic value of their expressions or have contradictory intuitions about it. Answer (b) is more promising.

---

[5]Note that this discussion is equally relevant for plurals. However, there is a general agreement that there are plurals in natural language.

As it happens, the mainstream view in linguistics suggests that there are indeed plural and HLP terms in ordinary English. Nevertheless, it suggests a mereological reading of them, which is in tension with the ontological innocence claim. There is, however, one dissonant voice in linguistics, as I mentioned above: that of Moltmann, who has argued that not only the plural-reference approach can do all the work the mereological approach does, but that it does it better. This is controversial, but I won't discuss it in this paper.

Be that as it may, one could deny that linguistic semantics is relevant in this respect. For example, if one believed that semantics is a useful tool to make predictions about ordinary inferences, but it does not reveal the logical form of natural language, which is what matters for the purposes of the philosopher.

Finally, option (c) has often been considered the right sort of strategy by philosophers and logicians.[6] The underlying principle is that the logical form and the ontology of ordinary language are revealed by the ineliminable expressions of our language. The expressions which cannot be paraphrased away are the ones that are significant for metaphysical conclusions. I will engage with this dialectic at the end of this paper.

2.2 The iteration principle and the metaphysical objection

We have considered the objections from natural language and from notational disguise in the section above. It remains to look into the metaphysical objection.[7] Here is one way of putting the argument: (i) A higher-level plurality is a plurality of pluralities, (ii) since plural language is ontologically innocent, there are no such things as pluralities. Therefore, there are no higher-level pluralities.

Note the use of 'plurality' and 'higher-level plurality'. The assumption behind this argument, sometimes made explicit, is that plural terms refer to pluralities and HLP terms refer to higher-level pluralities.

The objection is not compelling, because it is based on an unjustified jump from the ideological to the ontological domain: from the fact that some terms are HLP terms does not necessarily follow that there are some corresponding sort of objects to which they refer. Moreover, the argument is based on a tenet which, I argue as follows, should be rejected:

---

[6] See Oliver and Smiley (2013).
[7] This is raised, for example, in Russell (1903) and in McKay (2006).

Iteration Principle: HLP terms are the result of an iteration of the step from singular terms to plural terms.

Even though there is no widespread agreement as to how to understand higher-level plurals, most authors seem to share the intuition expressed in the following passage:[8]

'(...) a HLP (noun, pronoun, verb form...) is related to plurals as plurals are to singulars. As a semi-serious example, pretend our plural endings on nouns are iterable: then we could assert the existence of infinitely many cats by saying something like:

There are some catses such that for each cats among thems there are some cats among thems including at least one more cat.'
(Hazen, 1997, p. 247)

Even though, prima facie, the iteration principle might seem to be sufficiently clear, it is not, for there are different ways of cashing out the idea of iterating the step from the singular to the plural and neither of them appears to deliver the desired result.

The step from the singular to the plural may be seen as the fact that whereas a singular term refers to one thing, a HLP term refers to many things. By iterating this statement, that is, applying it in some way to itself, we obtain that whereas a plural term refers to many things, a HLP term refers to...

(1) ...(many many) things.

(2) ...many (many things).

The parentheses intend to indicate how the iteration is to be understood exactly. In (1), the use of the determiner 'many' has been iterated thus obtaining a new determiner, 'many many'. It seems that when we say that a HLP refers to many many things, that has to be understood as saying that higher-level plurals refer to very many things (meaning perhaps 'more than two things'). In (2), by contrast, the new occurrence of 'many' applies to the whole phrase 'many things', forcing a singularization[9] thereof. We obtain thus that whereas a plural term refers to many things, a HLP term refers to

---

[8] Also expressed in Uzquiano (2004, p. 438) and Linnebo and Nicolas (2008, p. 186).
[9] What linguists might describe as a case of type-shifting.

many many's[10], where 'many's' has to be understood as denoting more than one set, group or sum.

Neither of the two options results in a characterization of the higher-level plural. According to (1), HLP terms refer to very many things. But this variety of reference is subsumable into the general notion of plural reference.

According to (2), HLP terms refer to many sets, groups or sums, but this is simply plural reference to a special kind of individuals: complex entities. The latter path is what would indeed motivate the metaphysical objection, according to which higher-level plurals refer to pluralities of pluralities of individuals – just as plurals refer to pluralities of individuals.

The fact that it is prima facie difficult to make sense of the iteration principle gives us a reason to be wary of it. I submit that we should not assess the success of our investigation around higher-level plurals on its concordance with this principle.[11]

I hope to have shown that the main objections to the notion of HLP reference are not compelling. Moreover, at this point it should be clear why natural language analysis is relevant for the debate from the point of view of the philosopher. Finally, I have argued that the iteration principle should not guide our research for an adequate notion of HLP reference.

In the next section, I am going to propose a way of analyzing the HLP fragment of natural language which makes the right predictions about linguistic facts. My aim will be to give support to the claim that HLP logic cannot be criticized on the basis of the objection from natural language. In section 4, I will respond to the notational disguise objection by following the paraphrasing strategy and showing that there are HLP expressions which are ineliminable, in a certain sense.

3 Higher-level plurals as referential-qua terms

Even though most authors have discussed higher-level plurals without providing a precise definition of the notion, some of them seem to agree on a few examples of terms which should count as HLP. In this section, I will propose a precise characterization which is materially adequate. That is, which renders what others have considered to be HLP terms as such. The examples of HLP term given in the literature are lists of plurals (e.g. 'these

---

[10] I am borrowing Russell's terminology here. See, for example, Russell (1903, p. 489).

[11] Against Ben-Yami (2013), who believes the iteration principle describes an essential feature of higher-level plurals and dismisses the latter, partly, on this basis.

students and their professors'[12]), lists of lists of singulars (e.g. 'Whitehead and Russell, and Hilbert and Bernays'[13]), count noun phrases built from pseudo-singular[14] terms (e.g. 'the pairs of shoes'[15]) and count noun phrases built from collective predicates (e.g. 'the authors of multivolume classics in logic'[16]).

Higher-level plural language is more expressive than plural language. The way in which the analysis I'm advocating here will achieve this expressive power is by incorporating a new kind of reference – HLP reference. By using this language one will not be able to refer to any objects she could not refer to before, but will nonetheless be able to refer to old objects in a new way.

Roughly speaking, HLP reference allows us to pick out objects while giving certain information about them. What kind of information? HLP reference tells us about the way in which objects are related to each other. It refers to objects *qua* being related in a certain way. For example, when referring to some people as 'these people, those people and these other people' one may be highlighting the fact that some of them are at one end of a field, the others are in the middle and the rest are at the other end of the field. Or perhaps one intends to emphasize that the people picked out by each of the plural constituents of the list have different professions.

As will become clear in section 4, the main role of HLP language is that it allows us to express complex collective predications that would otherwise be very difficult to convey.

Objects can be considered as related to each other in many different ways. However, we will consider only equivalence relations. Roughly speaking, we will say that HLP terms pick out objects by highlighting that one or more equivalence relations hold between them. If it is only one, then the plural is of second level (first level plurals being ordinary plurals). If it's more than one, the plural is of a level higher than second.

Let's put this more precisely. First, a preliminary definition:

Definition 1 (Sub-relation)   *R is a strict sub-relation of S iff for any $\langle a, b \rangle \in R$, $\langle a, b \rangle \in S$ and there is $\langle a, b \rangle \in S$ such that $\langle a, b \rangle \notin R$.*

Now we can define the kind of reference carried by higher-level plurals

---

[12] See Link (1984); Landman (1989a); Linnebo and Nicolas (2008, p. 193).
[13] See Oliver and Smiley (2005, p. 1062).
[14] Pseudo-singulars are syntactically singular terms which function semantically as plurals.
[15] See Linnebo (2008).
[16] See Oliver and Smiley (2013, p. 28).

in a more precise way.

Definition 2 (*n*th-level plural term)   *A nth-level plural term (where n>1) is a term that refers to some objects qua bearing $n-1$ equivalence relations, $R_1, ..., R_{n-1}$, to one another, where for $1 \leqslant k < n-1$, R_k is a strict sub-relation of R_{k+1}.*[17][18]

This is only a preliminary definition. The notion of reference-qua is still in need of clarification. We will say more about it below.

Naturally, I do not maintain that higher-level plurals of any level can actually be found in natural languages, but simply that it is possible to characterize a hierarchy of them so as to make sense of (artificial) expressions of any such level.

3.1   A hierarchy of collective and distributive predicates

The relational information carried by HLP terms determines how different predicates distribute over the individuals picked out by them.

Collective predicates take plural arguments and apply to their referents collectively.[19] These are what I will call 'collective$_1$ predicates'. On the other hand, with the notion of second-level plural at hand, we can define collective$_2$ predicates as predicates which apply collectively to the referents of second-level plural terms. For example, whereas surrounding a building is usually understood as applying collectively to some things simpliciter, there is an interpretation of competing in a 3-way game in which it applies to some objects qua being in an equivalence relation. For example, the interpretation it would receive in 'These people, those people and these other people competed in a 3-way game' if what one meant is that all of those people were playing a single game as members of three different teams.

More generally, we can classify predicates according to the kind of subjects for which they can be interpreted as being collective. Accordingly,

---

[17] Note that this definition allows for mixed higher-level plural terms. That is, it renders terms such as 'Barbara and the other students' and 'The students and their teachers, and the firefighters' HLP.

[18] Since arriving at these views, I have discovered the recent article (Moltmann, 2016) which presents a position similar to mine. The article contains a number of arguments from the point of view of natural language semantics which give further support to the view presented here. However, one of the crucial differences between our accounts is that it is not clear that (and, if so, how) Moltmann's notion of second-level plural can be iterated.

[19] A predicate is collective iff it is not the case that it holds of some objects iff it holds of every one of them.

since a predicate is collective$_n$ iff it is distributive$_{n+1}$, we can see the hierarchy as a hierarchy of distributive predicates.

3.2 Higher-level plurals and intensionality

The idea of considering some objects qua bearing some relations to each other needs to be further elucidated.

HLP terms are a special sort of referential-qua term. Referential-qua terms are what Landman[20] has called 'atemporal/non-modal' intensional terms. According to Landman, there are two sorts of intensional terms. The first one is what he has called 'temporal/modal intensional' terms. In this sense, a term is intensional iff it picks out different objects at different times or in different possible worlds. The second one is the atemporal/non-modal sense. According to it, two terms are intensional iff they both denote the same individual(s) and yet they are not inter-substitutable (in a variety of contexts – not only in opaque contexts – we will see an example below). Note that terms can be either intensional in one sense or the other, in both or in neither (i.e. be referential – e.g. demonstrative noun phrases, proper names and non-nested lists of them). In what follows, I will focus on atemporal/non-modal intensionality and ignore the mixed cases, for ease of exposition.

My claim is that HLP terms are a special kind of referential-qua term, in the sense that substitutivity fails for them in certain sorts of contexts. The contexts in question are collective and distributive predications of the right level. The examples I'll give in what follows all involve second-level plurals, thus the relevant sort of contexts will involve collective$_2$ and distributive$_2$ predicates.

For example, suppose 'these students and these professors' and 'these women and these men' (where 'these' is used in a demonstrative sense) pick out the same individuals (in the sense that the totality of individuals they denote is the same). The following sentences, involving the same collective$_2$ predicate are not equivalent, since the two statements may have different truth-conditions: 'These students and these professors played against each other' and 'These women and these men played against each other'.[21]

---

[20]In Landman (1989b).

[21]Here the intended reading of 'playing against each other' is that according to which it is two teams that play against each other. The reading is certainly available. So, in particular, the equivalence fails when 'these students' does not pick out the same individuals as 'these women', and 'these professors' does not pick out the same individuals as 'these men' (or the other way around).

Making Sense of Higher-Level Plurals

Now let us look at an example where the predication involved is distributive$_2$. Consider the statement 'Barbara admires these people and these other people', where the intended reading is that Barbara admires these people collectively (not each of them individually) and she also admires these other people collectively. From this does not follow that she admires these men and these women, even if 'these people and these other people' picks out the same individuals as 'these men and these women'.

And the reason for these failures is that, in the case of statements involving HLP term, it is not only the denotation of the term, but also certain equivalence relations that enter into their truth-conditions.

By contrast, plural terms aren't referential-qua in this sense, since they generally enjoy substitutivity in collective$_1$ and distributive$_1$ contexts. Think of 'All those people played against each other' and 'A, B, C, D... played against each other' (where 'A, B, C, D...' is a list of all the proper names of the people denoted by 'all of those people' in the previous statement). They have the same truth-conditions. Similarly for 'All those people are older than 20 years old' and 'A, B, C, D... are older than 20 years old', where the predicate involved is distributive.

This is not to say that plural terms are always substitutable in these sorts of contexts. In fact, Landman has shown that, often, they aren't. Suppose the judges are the same people as the hangmen (i.e. they happen two have two jobs). In such a situation we can't infer from 'The judges are on strike' that the hangmen are on strike or vice-versa. At this point, one might object that being on strike is a distributive predicate and thus we haven't really managed to distinguish HLP from plural terms in terms of substitutivity failures. Our claim is that what is distinctive of HLP terms is that substitutivity always fails for them in contexts which are sensitive to the highlighted equivalence relations, whereas that is not always the case for plurals. This is because there is only one equivalence relation qua which plurals may refer to some objects – identity. However, plurals are referential-qua in other ways, as the example of the judges and the hangmen shows. I will not discuss these other ways in this article.

Now it should be clear that HLP terms are a particular case of a more general phenomenon: atemporal/non-modal intensional terms. This should provide some plausibility to my proposal by disuading concerns about ad-hocness.

3.3 Types of HLP terms

Recall the examples of higher-level plurals presented above: 'Whitehead and Russell, and Hilbert and Bernays'; 'these students and their professors'; 'those pairs of shoes'; 'the authors of multivolume classics in logic'. I submit that all of them are capable of referring to some objects higher-level plurally. Let's have a quick look at each of these types of expression.

Lists of plurals The first two are lists of plural terms. One, in the form of a nested list of singular terms; the other one, in the form of a list of plural demonstrative phrases. In either case, it is assumed that lists of referring expressions themselves are referring expressions.[22] We are also restricting the case to non-ordered uses of lists.[23]

The way in which these terms manage to refer higher-level plurally is straightforward: speakers read off the relevant equivalence relations directly from the form of the list.[24]

The fact that lists convey HLP reference shows that, given the easiness with which we can form nested lists and against what other authors have claimed, in ordinary English there are plurals of a level higher than second.

Terms formed from pseudo-singulars There are at least two kinds of higher-level plurals which can be formed from pseudo-singulars: those in the form of a list of pseudo-singulars and those in the form of a noun phrase. An example falling within the first category is 'this pair of shoes, that pair of shoes and this other pair of shoes'. We touched on this kind of terms above.

---

[22] We are following Oliver and Smiley (2013) on this.

[23] In fact, there are reasons to think that lists are always non-ordered. For one such view, see Florio and Nicolas (2015). According to their view, order does not come engrained in the mode of reference, but is given by modifers such as 'in that order' and 'respectively'.

[24] Linnebo and Nicolas (2008) characterize HLP terms in a way that renders lists of plurals the only kind of HLP term. For them, HLP predicates (which are more fundamental than HLP terms) are multigrade predicates which take a variable number of plural terms as subjects. Thus, they are predicates that take lists of plurals of different length as subjects. There is some overlap between their proposal and mine, however I believe theirs has problems that mine overcomes. One of the issues is that their account does not deem a predicate like 'being arch-rivals' in 'Barcelona's supporters and Real Madrid's supporters are arch-rivals' as a HLP predicate, since it is not multigrade. The same holds of number-related predicates. Moreover, if multigradedness is a sufficient condition for being a higher-level plural, then the predicate in 'These people and those people weigh 500 kilos' has to be HLP. But it needs not be. In fact, a more natural reading of this sentence is one in which the subject is a plural one, because the predicate is not sensitive to the relational information conveyed by the subject.

To the second category belong 'the most expensive pairs of shoes in this shop', 'my favorite teams' and 'those couples over there'. The existence of these higher-level plurals suggests that Hazen's semi-serious example (i.e. 'catses') may not be so misguided after all: just as 'catses' would be a pluralization of 'cats', 'pairs' is a pluralization of 'pair'. The fact that some semantically plural terms happen to be syntactically singular, allows for the process of syntactic pluralization to be applied to plural terms.

Plural noun phrases restricted by a collective relative clause Finally, we have plural count noun phrases restricted by a collective relative clause: 'the numbers whose product is larger than 25', 'the specialists competing for the same jobs', 'the authors of multi-volume classics in logic'.

Oliver and Smiley[25] have called this kind of terms 'plurally exhaustive descriptions'. Plurally exhaustive descriptions are interpreted as picking out all the objects that jointly satisfy the predicate in question. The examples above have also what Oliver and Smiley have called a 'plurally unique description interpretation', according to which the expressions would pick out the only joint satisfiers of the relevant predicate in one occasion. Under this reading, the terms would refer merely plurally.

4 Indispensability of higher-level plurals

In this final section, I will show that statements involving higher-level plurals are not in general reducible to statements which only involve singular or plural terms. Recall that the main motivation for doing this is given by the assumption that it is the ineliminable expressions of our language which reveal what the logical form and the ontological commitments of the language are. In the philosophical debate around plurals, this is often common ground.

In a certain sense, it is always possible in principle to turn any expression of a certain language into an expression of another language by using an adequate translation scheme. However, our claim is that there are HLP expressions whose plural or singular paraphrases are not adequate, in the sense that they involve too radical a departure from the original expression and its ordinary interpretation.

We will assume that the following minimal requirement for difference in sentence-meaning holds: if two sentences have different epistemic status

---

[25] In Oliver and Smiley (2013, chapter 8).

(i.e. a speaker may believe one and not the other), then they have different meanings. Difference in epistemic status may arise for different reasons. The examples I will give fall within one of these two categories:

(i) The paraphrases have (very) different lengths.

(ii) The paraphrases involve different predicates.

An example falling both within (i) and (ii) is given by the sentence mentioned above: 'These people, those people and these other people played against each other in a 3-way game'.

It is easy to check that this proposition cannot be expressed in terms of two-way games, that is, as something like 'These people played against those other people, and the latter played against these other people...'. However, perhaps it could be expressed as an extremely long disjunction of sentences, each of which being a long conjunction describing a possible role played by each of the players. In this case, it is quite clear that the sentences would have different epistemic status. The second sentence would be extremely long and convoluted. Moreover, note that this example would also fall within the second category of examples given above: the new paraphrase would involve loads of new predicates describing the way in which each of the players participates in the game.

Another example is given by 'Some couples admire only one another', a HLP version of the Geach-Kaplan sentence (if 'couple' is a pseudo-singular). It seems that the only way of paraphrasing the HLP quantifier 'some couples' away is as follows: 'There is a set such that if a couple belongs to it, then, if another couple belongs to it, then the former admires the latter, and any couple which is admired by a couple that belongs to it belongs to it.'[26] However, this paraphrase appeals crucially to sets; whereas someone might be perfectly familiar with the relations of being a couple and admiring one another, they may lack a good command of the logical operators and quantifiers employed in the new sentence and even of the notion of set.

Finally, sentences involving reference to an infinite number of objects, like 'The twin primes decrease in density', provide good instances of (ii). That's because the impossibility for giving infinitely long lists of terms will force a paraphrase which resorts to new predicates.

***

---

[26] See Oliver and Smiley (2013, p. 39).

Making Sense of Higher-Level Plurals

In this article, I have made a case for a characterization of HLP terms as being one particular sort of atemporal/non-modal intensional terms. Terms are intensional in this sense iff not only their denotation, but also some relation or property enters into the contribution they make to the truth-conditions of statements involving them. Define 'pluralism' as the view that only appeal to objects from the first-order domain is needed in order to state the truth-conditional contribution of referring terms. Then, although HLP reference is not a form of singular reference, our proposal is not a pluralist one, since one needs to appeal to something other than the individuals picked out in order to lay down adequate truth-conditions. This does not necessarily imply that these additional elements are to be counted amongst the ontological commitments of speakers, since we can distinguish the fact that such and such need to be appealed to in the truth-conditions from the further conclusion that one is thereby committed to the existence of such and such.

Although the abandoning of pluralism may seem to be an unwelcome consequence of our account, we submit it is an inevitable one. Our hypothesis is that any tenable characterization of the higher-level plural will have to give up pluralism in the sense described above. However, we believe that our proposal, which takes higher-level plural reference to be an idiosyncratic linguistic phenomenon and demands it to be distinguished from mere singular and plural reference, leaves the door open to the nominalist. We are left with a trilemma: (i) drop pluralism while providing an argument for ontological innocence, (ii) give up ontological innocence altogether or (iii) resist the motivations for going up the hierarchy of higher-level plurals.

References

Ben-Yami, H. (2013). Higher-level plurals versus articulated reference, and an elaboration of salva veritate. *Dialectica*, *67*(1), 81–102.
Boccuni, F. (2013). Plural logicism. *Erkenntnis*, *78*(5), 1051–1067.
Boolos, G. (1984). To be is to be a value of a variable (or to be some values of some variables). *Journal of Philosophy*, *81*(8), 430–449.
Boolos, G. (1985). Nominalist platonism. *Philosophical Review*, *94*(3), 327–344.
Florio, S., & Nicolas, D. (2015). Plural logic and sensitivity to order. *Australasian Journal of Philosophy*, *93*(3), 444–464.
Hazen, A. P. (1997). Relations in Lewis' framework without atoms. *Analysis*, *57*(4), 243–248.

Hossack, K. (2000). Plurals and complexes. *British Journal for the Philosophy of Science, 51*(3), 411–443.
Landman, F. (1989a). Groups, I. *Linguistics and Philosophy, 12*(5), 559–605.
Landman, F. (1989b). Groups, II. *Linguistics and Philosophy, 12*(6), 723–744.
Lewis, D. (1991). *Parts of Classes*. Blackwell.
Link, G. (1984). Hydras: On the logic of relative clause constructions with multiple heads. In F. Landman & F. Veltman (Eds.), *Varieties of Formal Semantics: Proceedings of the Fourth Amsterdam Colloquium*. Foris Publications.
Linnebo, O. (2008). Plural quantification. In E. N. Zalta (Ed.), *Stanford Encyclopedia of Philosophy*. The Metaphysics Research Lab, Stanford University.
Linnebo, O., & Nicolas, D. (2008). Superplurals in English. *Analysis*(68), 186–197.
McKay, T. J. (2006). *Plural Predication*. Oxford University Press.
Moltmann, F. (2016). Plural reference and reference to a plurality. In M. Carrara, A. Arapinis, & F. Moltmann (Eds.), *Unity and Plurality. Philosophy, Logic, and Semantics*. Oxford University Press.
Oliver, A., & Smiley, T. (2005). Plural descriptions and many-valued functions. *Mind, 114*(456), 1039–1068.
Oliver, A., & Smiley, T. (2013). *Plural Logic*. Oxford University Press.
Rayo, A. (2006). Beyond plurals. In A. Rayo & G. Uzquiano (Eds.), *Absolute Generality* (pp. 220–254). Oxford University Press.
Russell, B. (1903). *The Principles of Mathematics*. Cambridge University Press.
Uzquiano, G. (2004). Plurals and simples. *The Monist, 87*(3), 429–451.

Berta Grimau
University of Glasgow
United Kingdom
E-mail: b.grimau.1@research.gla.ac.uk

When Structural Principles Hold Merely Locally

Ulf Hlobil[1]

Abstract: In substructural logics, structural principles may hold in some fragments of a consequence relation without holding globally. I look at this phenomenon in my preferred substructural logic, in which Weakening and Cut fail but which is supra-intuitionistic. I introduce object language operators that keep track of the admissibility of Weakening and of intuitionistic implications. I end with some ideas about local transitivity.

Keywords: Structural Principles, Nonmonotonic Logic, Nontransitive Logic, Logical Expressivism, Inferentialism, Atomic Systems

1 Introduction

Consequence relations are traditionally thought to obey the following—so-called structural—principles:[2]

WEAKENING   If $\Gamma \vdash A$ and $\Delta \supseteq \Gamma$, then $\Delta \vdash A$.

CUT   If $\Gamma \vdash A$ and $\Gamma, A \vdash B$, then $\Gamma \vdash B$.

CONTRACTION   If $\Gamma, A, A \vdash B$, then $\Gamma, A \vdash B$.

PERMUTATION   If $\Gamma, A, B, \Delta \vdash C$, then $\Gamma, B, A, \Delta \vdash C$.

REFLEXIVITY   If $A \in \Gamma$, then $\Gamma \vdash A$.

---

[1] Thanks goes to Robert Brandom, Daniel Kaplan, Shuhei Shimamura, Rea Golan and the audience at the Logica 2016 conference for invaluable comments, criticism and feedback. The work I am presenting here is part of my contribution to a joint project of Robert Brandom's research group. I also thank Graham Priest, who got me started on this (sub)project by asking me whether I had anything like Girard's shriek for my preferred logic.

[2] The first two are also known as (i) "monotonicity" or "monotony" or "MO," and (ii) "cumulative transitivity" or "CT." Reflexivity is also known as "identity" or "containment" or "CO."

Substructural logics are logics in which at least one of these five conditions fails. A nonmonotonic logic (denoted by "$\hspace{0.5pt}\mid\sim$") is one in which Weakening fails. Even in a nonmonotonic logic, however, there can be a set Θ such that for all $\Delta \supseteq \Theta$ we have $\Delta \mid\sim A$. In such a case, we may say that Θ implies A "monotonically" or "persistently." We can think of this as monotonicity holding locally at the sequent $\Theta \mid\sim A$.

Suppose we could say in the object language that A is a monotonic consequence of Θ, e.g., by having an operator $\mathfrak{M}$ such that Θ implies $\mathfrak{M}A$ iff Θ persistently implies A. We could then view monotonic logic as merely a restricted perspective on a shared topic: Monotonic logic looks at the turnstile as if it were always followed by a silent $\mathfrak{M}$. Such an operator lets us say what the monotonic logician is getting right and which consequences she simply ignores. Analogous ideas apply to the other structural rules.

When we look at the matter in this way, a new task for logicians becomes visible, namely the task of investigating and developing the expressive resources that are needed to think about local structural features. If logicians study and develop such expressive resources, they give us the expressive tools to be explicit about the structural features that we think our inferences obey and to disagree about whether we are right about this in a particular case. Acquiring these expressive resources will allow us to see structural logics as being blind to everything but a very special class of consequences and lacking the expressive power to see how special these consequences really are.

These ideas bear some resemblance, e.g., to Girard's (1987) recovery of intuitionistic logic within linear logic. In linear logic Weakening and Contraction fail. Girard introduces an operator "!" (shriek) that, in effect, allows applications of Weakening and Contraction. You can weaken any sequent with $!A$ on the left, and you can contract $!A\,!A$ to $!A$ on the left. With this operator in hand, Girard can define a translation function, θ, such that $A_1, ..., A_n \vdash_{int} B$ holds in intuitionistic logic just in case $!\theta(A_1), ..., !\theta(A_n) \vdash_{linear} \theta(B)$ holds in linear logic. Thus, we have an operator that allows us to use the structural rules that distinguish linear logic from intuitionistic logic in a controlled fashion, and we can thus use the operator to recover intuitionistic logic.

There are three main differences between Girard's recovery of intuitionistic logic and my aims in the remainder of this paper. First and most importantly, I am looking for ways of making something explicit that is there anyway. I start with a material consequence relation over atomic sentences, and I want to keep track of the structural principles that hold locally in this material consequence relation and its logical extension. I don't want to simply create a new part of the language in which certain structural principles hold by introducing a new operator. That is what Girard does by allowing us to weaken any sequent with !A irrespective of the weakening behavior of the sequent before the introduction of shriek. In contrast to this, I want to find ways to express facts about where structural principles hold locally even before we introduce any new expressions. Second, the structural principles with which I will be concerned are Weakening and Cut, rather than Weakening and Contraction. Third, the substructural logic I shall work with is supra-intuitionistic. Hence, the problem of "recovering" intuitionistic logic is really just the problem of singling out all and only the implications that this logic shares with intuitionistic logic.

The paper is structured as follows: In Section 2, I introduce the substructural logic with which I will be working. I then show how to introduce an operator that says, in the object language, that Weakening holds at a sequent (Section 3). In Section 4, I introduce an object language expression that allows us to keep track of intuitionistic implications. I end, in Section 5, with some ideas about the local admissibility of Cut.

2 A Nonmonotonic and nontransitive Logic

In this section, I introduce the reader to the substructural logic within which I want to keep track of the admissibility of some structural principles and of intuitionistic logic. I will do so very briefly, as I have laid out a very similar logic in more detail elsewhere (Hlobil, 2016).

2.1 Philosophical background

The logic that I will be concerned with is motivated by semantic inferentialism, i.e. the view (very roughly) that the meaning of a sentence is settled once it is settled what follows from it, what it follows from, and what is incompatible with it.[3] For our current purposes, we can think of inferentialism as the view that the meaning of a sentence is given by its place in a consequence relation and incoherence property defined over the language to which it belongs.

---

[3] I am ignoring here how causal relations—especially perception and action—play a role in the fixation of meaning. Of course, a plausible semantic inferentialism must ultimately include what Sellars (1954) calls "language-entry transitions" and "language-exit transitions."

Ulf Hlobil

Inferentialism is a general claim about content, not just a claim about the meaning of logical vocabulary (Brandom, 1994; Peregrin, 2014). The claim applies to atomic sentences just as much as it applies to logically complex sentences. Now, formal consequence relations are closed under uniform substitutions of atomic sentences. In other words, formal consequence relations assign to all atomic sentences symmetrical inferential roles. Any set of atoms implies all its members and no other atoms. These only minimally different inferential roles can hardly suffice to confer genuinely distinct meanings onto the different atomic sentences. So if we want to capture the inferential roles of atomic sentences, we must reject the idea that all consequence relations are formal. Rather, we must allow that a set of atomic sentences implies another atomic sentence that is not in the set.

Once we accept such material implications, it becomes natural to think of logic as extending a given material consequence relation (and incoherence property) to logically complex sentences.[4] Logic introduces logical vocabulary into an atomic language whose sentences already stand in rich inferential relations.

Now, it seems that the inferential and incompatibility relations between atomic sentences are virtually always defeasible. After all, we can normally infer that the streets are wet from the fact that it is raining, but it can happen that the authorities have covered up the streets to protect it from the rain, etc. Hence, the consequence relation over the atomic language with which we begin must be nonmonotonic. Given that a logic should extend this relation in a conservative fashion, the consequence relation that results from introducing logical vocabulary must also be nonmonotonic.

At this point a second motivating idea behind this logic becomes relevant: logical expressivism (Brandom, 2008). According to logical expressivism, it is the expressive job of logical vocabulary to allow us to express claims about the inferential and incoherence roles of non-logical expressions (where the "about" must not be understood in a representationalist fashion). The paradigm case is the conditional. According to logical expressivism, the conditional lets us claim that the antecedent implies the consequent (in a given context). If we spell this out in a non-representationalist way, we should say something like this: You have reasons to assert a conditional just in case these reasons together with the antecedent are reasons to assert the

---

[4]Similar ideas have recently been discussed under the heading of "atomic systems" or "atomic bases" and extensions thereof (Sandqvist, 2015). The present approach differs from familiar approaches in that it doesn't enforce transitivity in the atomic base. Thus, we reject what is sometimes called "definitional reflection."

consequent. In other words, a deduction-detachment theorem (DDT) holds:

DDT $\qquad \Gamma \mathrel{\smash{\vert\!\sim}} A \to B$ iff $\Gamma, A \mathrel{\smash{\vert\!\sim}} B$.

This is a substantive constraint on the inferential behavior of the conditional. It is justified by the claim that it is the job of the conditional to let us express what follows from what. The conditional makes explicit the consequence relation. Together with Reflexivity, DDT implies that our logic must be nontransitive. For a reflexive consequence relation in which DDT holds and that obeys Cut must be monotonic.[5]

The upshot of all this is that if you accept the inferentialism and expressivism I am endorsing, then you need rules for introducing logical connectives that extend a nonmonotonic, material consequence relation in such a way that the resulting consequence relation is not only nonmonotonic but also nontransitive and obeys DDT. Furthermore, the rules governing the logical connectives should be plausible, in the sense that they can naturally be thought to capture the meanings of the logical connectives.

2.2 A sequent calculus

In order to construct a system that has the properties just mentioned, we start with a consequence relation and incoherence property over a finite atomic language, $\mathfrak{L}_{0-}$. In order to represent the consequence relation and the incoherence property in a unified fashion, we first introduce the symbol "$\perp$," which functions like an empty right side in Gentzen's sequent calculus for intuitionistic logic in that it cannot be embedded and can only occur on the right. Let $\mathfrak{L}_0 = \mathfrak{L}_{0-} \cup \{\perp\}$. We define the relation $\mathrel{\smash{\vert\!\sim}}_0 \subseteq \mathcal{P}(\mathfrak{L}_{0-}) \times \mathfrak{L}_0$ by saying that $\Gamma \mathrel{\smash{\vert\!\sim}}_0 p$ holds just in case either (i) Γ is materially incoherent and $p = \perp$ or (ii) Γ materially implies p. We require that $\mathrel{\smash{\vert\!\sim}}_0$ obeys Reflexivity and that if $\forall \Delta \subseteq \mathfrak{L}_{0-} \, (\Gamma, \Delta \mathrel{\smash{\vert\!\sim}}_0 \perp)$, then $\Gamma \mathrel{\smash{\vert\!\sim}} p$ for all atoms, p, in $\mathfrak{L}_{0-}$. The second requirement, which I call *ex falso fixo quodlibet* (ExFF), is a restricted version of explosion, which applies only if a set is not only incoherent but is, as I shall say, persistently incoherent, i.e., all of its supersets are incoherent.

We now give a sequent calculus formulation of the extended consequence relation. I call the resulting system NM (for Non-Monotonic). Our

---

[5] A mixed-context version of Cut together with Reflexivity implies Weakening even without DDT. After all, suppose that $\Gamma \mathrel{\smash{\vert\!\sim}} A$. By Reflexivity, $\Gamma, \Delta, A \mathrel{\smash{\vert\!\sim}} A$. By mixed-context Cut, $\Gamma, \Delta \mathrel{\smash{\vert\!\sim}} A$. The deduction-detachment theorem is just needed to show that the same holds for shared-context versions of Cut.

axioms are all the sequents in the underlying material consequence relation, $\mathrel{\mid\hspace{-.3em}\sim}_0$. Moreover, we define not only a single snake-turnstile but one for every $X \subseteq \mathcal{P}(\mathcal{L}_{0-})$. The idea is that $\Gamma \mathrel{\mid\hspace{-.3em}\sim}^{\uparrow X} p$ holds just in case $\forall \Delta \in X\, (\Delta, \Gamma \mathrel{\mid\hspace{-.3em}\sim} p)$. By convention we can write $\Gamma \mathrel{\mid\hspace{-.3em}\sim}^{\uparrow \mathcal{P}(\mathcal{L}_{0-})} p$ as $\Gamma \mathrel{\mid\hspace{-.3em}\sim}^{\uparrow} p$. Our axioms are:

Axioms of NM:

Ax1: If $\Gamma \mathrel{\mid\hspace{-.3em}\sim}_0 p$, then $\Gamma \mathrel{\mid\hspace{-.3em}\sim} p$ is an axiom.

Ax2: If $X \subseteq \mathcal{P}(\mathcal{L}_{0-})$ and $\forall \Delta \in X\, (\Delta, \Gamma \mathrel{\mid\hspace{-.3em}\sim}_0 p)$, then $\Gamma \mathrel{\mid\hspace{-.3em}\sim}^{\uparrow X} p$ is an axiom.

We define the extended consequence relation, $\mathrel{\mid\hspace{-.3em}\sim}$, as the smallest relation that closes these axioms under the following rules.

Rules of NM:

Notation: The square brackets mean that what is inside the brackets is optional or, if preceded by a "/", that it is an alternative. The rules are systematically ambiguous: unless a rule explicitly contains a sequent with an upward arrow, the rule applies if we uniformly replace the plain snake-turnstile by a snake-turnstile with an upward arrow with a particular set of atomic sets throughout the rule. In LC, we can uniformly substitute the right top-sequent and the root-sequent for ones with the same upward arrow. In PushUp, the top sequent cannot have an upward arrow.

$$\frac{\Gamma, A \mathrel{\mid\hspace{-.3em}\sim} B}{\Gamma \mathrel{\mid\hspace{-.3em}\sim} A \to B} \text{ RC} \qquad \frac{\Gamma \mathrel{\mid\hspace{-.3em}\sim}^{\uparrow} A \quad \Gamma, B \mathrel{\mid\hspace{-.3em}\sim} C}{\Gamma, A \to B \mathrel{\mid\hspace{-.3em}\sim} C} \text{ LC}$$

$$\frac{\Gamma, A \mathrel{\mid\hspace{-.3em}\sim} \bot}{\Gamma \mathrel{\mid\hspace{-.3em}\sim} \neg A} \text{ RN} \qquad \frac{\Gamma \mathrel{\mid\hspace{-.3em}\sim} A}{\Gamma, \neg A \mathrel{\mid\hspace{-.3em}\sim} \bot} \text{ LN}$$

$$\frac{\Gamma \mathrel{\mid\hspace{-.3em}\sim} A \quad \Gamma \mathrel{\mid\hspace{-.3em}\sim} B}{\Gamma \mathrel{\mid\hspace{-.3em}\sim} A \& B} \text{ R\&} \qquad \frac{\Gamma, A, B \mathrel{\mid\hspace{-.3em}\sim} C}{\Gamma, A \& B \mathrel{\mid\hspace{-.3em}\sim} C} \text{ L\&}$$

When Structural Principles Hold Merely Locally

$$\frac{\Gamma \mathrel{\vert\!\sim} A\,[/B]}{\Gamma \mathrel{\vert\!\sim} A \vee B}\,\text{Rv} \qquad \frac{\Gamma, A \mathrel{\vert\!\sim} C \quad \Gamma, B \mathrel{\vert\!\sim} C}{\Gamma, A \vee B, [A], [B] \mathrel{\vert\!\sim} C}\,\text{Lv}$$

$$\frac{\Gamma \mathrel{\vert\!\sim} \uparrow A}{\Gamma, B \to C \mathrel{\vert\!\sim} {}^{[\uparrow]}A}\,\text{pCW} \qquad \frac{\Gamma \mathrel{\vert\!\sim} \uparrow A}{\Gamma, \neg B \mathrel{\vert\!\sim} {}^{[\uparrow]}A}\,\text{pNW}$$

$$\frac{\Gamma \mathrel{\vert\!\sim} \uparrow^X A \quad \Gamma \mathrel{\vert\!\sim} \uparrow^Y A}{\Gamma \mathrel{\vert\!\sim} \uparrow^{X \cup Y} A}\,\text{UN} \qquad \frac{p_1...p_n, \Gamma \mathrel{\vert\!\sim} A}{\Gamma \mathrel{\vert\!\sim} \uparrow^{\{\{p_1...p_n\}\}} A}\,\text{PushUp}$$

$$\frac{\Gamma \mathrel{\vert\!\sim} \uparrow \bot}{\Gamma \mathrel{\vert\!\sim} {}^{[\uparrow]}A}\,\text{ExFF}$$

This gives us a nonmonotonic consequence relation over a language, $\mathfrak{L}_-$, that contains conditionals, negations, conjunctions, and disjunctions. All but the last five rules should look more or less familiar from standard sequent calculi. The differences to familiar versions of the connective rules—e.g. that LC requires a persistent sequent as the left premise-sequent—are all such that they cannot even be formulated in a monotonic setting. Note that we are building permutation and contraction in by working with sets on the left.

Among the unfamiliar rules, pCW and pNW allow us to weaken with conditionals and negations respectively if we can weaken with arbitrary sets of atoms. And UN and PushUp allow us to form new upward arrows. ExFF can be viewed as a restricted version of Gentzen's right weakening rule.

Since our axioms contain no logically complex sentences, Reflexivity isn't true by stipulation in NM. It can be shown, however, that the rules preserve Reflexivity. Moreover, it can be shown that the extension is conservative, i.e., if $\Gamma \subseteq \mathfrak{L}_{0-}$ and $p \in \mathfrak{L}_0$, then $\Gamma \mathrel{\vert\!\sim} p$ iff $\Gamma \mathrel{\vert\!\sim}_0 p$. This ensures not only that our connectives don't trivialize the consequence relation in a "tonk-like" fashion; it also shows that they don't change the facts that they are meant to make explicit and that they don't force monotonicity. Furthermore, the connectives are well-behaved. The conditional obeys the deduction-detachment theorem. A negation is implied just in case the premises together with the negated sentences are incoherent. A conjunction is implied just in case both conjuncts are implied. And a disjunction is implied just in case at least one of the disjuncts is implied.

In sum, the NM system gives us a surprisingly well-behaved nonmonotonic logic that conservatively extends a nonmonotonic material consequence relation. Moreover, NM is supra-intuitionistic.

Theorem 1 *NM is supra-intuitionistic, i.e., if* $\Gamma \vdash A$ *holds in intuitionistic logic, then* $\Gamma \mid\!\sim A$ *holds in NM.*

Proof. Due to limitations of space, I just give a sketch of the proof. First, we notice that Cut and Weakening can both be eliminated in Gentzen's sequent calculus formulation of intuitionistic logic, LJ, if we are allowed to use as our axioms atomic instances of Reflexivity, i.e., if the leaves of the proof-trees can be of the form $p_0, ..., p_n \vdash p_n$.[6] Next, we notice that Weakening is admissible in NM proof-trees all of whose leaves are instances of Reflexivity. This is because we can weaken such leaves with arbitrary atoms. Given these facts, we can show that every proof-tree that uses the rules of LJ but not Cut or Weakening and whose leaves are instances of Reflexivity (with context) can be translated into NM because we can translate all the remaining rules of LJ into NM.

The translation from LJ works as follows: KA and Cut are eliminated. KS translates into ExFF after making the premise-sequent persistent via PushUp and UN. W and C are given by working with sets. FES translates into RC. FEA can be derived from LC after making the left premise-sequent persistent. The translations of the rules for conjunction, disjunction and negation are similarly straightforward. ∎

Although NM is supra-intuitionistic, Weakening and Cut are not globally admissible in NM. Outside of the intuitionistic fragment of the consequence relation they may fail. Can we introduce operators that allow us to keep track—in the object language—of the admissibility of Weakening and Cut? And can we introduce an operator that allows us to make explicit, in the object language, that an implication of NM also holds in intuitionistic logic? Those are the questions that I will address in the remainder of this paper.

---

[6]Here, as elsewhere in this paper, I am assuming that we are only concerned with finite premise sets.

3 Making monotonicity explicit

Weakening is not always admissible in NM. However, it is sometimes admissible; e.g., it is admissible to weaken instances of Reflexivity. We want to introduce an operator that allows us to keep track of where Weakening is admissible. In order to do this, it is helpful to think of monotonicity as analogous to a modality. We can think of premise sets as points of evaluation. From each premise set all and only its supersets are accessible. A sentence is verified at a premise set just in case the premise set implies the sentence. A set, Γ, monotonically implies a sentence, A, just in case all points of evaluation accessible from Γ verify A. Looking at monotonicity in this way, it is natural to understand it as a kind of necessity. Hence, it is natural to express it in a similar way in the object language, namely by introducing an operator, $\Box$, such that $\Gamma \mathrel{\mid\!\sim} \Box A$ just in case $\forall \Delta \subseteq \mathfrak{L}_- \ (\Delta, \Gamma \mathrel{\mid\!\sim} A)$.

Fortunately, such an operator can easily be introduced. That is in part because it can be shown that weakening with arbitrary sets of sentences is admissible for a particular sequent just in case the sequent can be weakened with arbitrary sets of atomic sentences. That is, $\forall \Delta \subseteq \mathfrak{L}_- \ (\Delta, \Gamma \mathrel{\mid\!\sim} A)$ iff $\forall \Delta_0 \subseteq \mathfrak{L}_{0-} \ (\Delta_0, \Gamma \mathrel{\mid\!\sim} A)$. Hence, it suffices to have an operator that keeps track of when we can weaken with arbitrary sets of atoms. But we already have a device that does that in the meta-language, namely the upward arrow. All we need to do is to find rules that use the upward arrow in the right way to introduce the box. It can be shown that the following rules do that:

$$\frac{\Gamma \mathrel{\mid\!\sim} {\uparrow} A}{\Gamma \mathrel{\mid\!\sim} {}^{[\uparrow]} \Box A} \text{ RB} \qquad\qquad \frac{\Gamma, A \mathrel{\mid\!\sim} B}{\Gamma, \Box A \mathrel{\mid\!\sim} B} \text{ LB}$$

In other words, if we introduce the box into NM by adding these two rules, then the box obeys the following principle.[7]

BOX $\quad \Gamma \mathrel{\mid\!\sim} \Box A$ iff $\forall \Delta \subseteq \mathfrak{L}_- \ (\Gamma, \Delta \mathrel{\mid\!\sim} A)$.

This means that the box makes explicit where monotonicity holds locally. We can thus keep track of the local admissibility of Weakening in the object language.

---

[7]I omit the proof for reasons of space. The strategy is to show that $\Gamma \mathrel{\mid\!\sim} \Box A$ iff $\Gamma \mathrel{\mid\!\sim}^{\uparrow} A$ iff $\forall \Delta \subseteq \mathfrak{L}_- \ (\Gamma, \Delta \mathrel{\mid\!\sim} A)$. The left-to-right direction of the first biconditional is shown via induction on proof-height. The left-to-right direction of the second biconditional is shown by induction on the complexity of A. The other direction is straightforward via PushUp, UN, and RB.

4 Making intuitionism explicit

Keeping control over Weakening turned out to be relatively easy in NM. Unfortunately, keeping control over Cut is more complicated. However, it is easy to introduce an operator that keeps track of intuitionistic implications. Hence, I will start with that.

In this section, I will introduce an operator, ⊡ (pronounced "I-box"), that allows us to keep track of intuitionistic logic. More precisely, $A_0, ..., A_n \vdash B$ holds in intuitionistic logic just in case $⊡A_0, ..., ⊡A_n \mathrel{|\!\sim} ⊡B$ is derivable in NM plus the I-box, a system which I will call NM⊡. Thus, the I-box is our analog to Girard's shriek.

For ease of exposition, let us write ⊡Γ for a set of sentences that is like Γ except that every sentence is prefaced by an I-box.

How can we introduce the I-box? It can be shown that the sequents of intuitionistic logic are all and only the sequents of NM that can be derived in proof-trees all of whose leaves are atomic instances of Reflexivity. This is easy to see if we realize that the rules of NM and the rules of LJ are intertranslatable if Weakening and Cut hold for the leaves of proof-trees. For in this case, we can push all applications of Weakening and Cut into the leaves. And the rest of the translation is straightforward.

Since intuitionistic sequents are exactly those sequents that can be derived in NM from leaves that are all instances of Reflexivity, we can keep track of intuitionistic logic by keeping track of such sequents. The obvious way to do that is to mark atomic instances of Reflexivity and then allow the rules to transfer the mark from the premise sequent(s) to the conclusion sequent. We do this in the meta-language by introducing a new kind of snake-turnstile: $\mathrel{|\!\sim}_{\mathfrak{I}}$. Since this turnstile can be combined with the upward arrow turnstile to yield $\mathrel{|\!\sim}_{\mathfrak{I}}^{\uparrow X}$ via PushUp and UN, we double the number of our turnstiles. We introduce $\mathrel{|\!\sim}_{\mathfrak{I}}$ at the ground level by adding the following axioms:

Additional Axioms of NM⊡

Ax3: If $p \in \Gamma \subseteq \mathfrak{L}_{0-}$, then $\Gamma \mathrel{|\!\sim}_{\mathfrak{I}} p$ is an axiom.

We allow all of the rules of NM to be applied to sequents with our new turnstiles, i.e. to sequents of the form $\Gamma \mathrel{|\!\sim}_{\mathfrak{I}}^{[\uparrow X]} A$. Next we need to introduce ⊡ in such a way that it expresses our new turnstile in much the same way in which the box expresses the plain upward arrow. The following rules do that:

When Structural Principles Hold Merely Locally

Additional Rules of NM⊡

$$\frac{\Gamma, A \mathrel{\mid\!\sim}_{\supset} B}{\Gamma, \boxdot A \mathrel{\mid\!\sim}_{[\supset]} B} \text{ LIB} \qquad \frac{\Gamma, A \mathrel{\mid\!\sim}_{\supset} B}{\Gamma, A \mathrel{\mid\!\sim}_{[\supset]} \boxdot B} \text{ RIB}$$

It is easy to show that NM⊡ preserves Reflexivity and is a conservative extension of any nonmonotonic material consequence relation that obeys ExFF and Reflexivity.[8] Most importantly, however, the I-box lets us keep track—in the object language—of intuitionistic logic.

Theorem 2 $\boxdot A_0, ..., \boxdot A_n \mathrel{\mid\!\sim} \boxdot B$ *holds in NM⊡, where* $A_0, ..., A_n$ *and* B *don't contain any I-boxes, iff* $A_0, ..., A_n \vdash B$ *holds in intuitionistic logic.*

Proof. ($\Leftarrow$) If $A_0, ..., A_n \vdash B$ is derivable in intuitionistic logic, it can be derived in Gentzen's LJ (Gentzen, 1934). Given a proof-tree in LJ, we can translate it into a proof-tree in NM. In order to do this, we push Cut and Weakening into the leaves. Then we translate all rule applications in accordance with the translation manual given above. All the leaves of the resulting tree are instances of Reflexivity, which are derivable in NM. Moving to NM⊡, we use $\mathrel{\mid\!\sim}_{\supset}^{[\uparrow X]}$-turnstiles in the leaves. The root of the resulting tree is $A_0, ..., A_n \mathrel{\mid\!\sim}_{\supset} B$. By RIB and repeated applications of LIB, we get $\boxdot A_0, ..., \boxdot A_n \mathrel{\mid\!\sim} \boxdot B$.

($\Rightarrow$) Suppose $\boxdot A_0, ..., \boxdot A_n \mathrel{\mid\!\sim} \boxdot B$. This must come by LIB, RIB, or ExFF. If it comes by LIB, the premise is $\boxdot A_0, ..., \boxdot A_{n-1}, A_n \mathrel{\mid\!\sim}_{\supset} \boxdot B$. We can show by induction on proof-height that having at least one I-box on the left as a principal operator suffices to ensure that $A_0, ..., A_n \mathrel{\mid\!\sim}_{\supset} \boxdot B$. Moreover, $\boxdot B$ on the right can only come from ExFF or RIB. In either case, we can get B instead. Hence, $A_0, ..., A_n \mathrel{\mid\!\sim}_{\supset} B$. The same reasoning applies to the cases of RIB and ExFF. We can also show that if there are no I-boxes in $A_0, ..., A_n, B$, then we can derive $A_0, ..., A_n \mathrel{\mid\!\sim}_{\supset} B$ just in case $A_0, ..., A_n \vdash B$ holds in intuitionistic logic. To show this, we simply reverse the translation scheme provided above. ∎

With this result in hand, we can say that the intuitionist is reasoning as if all sentences were prefaced by a silent "⊡". From the perspective of NM⊡, the problem with intuitionistic logic, is that it is blind to the rich structure

---

[8]For reasons of space, I am omitting the proofs. They work similarly to those for NM, which are sketched in (Hlobil, 2016).

consequence and incoherence that does not derive from Reflexivity. The I-box allows us to restrict our vision 'artificially,' as it were, to single out all and only the sequents that the intuitionist can see.

There is a general strategy behind the techniques that we have used in the last two sections: (a) We want to keep track of a local structural feature of our consequence relation within the object language; this structural feature was the admissibility of Weakening in the first case and the derivability in intuitionistic logic in the second case. (b) We do this by first introducing a new kind of turnstile that filters out all and only the sequents at which the feature in question holds. (c) Next we introduce an operator that marks the result of this filtration in the object language. The result is that we have an object language expression that allows us to mark precisely the region of our logic within which the structural feature in question holds.

5   Making some local transitivity explicit

Turning to transitivity, everything would be plain sailing if we were able to keep track of when Cut is admissible in a way that is analogous to what we did for Weakening. However, keeping track of local Cut admissibility turns out to be far more difficult than keeping track of local Weakening admissibility or intuitionistic logic.

The problem is that the admissibility of Cut is not preserved by our rules. To see this, let us define $Cn_0(\Gamma)$ as the set of atomic consequences of Γ and $Cn(\Gamma)$ as the consequences of Γ in NM. Suppose Γ obeys Cut in the base consequence relation, i.e., if $\Delta \subseteq Cn_0(\Gamma)$, then $Cn_0(\Gamma) \supseteq Cn_0(\Gamma \cup \Delta)$. Now, suppose that $\Gamma \not\mid\!\sim_0 p$ and $\Gamma, p \mid\!\sim_0 \bot$ and $\Gamma \not\mid\!\sim_0 \bot$. So, $\Gamma \mid\!\sim \neg p$. However, suppose also that $\Gamma \not\mid\!\sim_0 q$ and $\Gamma, p, q \mid\!\sim_0 \bot$, which doesn't violate the assumption that Γ obeys Cut because p is not a consequence of Γ. By RN, $\Gamma, q \mid\!\sim \neg p$. Hence, $q \in Cn(\Gamma)$ and $\neg p \notin Cn(\Gamma)$ but $\neg p \in Cn(\Gamma \cup \{q\})$.

Unfortunately, I don't know how to solve this problem in full generality. Here I can only make a first step towards a solution. The step consists in a generalization of the idea behind the I-box. In the case of the I-box, we exploited the fact that the atomic instances of Reflexivity are closed under Cut and Weakening. That makes Cut and Weakening admissible in proof-trees all of whose leaves are instances of Reflexivity. The general lesson that we should learn from this special case is that a set of sequents is closed under Cut if they can all be derived in proof-trees all of whose leaves belong

When Structural Principles Hold Merely Locally

to a set of atomic sequents that are closed under Cut and Weakening. Let us define sets of atomic sequents that have this property:

Definition 1 *A set, $\mathfrak{S}$, of atomic sequents has the T+W property iff (a) if $\Gamma \mathrel{\mid\!\sim}_0 p \in \mathfrak{S}$, then $\forall \Delta \subseteq \mathfrak{L}_0 \, (\Delta, \Gamma \mathrel{\mid\!\sim}_0 p \in \mathfrak{S})$; and (b) if $\Gamma \mathrel{\mid\!\sim}_0 p \in \mathfrak{S}$ and $\Gamma, p \mathrel{\mid\!\sim}_0 q \in \mathfrak{S}$, then $\Gamma, \mathrel{\mid\!\sim}_0 q \in \mathfrak{S}$.*

Let us focus on maximal sets that have the T+W property, i.e., T+W sets such that no proper superset of them has the property. Let us enumerate these sets and call them $\mathfrak{S}_0, ..., \mathfrak{S}_n$. We can now treat each of these sets like we have treated the atomic instances of Reflexivity when we introduced the I-box. For each $\mathfrak{S}_i$ we can introduce a new kind of turnstile, $\mathrel{\mid\!\sim}_i^{[\uparrow X]}$, and a new operator, $\mathbb{T}^i$ (pronounced "T-box", for transitivity-box). For the new turnstiles, we add axioms with those turnstiles.

Ax3: If $\Gamma \mathrel{\mid\!\sim}_0 p$ is in $\mathfrak{S}_i$, then $\Gamma \mathrel{\mid\!\sim}_i p$ is an axiom.

We apply the usual NM rules to our new kinds of sequents if they are of the same kind, i.e. if the same subscript on the turnstile occurs in all the sequents in the application of the rule. Moreover, we add rules that introduce an operator $\mathbb{T}^i$ for each $\mathfrak{S}_i$.

$$\frac{\Gamma, A \mathrel{\mid\!\sim}_i B}{\Gamma, \mathbb{T}^i A \mathrel{\mid\!\sim}_{[i]} B} \text{LTB} \qquad \frac{\Gamma \mathrel{\mid\!\sim}_i A}{\Gamma \mathrel{\mid\!\sim}_{[i]} \mathbb{T}^i A} \text{RTB}$$

Let us call the resulting system NM$\mathbb{T}$. In this system we can keep track of some regions in which Cut is admissible, namely regions that are extensions of maximal T+W sets in the base. More precisely, we can show the following:

Theorem 3 *Sequents of the form $\Gamma \mathrel{\mid\!\sim}_i A$ are closed under Cut.*

Proof. First, we note that in order to derive a sequent of the form $\Gamma \mathrel{\mid\!\sim}_i A$, all the sequents in the proof-tree must be of that form. The rest of the proof follows Gentzen's original Cut-elimination proof very closely. Just as in Gentzen, it is a double induction on rank and degree; and we can divide the cases in the way he did. In effect, we push Cut up into the leaves; and we know that Cut holds among the leaves. The proof is tedious but straightforward. ∎

So we have a way to mark off some regions in which Cut holds locally in the metalanguage of the sequent calculus. But we also have object language operators that allow us to keep track of these regions.

Theorem 4 *If we can derive $\mathbb{T}^i A_0, ..., \mathbb{T}^i A_m \mid\!\sim \mathbb{T}^i B$ in NM$\mathbb{T}$ and we can derive $\mathbb{T}^i A_0, ..., \mathbb{T}^i A_m, \mathbb{T}^i B \mid\!\sim \mathbb{T}^i C$, then $\mathbb{T}^i A_0, ..., \mathbb{T}^i A_m \mid\!\sim \mathbb{T}^i C$ also holds.*

Proof. It suffices to show that if $\mathbb{T}^i \alpha_0, ..., \mathbb{T}^i \alpha_k \mid\!\sim \mathbb{T}^i \beta$, then we also have $\alpha_0, ... \alpha_k \mid\!\sim_i \beta$. The proof is parallel to the one for the I-box. We argue by induction on proof-height that if $\mathbb{T}^i \alpha_0, ..., \mathbb{T}^i \alpha_k \mid\!\sim \mathbb{T}^i \beta$, this must come from $\alpha_0, ... \alpha_k \mid\!\sim_i \mathbb{T}^i \beta$. Then we argue that wherever $\mathbb{T}^i \beta$ was first introduced into the proof-tree, we can replace it with β. ∎

This technique gives us a family of object language operators each of which keeps track of a particular region of our consequence relation that is closed under Cut. The regions we keep track of are extensions of atomic sequents that are closed under Weakening and Cut. It may be worth mentioning that these regions behave intuitionistically, in the sense that they are equivalent to extensions of the underlying atomic sequents via the rules of Gentzen's LJ.

Let me point out some limitations of this technique. (1) It does not give us a way to say that a set, Γ, obeys Cut in the sense that $\Delta \subseteq Cn(\Gamma)$, then $Cn(\Gamma) \supseteq Cn(\Gamma \cup \Delta)$. After all, it can happen that $\Gamma \mid\!\sim_i B$ and $\Gamma, B \mid\!\sim A$ but not $\Gamma \mid\!\sim A$. (2) The technique does not allow us to keep track of regions that are closed under Cut but where Weakening fails. (3) There is no guarantee that the technique allows us to keep track of all the sets of sequents that are closed under Cut and Weakening. For some such regions may not be traceable to monotonic and transitive regions in the atomic consequence relation.

Despite these limitations, the T-boxes allow us to make explicit in the object language a particular class of cases in which Cut is admissible. From the perspective of NM$\mathbb{T}$, to insist that Cut holds globally is to insist that we should always think of our premises and conclusions as if they were prefaced by silent T-boxes (with the same superscript). In NM$\mathbb{T}$ we can talk about such implications without being blind to all other implications.

6 Conclusion

Let us take stock. The sequent calculus NM extends a nonmonotonic, non-transitive material consequence relation over atomic sentences to the language of propositional logic. The resulting consequence relation is supra-intuitionistic. We can introduce an operator, $\Box$, that keeps track of where monotonicity holds locally. Moreover, we can introduce another operator, $\boxdot$, that keeps track of intuitionistic logic within NM. Finally, we can introduce a family of operators, $\boxdot^i$, that keep track of some of the regions of our consequence relation that are closed under Cut and Weakening.

The results I have presented here are limited in various respects. In particular, we still need better ways to keep track of where Cut is admissible. If we can develop such techniques and do the same for the other structural principles, we will have a general framework in which what used to look like disagreements about the foundations of logic will emerge as disagreements about particular claims that can be formulated in a common logical framework.

References

Brandom, R. B. (1994). *Making It Explicit: Reasoning, Representing, and Discursive Commitment.* Cambridge, Mass.: Harvard University Press.

Brandom, R. B. (2008). *Between saying and doing: Towards an analytic pragmatism.* Oxford: Oxford University Press.

Gentzen, G. (1934). Untersuchungen über das logische Schließen: I. *Mathematische Zeitschrift, 39*(2), 176–210.

Girard, J.-Y. (1987). Linear logic. *Theoretical Computer Science, 50*(1), 1–101.

Hlobil, U. (2016). A nonmonotonic sequent calculus for inferentialist expressivists. In P. Arazim & M. Dančák (Eds.), *The Logica Yearbook 2015* (pp. 87–105). London: College Publications.

Peregrin, J. (2014). *Inferentialism: Why Rules Matter.* New York: Palgrave MacMillan.

Sandqvist, T. (2015). Base-extension semantics for intuitionistic sentential logic. *Logic Journal of the IGPL, 23*(5), 719–731.

Sellars, W. (1954). Some reflections on language games. *Philosophy of Science, 21*, 204–228.

Ulf Hlobil
Concordia University
Canada
E-mail: ulf.hlobil@concordia.ca

Four Remarks on Relations and Predication

HAYTHEM O. ISMAIL

Abstract: In four brief, but rather technical, remarks, this paper presents an account of relations and predication which purports to address several of the issues that have been the source of nagging problems over the years. An account of variably polyadic relations is motivated and laid out, with properties such as reflexivity, symmetry, and transitivity generalized from the classical dyadic case. Said account is then employed to give formal semantics for a language with a flexible and parsimonious predication apparatus.

Keywords: Relations, Predication, Knowledge Representation

1 Introduction

The classical account of relations and predication in predicate logic, though mostly successful and sufficiently intuitive, has raised a number of worries and objections over the years. In this preliminary report, I provide an account of relations and predication which is intended to sooth the worries and address the objections. My account is presented in four, rather technical, remarks. The first remark motivates the account by revisiting the disappointing aspects of the classical account. The second remark presents my account of relations as structured sets of structured tuples, referred to as *covered relations*. The third remark extends common relational properties (e.g., reflexivity, symmetry, and transitivity) to covered relations. The fourth, and final, remark addresses predication by outlining a formal language for representing and reasoning about covered relations.

2 Remark 1: On classical relations and predication

The classical view of relations can be stated thus:

CR. A relation R on a set D is a set of k-tuples of members of D ($R \subseteq D^k$) for some $k \in \mathbb{N}$, where k is referred to as the adicity of the relation.

Haythem O. Ismail

This view has been challenged by several philosophers over the years. Perhaps the main challenge is the one raised against the assumption (encapsulated in **(CR)**) that relations hold among their relata in some *order* (Diro, 2008; Fine, 2000; Johansson, 2011, 2014; Leo, 2008a, 2008b, for example). The problem with this assumption is that it entails that every relation has a converse relation, and the converse is, for non-symmetric relations, distinct from the relation if both are construed *à la* **(CR)**. But this raises a problem for the meta-physicist who cannot find an acceptable reason for making an ontological distinction between a relation and its, necessarily companying, converse. As a response to such difficulties, other accounts of relations have been proposed: the positionalist account views relations as sets of unordered, distinct *argument places* (see Fine, 2000; Leo, 2008a and particularly Gilmore, 2013); the anti-positionalist account takes relations to be networks of states of affairs linked by substitutions (see Fine, 2000; Leo, 2008a, 2008b).

But my primary interest here is the *logic* of relations. And while the ontological considerations highlighted above may, in the final analysis, prove indispensable to the practice of logic, the issues I address in this report are somewhat orthogonal to these considerations. Inasmuch as the focus is mainly on logic, we need only consider *formal models* of relations to serve as denotations of predicate symbols; said models are just what they are—models—and I by no means claim that the particular model I present here is a faithful account of the reality of relations. This being said, consider the classical account of predication:

CP1. Each predicate symbol p has a unique associated arity $a(p) \in \mathbb{N}$.[1]

CP2. Predication is the formal combination of a single predicate symbol p with a single sequence of $a(p)$, individual-denoting, terms. Classically, this is displayed as $p(t_1, \ldots, t_{a(p)})$, where t_i is a term (for every $1 \leq i \leq a(p)$).

CP3. A predicate symbol p denotes an $a(p)$-adic relation, and the predication $p(t_1, \ldots, t_{a(p)})$ is true just in case $(\llbracket t_1 \rrbracket, \ldots, \llbracket t_{a(p)} \rrbracket) \in \llbracket p \rrbracket$, where $\llbracket s \rrbracket$ is the referent of the symbol s.

First off, there are at least two classical objections against **(CP1)**. The first one is based on the apparently unbounded number of arguments that action verbs admit (Davidson, 1967; Kenny, 1963). This may be attested to by examples such as the following Kenny-sequence, where not only do the arguments vary in number, but they also vary in the *roles* they play even if they have the same number (cf. (1c) and (1d)).

---

[1] I reserve "adicity" to relations and "arity" to predicate symbols.

Relations and Predication

(1) a. Brutus killed Caesar.

b. Brutus killed Caesar with a knife.

c. Brutus killed Caesar with a knife at noon.

d. Brutus killed Caesar with a knife at Pompey's theater.

This classical objection has well-known classical responses (Davidson, 1967; Parsons, 1990). But such responses cannot address the second objection which is based on the observation that some examples, not necessarily involving actions, appear to include *multigrade* predicates (Morton, 1975; Oliver & Smiley, 2004):

(2) a. Tom and Dick cooked dinner.

b. Tom, Dick, and Harry cooked dinner.

c. Sue, Molly, and Lilly live together.

d. Adam and Bill fought with Yuri and Zero.

Although sequences (1) and (2) both provide evidence against **(CP1)**, they do so differently. Whereas sequence (1) alludes to the unbounded number of *roles* that arguments of a predicate may play, sequence (2), while maintaining a fixed number of roles, suggests that an unbounded number of individuals may play (what intuitively is) the same role. Consequently, sentences like those in (2) are susceptible to distributive/collective ambiguities. Again, responses to *some* of the questions raised by (2) have been proposed, alluding to logics of plurals (Morton, 1975; Oliver & Smiley, 2004), logics with flexible predicates (Taylor & Hazen, 1992), or logics with "set arguments" (Shapiro, 1986, 2000). The account which I put forward in this report is both less ambitious and more ambitious than these proposals.

Now, rejecting **(CP1)** requires revising **(CP2)**. For instance, **(CP2)** may, in principle, be revised by simply relaxing the constraint that p combines with a fixed number of terms. Nevertheless, proposals that seriously attempt to address objections to **(CP1)** typically require more drastic revisions of **(CP2)** (Morton, 1975; Oliver & Smiley, 2004; Taylor & Hazen, 1992). The case of SNePS (Shapiro, 1986, 2000; Shapiro & Rapaport, 1987) is particularly interesting since it (i) exemplifies a non-classical mode of "combining" predicates and arguments and (ii) is a major motivation of the work presented in this report.

SNePS is a computational knowledge representation and reasoning system based on semantic network processing. For example, a typical SNePS representation of (2c) is shown in Figure 1. Here predication is displayed

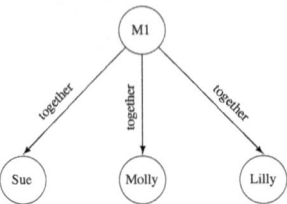

Figure 1: A SNePS representation of "Sue, Molly, and Lilly live together".

graphically, and one cannot identify a single "predicate symbol" in the classical sense; rather, it is the collection of arc labels that plays the role a predicate symbol classically plays in predication (cf. Shapiro & Rapaport, 1987). A linear representation of the network of Figure 1 employs what Shapiro refers to as "set arguments" (Shapiro, 1986, 2000):

(3) LiveTogether({Sue, Molly, Lilly})

Interestingly, for expressions with set arguments like (3), the SNePS inference system allows us to infer similar expressions with a set argument replaced by an appropriately-sized subset thereof. Thus, from (3), we can infer LiveTogether({Sue, Molly}), LiveTogether({Sue, Lilly}), or any similar expression with a size-2 set argument.

We now come to **CP3**. Clearly, rejecting **CP1**, revising **CP2**, and endorsing multigrade predicates forces us to consider relations which are variably polyadic. This in turn calls for, at least, a revision of **CR** and, hence, **CP3**. We are, thus, committed to the following agenda:

1. We need to come up with a theory of variably polyadic relations. Said theory should account for relations where an unbounded number of individuals play the same role (or occupy the same "argument place" (Leo, 2008a; Taylor & Hazen, 1992)) and, while we are at it, the theory might as well account for relations with an unbounded number of roles without heeding the (neo-)Davidsonian advice on logical form.

2. It should be meaningful, in our theory, to talk about a relation's being, for example, reflexive, symmetric, or transitive just as it is meaningful to do so with classical dyadic relations. One motivation for this is giving precise semantics for SNePS expressions such as (3). In (Shapiro,

Relations and Predication

1986) and (Shapiro, 2000), Shapiro advises that the set-argument device should only be used when representing symmetric and (almost) transitive relations. In examples like (3), where the underlying relation of "living together" is basically dyadic, this is well-defined. It is not clear, however, what that exactly means when said relation is not dyadic or, even worse, when it is variably polyadic.

3. A fresh view of predication and a formal semantics grounded in our theory of variably polyadic relations is to be laid out.

The above three points will be addressed, in order, in the following sections. The outcome, which is still preliminary in some respects, serves as a foundation for an extension of predication in SNePS with precise syntax and formal semantics.

3 Remark 2: On the nature of relations

We start as close as possible to **CR**: Relations are sets of tuples over some carrier set. Contra the standard model, we allow relations to include tuples which are (i) structured and (ii) of varying lengths. In the sequel, $[n]$ is used to refer to the set $\{1, 2, \ldots, n\}$.

Definition 1 *A **relation** R on a non-empty set D is a set of pairs (t, p), where $t \in D^j$, for some $j \in \mathbb{Z}^+$, and p is an ordered **partition** of j. R is $k^\uparrow$-**adic** if $k = \min\{|t| \mid (t, p) \in R\}$. Likewise, R is $k^\downarrow$-**adic** if $\{|t| \mid (t, p) \in R\}$ is finite and $k = \max\{|t| \mid (t, p) \in R\}$. R is k-**adic** if it is both $k^\uparrow$-adic and $k^\downarrow$-adic. R is $k^\uparrow$-**valent** if $k = \min\{|p| \mid (t, p) \in R\}$, it is $k^\downarrow$-**valent** if $\{|p| \mid (t, p) \in R\}$ is finite and $k = \max\{|p| \mid (t, p) \in R\}$, and it is k-**valent** if it is both $k^\uparrow$-valent and $k^\downarrow$-valent.*

The following notions will later prove useful for a succinct presentation of properties of relations.

Definition 2 *Let R be relation with $(t, p) \in R$. The **lub sequence of** p, denoted $\mathrm{lub}(p)$, is the sequence of prefix sums of p: $\mathrm{lub}(p)_i = \sum_{j=1}^{i} p_j$. For $i \in [|p|]$, $\mathrm{cell}_p(i) = \{\mathrm{lub}(p)_{i-1} + 1, \ldots, \mathrm{lub}(p)_i\}$, and $t|_{p,i}$ denotes the sub-sequence $(t_k)_{k \in \mathrm{cell}_p(i)}$ of t. (For brevity, the subscript p will be dropped whenever clear from context and we write $\mathrm{cell}(i)$ and $t|_i$, respectively.)*

Haythem O. Ismail

How does this account of relations respond to the objections raised in Section 2? As already declared, it is not my intention to account for the *reality* of relations; thus, Definition 1 is still vulnerable to attacks on the use of tuples. But the definition is a reasonable revision of **CR** as a response to the need for revising **CP1–CP3**. The partition of a tuple divides it into *places* each having multiple *positions* (Oliver & Smiley, 2004): Members of a partition identify the *sizes* of the different places, indicating different ways whereby a tuple of entities may be related by the relation. For example, a pair $((a, b, c, d, e, f), (1, 3, 2)) \in R$ indicates that the tuple (a, b, c, d, e, f) has three places, the first having only position 1, the second extending from position 2 through position 4, and the third having positions 5 and 6. The pair $((a, b, c, d, e, f), (2, 1, 2, 1)) \in R$ indicates that the same tuple of entities are related by R, albeit in a quite different way. The number of places (i.e. $|p|$) indicates the number of roles played in the relation (which varies from one pair (t, p) to another), and the number of positions within the ith place (i.e., p_i) indicates the number of individuals playing the corresponding role (which also varies from one pair to another).

A standard k-adic relation may be conceived as a k-adic, k-valent relation as per our definition. Thus, the standard dyadic identity relation, for instance, is given by $\{((a, a), (1, 1)) | a \in D\}$. For ease of reference, Table 1 gives examples of relations (to some of which we shall refer frequently) construed as per Definition 1.[2]

But, if we are motivated by capturing the intuition of a single relation's holding in different ways among the same tuple of individuals, Definition 1 does the job only partially. According to this definition, sentences such as (4a) and (4b) correspond to a single pair (mnemonically $((i, t, P), (1, 1, 1))$) in a formal travelling relation:

(4) a. I travelled by train to Paris.

b. I travelled by train from Paris.

Nothing in the partitioned tuple device signals the different roles played by Paris in the travelling relation. Something similar is going on with †-examples like (1c) and (1d), where nothing in our formal device may identify the kinship that (1c), but not (1d), has to sentences such as "Brutus killed Caesar with a knife at midnight". I believe that, ultimately, no purely formal property of the tuples may account for the kinds of distinction we would

---

[2]Examples similar to ⟨**Arr**⟩ can be generated by considering any preorder.

Relations and Predication

Name	Description/Example	Adicity	Valency
$\langle \cdot \cdot \cdot \rangle$	Holds among a tuple of collinear points	$2^\uparrow$-adic	univalent
⋒	Holds among two tuples of sets, $(A_1, \ldots, A_m)$ and $(B_1, \ldots, B_n)$, and a set C whenever the union of the As and the union of the Bs have the same intersection with C	$3^\uparrow$-adic	trivalent
⋘	Holds among a tuple of numbers if the tuple is monotonically increasing (but not strictly so)	$2^\uparrow$-adic	univalent
$\overrightarrow{\lozenge 3}$	An oriented between-ness relation: "Object b is after object a and before object c"	triadic	trivalent
$\overline{\lozenge 2}$	An un-oriented variant of $\overrightarrow{\lozenge 3}$ construed as bivalent: "Object b is between objects a and c"	triadic	bivalent
$\overline{\lozenge 3}$	An un-oriented variant of $\overrightarrow{\lozenge 3}$ construed as univalent	triadic	univalent
$\langle 6 \rangle$	Holds among a partitioned tuple of six natural numbers if the parts have equal sums	6-adic	$2^\uparrow$- and $6^\downarrow$-valent
ΣΣΣ	$\sum_{i=1}^{3} a_i + \sum_{i=1}^{3} b_i = \sum_{i=1}^{3} c_i$	9-adic	trivalent
♡	The love circle relation construed as univalent: "A loves B, B loves C, and C loves A"	triadic	univalent
†	The killing relation, with example sentences in sequence (1)	$1^\uparrow$-adic	$1^\uparrow$-valent
$\langle \mathbf{Arr} \rangle$	"Tim and Joe and then Dick, Sue, and Sally arrived in that order"	$2^\uparrow$-adic	$2^\uparrow$-valent

Table 1: Examples of relations à la Definition 1

like to capture. Rather, imposing some extrinsic structure on the pairs of a relation is needed.

Definition 3 *Let R be a relation. A cover $\mathfrak{n}$ of R is a **normal cover** of R if (i) it is a Sperner family and (ii) whenever (t_1, p_1) and (t_2, p_2) are in the same $\mathfrak{n}$-block we have $p_1 = p_2$. We refer to an $\mathfrak{n}$-block containing (t, p) as an $\mathfrak{n}(t, p)$-**block**.*

Each block in $\mathfrak{n}$ intuitively corresponds to one way the relation may hold among a tuple of individuals. An important example of a normal cover is the *partition* $\mathfrak{n}_0$ where $(t', p') \in \mathfrak{n}_0(t, p)$ if and only if $p' = p$. All normal covers of R are refinements of $\mathfrak{n}_0$. It is the liberty of choosing such refinements that allows us to distinguish the pairs corresponding to (4a) and (4b) and to differentiate between the roles involved in pairs corresponding to (1c) and (1d). Normal covers also carry a flavor of Fine's anti-positionalist state-similarity relation (Fine, 2000).

But normal covers alone do not exactly capture the intuition we are after. How would we relate two distinct blocks $\mathfrak{n}(t_1, p_1)$ and $\mathfrak{n}(t_2, p_2)$, of a normal cover where $|p_1| = |p_2|$, intuitively corresponding to the same set of roles played in the relation. For this, we need one more level of structure on R.

Definition 4 *Let R be a relation and $\mathfrak{n}$ a normal cover thereof. A triple $\mathfrak{R} = (R, \mathfrak{n}, \mathfrak{p})$ is called a **covered relation** if $\mathfrak{p}$ is a partition of $\mathfrak{n}$ where* whenever $\mathfrak{n}(t_1, p_1)$ and $\mathfrak{n}(t_2, p_2)$ are in the same $\mathfrak{p}$-block, then $|p_1| = |p_2|$ and $p_1 \neq p_2$.

For $\mathfrak{n} = \mathfrak{n}_0$, we refer to the coarsest partition by $\mathfrak{p}_0$. Given this third level of structure on a relation, it is interesting to consider what reasonable relations should hold between the partitions of two normal covers where one refines the other. If cover refinement is done (as intended) to introduce more fine-grained distinctions among the different ways a relation may hold among individuals, then the corresponding partition modification should observe this intuition.

Definition 5 *If $\mathfrak{R}_1 = (R_1, \mathfrak{n}_1, \mathfrak{p}_1)$ and $\mathfrak{R}_2 = (R_2, \mathfrak{n}_2, \mathfrak{p}_2)$ are covered relations then $\mathfrak{R}_2$ **refines** $\mathfrak{R}_1$ if there is a function $\varrho : \mathfrak{p}_2 \longrightarrow \mathfrak{p}_1$ such that (1) for every $\mathfrak{B}_2 \in \mathfrak{p}_2$ and $B_2 \in \mathfrak{B}_2$, there is some $B_1 \in \varrho(\mathfrak{B}_2)$ with $B_2 \subseteq B_1$; and (2) for every $\mathfrak{B}_1 \in \mathfrak{p}_1$ and $\mathfrak{n}_1(t, p)$-block in $\mathfrak{B}_1$ there is some $\mathfrak{B}_2 \in \mathfrak{p}_2$ with $\varrho(\mathfrak{B}_2) = \mathfrak{B}_1$ and with some $\mathfrak{n}_2(t, p)$-block in $\mathfrak{B}_2$.*

Corollary 1 [3] *If $\mathfrak{R}_1 = (R_1, \mathfrak{n}_1, \mathfrak{p}_1)$ and $\mathfrak{R}_2 = (R_2, \mathfrak{n}_2, \mathfrak{p}_2)$ are covered relations with $\mathfrak{R}_2$ refining $\mathfrak{R}_1$, then (1) $R_2 = R_1$; (2) $\mathfrak{n}_2$ refines $\mathfrak{n}_1$; (3) for every $\mathfrak{B}_2 \in \mathfrak{p}_2$ and $B_2 \in \mathfrak{B}_2$, there is a unique $B_1 \in \varrho(\mathfrak{B}_2)$ with $B_2 \subseteq B_1$; and (4) for every $\mathfrak{B}_1 \in \mathfrak{p}_1$, every $B_1 \in \mathfrak{B}_1$, and every $\mathfrak{B}_2 \in \mathfrak{p}_2$ with $\varrho(\mathfrak{B}_2) = \mathfrak{B}_1$ there is at most one $B_2 \in \mathfrak{B}_2$ such that $B_2 \subseteq B_1$.*

4 Remark 3: On the properties of relations

While relational properties such as (ir)reflexivity, (a/anti-)symmetry, and transitivity are well-understood for classical dyadic relations, the same is far from true even for classical k-adic relations. Some few suggestions for generalizing such properties to the k-adic case do exist, but are rather hard to find (Pickett, 1966; Ušan & Šešelja, 1981; Žižović, 1990). The definitions to appear below are based on relations defined by Definition 1 and are generalizations of those found in the literature when restricted to the classical k-adic case.

Definition 6 *Let $(R, \mathfrak{n}, \mathfrak{p})$ be a covered relation. An $\mathfrak{n}$-**selector** on R is a function $\sigma : \mathfrak{n} \longrightarrow 2^{\mathbb{N}} \times 2^{\mathbb{N}}$, where if B is an $\mathfrak{n}(t, p)$-block, then $\sigma(B)_1 \subseteq [|t|]$ and $\sigma(B)_2 \subseteq [|p|]$. σ is **A-complete** if $\sigma(B)_1 = [|t|]$ and it is **V-complete** if $\sigma(B)_2 = [|p|]$.*

---

[3] Due to space limitations, all proofs (which are mostly straightforward) are not presented. A longer version of the paper contains all proofs and more definitions and discussions. The interested reader may contact the author who is happy to disseminate the long version.

Relations and Predication

Intuitively, a selector associates with every $\mathfrak{n}(t,p)$-block a selection of positions and argument places. The A and V refer to *adicity* and *valency*, respectively.

Definition 7 *Let $(R, \mathfrak{n}, \mathfrak{p})$ be a covered relation. If σ is an $\mathfrak{n}$-selector on R, then R is σ-**reflexive** if, for every $\mathfrak{n}(t,p)$-block B and every $a \in D$, (a) there is some $(t',p) \in B$ such that, if $i \in \sigma(B)_1$, then $t'_i = a$; and (b) there is some $(t'',p) \in B$ such that, if $i \in \sigma(B)_2$, $t''_j = a$, for some $j \in \text{cell}(i)$. R is $\mathfrak{n}$-**reflexive** if it is σ-reflexive for every $\mathfrak{n}$-selector σ.*

Observation 1 *Let $(R, \mathfrak{n}, \mathfrak{p})$ be a covered relation. R is $\mathfrak{n}$-reflexive if and only if it is σ-reflexive for some A-complete $\mathfrak{n}$-selector σ.*

How does the choice of a normal cover affect the reflexivity of a relation?

Proposition 1 *Let R be a relation and let $\mathfrak{n}_1$ and $\mathfrak{n}_2$ be normal covers thereof. If R is $\mathfrak{n}_2$-reflexive and $\mathfrak{n}_2$ is a refinement of $\mathfrak{n}_1$, then R is $\mathfrak{n}_1$-reflexive.*

No matter how much we try, there will always be covers of a relation with respect to which it is not reflexive—for example, any normal cover with blocks containing less than $|D|$ pairs.

If R is dyadic, the definitions of reflexivity reduce to the standard definitions.[4] According to the above definition, $\langle \cdot \cdot \rangle$ is n_0-reflexive. (Refer to Table 1.) It is also reflexive under a refinement that, say, distinguishes collinearity when the lines are vertical, horizontal, or slanted. ⌐ is also clearly n_0-reflexive, but is not $\mathfrak{n}$-reflexive for any *partition* $\mathfrak{n}$ refining n_0.

Definition 8 *Let $(R, \mathfrak{n}, \mathfrak{p})$ be a covered relation. An $\mathfrak{n}$-**permutation** on R is a function π which maps every $\mathfrak{n}(t,p)$-block to a pair of permutations of $[|t|]$ and $[|p|]$, respectively.*

For $(t,p) \in B$, we let $\pi(B)((t,p)) = (\pi(B)_1(t), \pi(B)_2(p))$. Moreover, we let $\pi(R) = \bigcup_{B \in \mathfrak{n}} \pi_B(R)$, where $\pi_B(R) = \{\pi(B)((t,p)) | (t,p) \in B\}$. It should be clear that $\pi(\mathfrak{n}) = \{\pi_B(R)\}_{B \in \mathfrak{n}}$ is a normal cover of $\pi(R)$. Hence, we let $\pi(R, \mathfrak{n}, \mathfrak{p}) = (\pi(R), \pi(\mathfrak{n}), \pi(\mathfrak{p}))$, where $\pi(\mathfrak{p}) = \{\{\pi_B(R) \mid B \in \mathfrak{B}\} \mid \mathfrak{B} \in \mathfrak{p}\}$. An $\mathfrak{n}$-permutation naturally gives rise to an associated $\mathfrak{n}$-selector. In particular, we let $\sigma_\pi(B) = (\{i | \pi(B)_1(i) \neq i\}, \{j | \pi(B)_2(j) \neq j\})$.

---
[4] Likewise, if R is k-adic and $|\sigma(B)_1| = 2$ for every $B \in \mathfrak{n}$, σ-reflexivity reduces to the (i,j)-reflexivity of Ušan and Šešelja (1981).

Definition 9 *Let $(R, \mathfrak{n}, \mathfrak{p})$ be a covered relation and let π be an $\mathfrak{n}$-permutation on R.*

1. *For every $k \in \mathbb{N}$, π is an $\mathfrak{n}^{A_k}$-permutation on R if for every $\mathfrak{n}(t,p)$-block B, (a) $\sigma_\pi(B)_2 = \varnothing$ and (b) if $|p| < k$, then $\sigma_\pi(B)_1 = \varnothing$; otherwise, $\sigma_\pi(B)_1 \subseteq \text{cell}_p(k)$.*

2. *For every $I \subseteq \mathbb{N}$, π is an $\mathfrak{n}^{V_I}$-permutation on R if for every $\mathfrak{n}(t,p)$-block B, if $I \not\subseteq [|p|]$, then $\sigma_\pi(B)_1 = \sigma_\pi(B)_2 = \varnothing$; otherwise $\sigma_\pi(B)_2 = I$, $\sigma_\pi(B)_1 = \bigcup_{j \in I} \text{cell}_p(j)$, and, for every $i \in I$, $t'|_j = t|_i$, where $t' = \pi(B)_1(t)$ and $j = \pi(B)_2(i)$.*

We say that an $\mathfrak{n}$-permutation is an $\mathfrak{n}$-**reversal** if, intuitively, it reverses the selected subsequences of both t and p. Likewise, an $\mathfrak{n}$-r-**rotation** is an $\mathfrak{n}$-permutation which rotates the selected subsequence of both t and p r steps to the right.

Definition 10 *Let $(R, \mathfrak{n}, \mathfrak{p})$ be a covered relation.*

1. *For every $\mathfrak{n}$-permutation π on R, R is π-**symmetric** if $\pi(R) \subseteq R$.*

2. *R is $\mathfrak{n}$-**symmetric** if it is π-symmetric, for every $\mathfrak{n}$-permutation π on R; it is $\mathfrak{n}$-**reversible** if it is π-symmetric for every $\mathfrak{n}$-reversal π on R; it is $\mathfrak{n}$-r-**rotary**, for $r \in \mathbb{N}$, if it is π-symmetric for every $\mathfrak{n}$-r-rotation π on R; and it is $\mathfrak{n}$-**rotary** if, for every $r \in \mathbb{N}$, it is π-symmetric for every $\mathfrak{n}$-r-rotation π on R.*

3. *For every $k \in \mathbb{N}$ and $I \subseteq \mathbb{N}$, R is $\mathfrak{n}^{A_k}$-**symmetric** ($\mathfrak{n}^{V_I}$-**symmetric**) if it is π-symmetric for every $\mathfrak{n}^{A_k}$-permutation ($\mathfrak{n}^{V_I}$-permutation) π on R.*

As defined above, the variants of symmetry collapse and reduce to the standard dyadic symmetry if R is a dyadic, bivalent relation. (For the more general case of standard n-ary relations, they reduce to the symmetry of Ušan & Šešelja, 1981.) Observations similar to Observation 1 can be made by overloading terminology from group theory.

Observation 2 [5] *Let $(R, \mathfrak{n}, \mathfrak{p})$ be a covered relation. R is $\mathfrak{n}$-symmetric if it is π-symmetric for every permutation π in some generator of the set of $\mathfrak{n}$-permutations.*

---

[5] Refer to the longer version of the paper for more details.

Relations and Predication

Proposition 2 *Let $\mathfrak{R}_1 = (R, \mathfrak{n}_1, \mathfrak{p}_1)$ and $\mathfrak{R}_2 = (R, \mathfrak{n}_2, \mathfrak{p}_2)$ be covered relations. R is $\mathfrak{n}_1$-symmetric if and only if it is $\mathfrak{n}_2$-symmetric.*

Hence, we can talk of R's being simply symmetric, without alluding to any particular cover. Similar results may likewise be established for the more specialized variants of symmetry given in Definition 10. Accordingly, it could be shown that $\langle\dot{\,\cdot\,}\rangle$ is symmetric; $\langle\!\!\langle\mathcal{Y}\rangle\!\!\rangle$ is rotary; $\langle\mathbf{6}\rangle$ is A_k-symmetric and V_I-symmetric, for any $k \in [6]$ and $I \subseteq [6]$; and both $\text{\tiny\textcircled{}}$ and $\mathbf{\Sigma\!\Sigma\!\Sigma}$ are A_1-symmetric, A_2-symmetric, and $V_{[2]}$-symmetric. The triadic, univalent un-oriented $\overline{\langle\mathbf{3}}$ is reversible; and the bivalent variant $\overline{\langle\mathbf{2}}$ is A_1-symmetric.

While the technical development of the generalizations of reflexivity and symmetry to our notion of relation (as per Definition 1) was quite involved, the intuitions driving it are (I hope) reasonable enough. This is not the case for transitivity. There is a wide array of possibilities when it comes to defining a generalized notion of transitivity (Pickett, 1966; Ušan & Šešelja, 1981, 2001; Žižović, 1990). Whatever methodology one may choose to follow in order to settle for a unique notion of transitivity is admittedly debatable. In this brief report, I present a generalized notions of transitivity which is arguably intuitive.[6]

Definition 11 *Let $(R, \mathfrak{n}, \mathfrak{p})$ be a covered relation and let $k \in \mathbb{N}$. R is A_k-transitive if for every $(t, p), (t', p') \in R$ such that (1) $k \leq |p|$, (2) $k \leq |p'|$, and (3) for some $m \geq 0$, $(t_i, t_{i+1}, \ldots, t_{i+m}) = (t'_j, t'_{j+1}, \ldots, t'_{j+m})$ where (a) $\{i, i+1, \ldots, i+m\} \subseteq \text{cell}_p(k)$, (b) $\{j, j+1, \ldots, j+m\} \subseteq \text{cell}_{p'}(k)$, (c) $i > \text{lub}(p)_{k-1} + 1$, and (d) $j + m < \text{lub}(p')_k$, it must be that $(\tilde{t}, \tilde{p}) \in R$, where $\tilde{t} = (t_1, \ldots, t_{i-1}, t'_{j+m+1}, \ldots, t'_{|t'|})$ and $\tilde{p} = (p_1, \ldots, p_{k-1}, i - \text{lub}(p)_{k-1} - 1 + \text{lub}(p')_k - j - m, p'_{k+1}, \ldots, p'_{|p'|})$.*

Construing the standard binary relation as a dyadic, univalent relation under $\mathfrak{n}_0$, A_1-transitivity reduces to the standard notion. Other simple examples include various univalent relations under $\mathfrak{n}_0$ and $\mathfrak{p}_0$: $\lll$ and $\overline{\langle\mathbf{3}}$ are both A_1-transitive, for instance.

The variable adicity and valency of our relations admit some new and interesting properties.

Definition 12 *Let $(R, \mathfrak{n}, \mathfrak{p})$ be a covered relation and $k \in \mathbb{N}$.*

1. *R is A_k-**concatenable** if for every $(t, p), (t', p') \in R$, with $k \leq |p|$ and $k \leq |p'|$, if there are $i \in \text{cell}_p(k)$ and $j \in \text{cell}_{p'}(k)$ such that*

---

[6] A different generalization of transitivity, along with definitions and results for irreflexivity, anti-symmetry, and asymmetry appear in the longer version of the paper.

$t_i = t'_j$ then $(\tilde{t}, \tilde{p}) \in R$, where $\tilde{t} = (t_1, \ldots, t_i, t'_{j+1}, \ldots, t'_{|t'|})$ and $\tilde{p} = (p_1, \ldots, p_{k-1}, i - \text{lub}(p)_{k-1} + \text{lub}(p')_k - j, p'_{k+1}, \ldots, p'_{|p'|})$.

2. *For every $m \in \mathbb{N}$, R is $(A_k{\downarrow}m)$-separable if for every $(t, p) \in R$, with $k \leq |p|$, and every $l \in \mathbb{N}$, with $m \leq l < p_k$, $(t', p') \in R$, where (a) $p' = (p_1, \ldots, p_{k-1}, l, p_{k+1}, \ldots, p_{|p|})$ (if $l = 0$, $p' = (p_1, \ldots, p_{k-1}, p_{k+1}, \ldots, p_{|p|})$); (b) $t'|_{p',j} = t|_{p,j}$ for $j \neq k$; and (c) if $l > 0$, $t'|_{p',k}$ is a size-l contiguous sub-sequence of $(t|\text{cell}_p(k))$. If R is $(A_k{\downarrow}m)$-separable for every $m \in \mathbb{N}$, then R is simply A_k-separable.*

3. *For every $m \in \mathbb{N}$, R is $(A_k{\uparrow}m)$-expandable if for every $(t, p) \in R$, with $k \leq |p|$, and every $l \in \mathbb{N}$, with $p_k < l \leq m$, $(t', p') \in R$ where (a) $p' = (p_1, \ldots, p_{k-1}, l, p_{k+1}, \ldots, p_{|p|})$; (b) $t'|_{p',j} = t|_{p,j}$ for $j \neq k$; and (c) $t|_{p,k}$ is a contiguous sub-sequence of $(t'|\text{cell}_p(k'))$. If R is $(A_k{\uparrow}m)$-expandable for every $m \in \mathbb{N}$, then R is simply A_k-expandable.*

Thus, $\lll$ and $\langle\cdot\cdot\rangle$ are both A_1-concatenable and, respectively, $(A_1{\downarrow}2)$-separable and $(A_1{\downarrow}1)$-separable; the $2^\uparrow$-adic, univalent relation which holds among a collection of positive integers when they do not have a common factor greater than 1 is A_1-expandable; and its *complement*, which holds among a collection of positive integers when they do have a common factor greater than 1 is $(A_1{\downarrow}2)$-separable.[7]

The properties of relations defined in previous sections do not always mesh well with the intuitive interpretation of covered relations. For example, Definition 10 of symmetry does not require a permutation of (t, p) to stay within the same p-block. Such a requirement is reasonable if we would like to capture the intuition that a set of argument places is permutable. A permutation's belonging to a different p-block means that we are talking about two possibly different sets of argument places. To remedy this, we introduce more *faithful* versions of the properties presented so far.

Definition 13 *Let $(R, \mathfrak{n}, \mathfrak{p})$ be a covered relation and $\mathcal{P}$ a property of relations. For every $\mathfrak{B} \in \mathfrak{p}$, we say that R is $\mathcal{P}$ in $\mathfrak{B}$ (with respect to $\mathfrak{n}$ and $\mathfrak{p}$) if, with respect to the covered relation $(\mathfrak{B}, \mathfrak{n}_0, \mathfrak{p}_0)$, $\mathfrak{B}$ is $\mathcal{P}$. We say that R is **pervasively** $\mathcal{P}$ if it is $\mathcal{P}$ in every $\mathfrak{B} \in \mathfrak{p}$.*

---

[7]This is always the case. It could be shown that the complement of any non-trivial expandable (separable) relation is separable (expandable).

Relations and Predication

Observation 3 *Let $(R, \mathfrak{n}, \mathfrak{p})$ be a covered relation, σ an $\mathfrak{n}$-selector, and π an $\mathfrak{n}$-permutation.*

1. *R is σ-reflexive ($\mathfrak{n}$-reflexive) if and only if it is pervasively σ-reflexive ($\mathfrak{n}$-reflexive).*

2. *If R is pervasively π-symmetric, then it is π-symmetric; and if $\mathfrak{n} = \mathfrak{n}_0$, $\mathfrak{p} = \mathfrak{p}_0$ and R is π-symmetric, then it is pervasively π-symmetric.*

3. *If $\mathfrak{n} = \mathfrak{n}_0$ and $\mathfrak{p} = \mathfrak{p}_0$, then for every $k \in \mathbb{N}$ it is the case that R is A_k-transitive (-concatenable) if and only if it is pervasively A_k-transitive (-concatenable).*

4. *For every $m \geq 1$, (i) if R is pervasively $(A{\downarrow}m)$-separable $((A{\uparrow}m)$-expandable), then it is $(A{\downarrow}m)$-separable $((A{\uparrow}m)$-expandable); and (ii) if $\mathfrak{n} = \mathfrak{n}_0$, $\mathfrak{p} = \mathfrak{p}_0$, and R is $(A{\downarrow}m)$-separable $((A{\uparrow}m)$-expandable), then it is pervasively $(A{\downarrow}m)$-separable $((A{\uparrow}m)$-expandable).*

5 Remark 4: On predication

Definition 14 *A predication structure is a tuple $\mathfrak{P} = \langle \mathbb{P}, \mathbb{R}, \mathbb{C}, \lfloor \cdot \rfloor, \lceil \cdot \rceil, \mathbb{D} \rangle$, where: (1) $\mathbb{P}$ is an alphabet, whose symbols are referred to as **predicates**; (2) $\mathbb{R}$ is a non-empty countable set of symbols referred to as **roles**; (3) $\mathbb{C} : \mathbb{P} \longrightarrow 2^{2^{\mathbb{R}}}$; (4) $\lfloor \cdot \rfloor : \mathbb{R} \longrightarrow \mathbb{N}$; (5) $\lceil \cdot \rceil : \mathbb{R} \dashrightarrow \mathbb{N}$; and (5) $\mathbb{D} : \mathbb{R} \longrightarrow 2^{\{\Downarrow, \Uparrow, \odot, \bowtie, \leftrightarrow, \circlearrowright\}}$, where $\mathbb{D}(r)$ is referred to as the **decoration** of r. When $x \in \mathbb{D}(r)$, we write $x(r)$ for brevity.*

Definition 15 *Let $\mathfrak{P}$ be a predication structure and D a non-empty set. For $P \in \mathbb{P}$, the interpretation of P, $[\![P]\!]^D$ with respect to D is a covered relation $(R, \mathfrak{n}, \mathfrak{p})$ on D where (1) R is $l^\uparrow$-valent with $l = \min\{|C| \mid C \in \mathbb{C}(P)\}$; (2) R is $l^\downarrow$-valent with $l = \max\{|C| \mid C \in \mathbb{C}(P)\}$, if the latter is defined; (3) R is $k^\uparrow$-adic with $k \geq \min\{\sum_{r \in C} \lfloor r \rfloor \mid C \in \mathbb{C}(P)\}$; (4) R is $k^\downarrow$-adic with $k \leq \max\{\sum_{r \in C} \lceil r \rceil \mid C \in \mathbb{C}(P)\}$, if the latter is defined; (5) there is a bijection $\varrho : \mathbb{C}(P) \longrightarrow \mathfrak{p}$ such that if $(t, p) \in B \in \varrho(C)$, then $|p| = |C|$; (6) for every $C \in \mathbb{C}(P)$, there is a bijection $\iota_C : C \longrightarrow [|C|]$; and (7) for every $C \in \mathbb{C}(P)$ and $r \in C$, if $\Downarrow (r)$ ($\Uparrow (r)$, $\odot(r)$, $\bowtie (r)$, $\leftrightarrow (r)$, $\circlearrowright (r)$), then R is $(\mathfrak{n}, \mathfrak{p})^{A_{\iota_C(r)}\downarrow \lfloor r \rfloor}$-separable (respectively expandable, concatenable, symmetric, reversible, rotary) in $\varrho(C)$.*

In what follows, let $\mathbb{S} = \langle \mathcal{V}, \mathcal{F} \rangle$ be a **signature**, where $\mathcal{V}$ is a countably-infinite set of variables and $\mathcal{F}$ is a finite set of function symbols each with an associated arity. (As usual, arities are indicated by superscripts; $f^0 \in \mathcal{F}$ is referred to as a constant.) The set of $\mathbb{S}$-terms is, as always, the set of variables and functional terms generated by $\mathbb{S}$.

Definition 16 *Let* $\mathbb{S} = \langle \mathcal{V}, \mathcal{F} \rangle$ *be a signature and* $\mathfrak{P} = \langle \mathbb{P}, \mathbb{R}, \mathbb{C}, \lfloor \cdot \rfloor, \lceil \cdot \rceil, \mathbb{D} \rangle$ *a predication structure. An* $\langle \mathbb{S}, \mathfrak{P} \rangle$-*generated atomic formula is a formula of the form*
$$P\{(r_1\ a_1), (r_2\ a_2), \ldots, (r_n\ a_n)\}$$
where $P \in \mathbb{P}, \{r_1, r_2, \ldots, r_n\} \in \mathbb{C}(P)$ *and* a_i, *for* $1 \leq i \leq n$, *is in one of the following forms (with* $t, t_1, t_2, \ldots, t_m$ $\mathbb{S}$-*terms).* (1) t. (2) $\{t_1, \ldots, t_m\}$, *if* $\bowtie \in \mathbb{D}(r_i)$. (3) $^\cup\{t_1, \ldots, t_m\}$, *if* $\{\bowtie, \odot\} \in \mathbb{D}(r_i)$. (4) $[t_1, \ldots, t_m]$. (5) $^\odot[t_1, \ldots, t_m]$, *if* $\odot \in \mathbb{D}(r_i)$. (6) $^\leftrightarrow[t_1, \ldots, t_m]$, *if* $\leftrightarrow \in \mathbb{D}(r_i)$. (7) $^\circlearrowleft[t_1, \ldots, t_m]$ *if* $\circlearrowleft \in \mathbb{D}(r_i)$.

Such atomic formulas are interpreted, as usual, with respect to a domain of interpretation D. A formula $P\{(r_1 a_1), (r_2 a_2), \ldots, (r_n a_n)\}$ is *true* with respect to an interpretation of P if there is some $\mathfrak{n}(t,p)$-block in $\varrho(\{r_1, r_2, \ldots, r_n\})$ satisfying the following properties: (1) $p_{\iota_C(r_i)} = |a_i|$, where $|a_i| = 1$ if a_i is an $\mathbb{S}$-term. (2) $t|_{\iota_C(r_i)} = [\![a_i]\!]$ if a_i is an $\mathbb{S}$-term, otherwise $t|_{\iota_C(r_i)} = ([\![t_1]\!], \ldots, [\![t_m]\!])$.

Together with the classical syntax of first-order logic, one may utilize the above definition of atomic formulas to construct parsimonious, semi-iconic logical formulas.

Example 1.

1. Colin$\{($points $^\cup\{a,b,c\})\}$: "a, b, and c are collinear."
2. Siblings$\{($sis $\{Mary, Sue, Sally\}), ($bro $John)\}$: "Mary, Sue, and Sally are sisters of John."
3. $\lll \{($num $^\odot[1,3,5])\}$: "1, 3, and 5 form an increasing sequence."
4. Unionsect$\{($C1 $\{A, B, C\}), ($C2 $\{D, E, F\}), ($S $G)\}$: "The union of the A, B, and C and the union of D, E, and F have the same intersection with G."

With this style of predication at our disposal, simple, yet useful, *analytical* inference rules are possible. Examples are displayed in Figure 2.

6 Conclusions

While the three-layered structure of covered relations allows us to accommodate variable polyadicity, it comes with a price. For the structure is admittedly much more complicated than the classical relational structure. But

$$\textbf{Reversal:} \quad \frac{P\{\ldots,(r \leftrightarrow [t_1,\ldots,t_n]),\ldots\}}{P\{\ldots,(r \leftrightarrow [t_n,\ldots,t_1]),\ldots\}}$$

$$\textbf{Rotation:} \quad \frac{P\{\ldots,(r^{\circlearrowleft}[t_1,\ldots,t_{n-1},t_n]),\ldots\}}{P\{\ldots,(r^{\circlearrowleft}[t_n,t_1,\ldots,t_{n-1}]),\ldots\}}$$

$$\textbf{Separation:} \quad \frac{P\{\ldots,(r\,\{t_1,\ldots,t_n\}),\ldots\},\ \Downarrow (r),\ \{s_1,\ldots,s_m\} \subseteq \{t_1,\ldots,t_n\}}{P\{\ldots,(r\,\{s_1,\ldots,s_m\}),\ldots\}}$$

$$\textbf{Expansion:} \quad \frac{P\{\ldots,(r\,\{t_1,\ldots,t_n\}),\ldots\},\ \Uparrow (r),\ \{t_1,\ldots,t_n\} \subseteq \{s_1,\ldots,s_m\}}{P\{\ldots,(r\,\{s_1,\ldots,s_m\}),\ldots\}}$$

$$\textbf{Concatenation:} \quad \frac{P\{(r_1\,a_1),\ldots,(r^{\circlearrowleft}[t_1,\ldots,t,\ldots,t_n]),\ldots,(r_n\,a_n)\},\ P\{(r_1\,b_1),\ldots,(r^{\circlearrowleft}[s_1,\ldots,t\ldots s_m]),\ldots,(r_n\,b_n)\}}{P\{(r_1\,a_1)\ldots,(r^{\circlearrowleft}[t_1,\ldots,t,\ldots,s_m]),\ldots,(r_n\,b_n)\}}$$

$$\textbf{Union:} \quad \frac{P\{(r_1\,a_1),\ldots,(r^{\cup}\{t_1,\ldots,t_n\}),\ldots,(r_n\,a_n)\},\ P\{(r_1\,b_1),\ldots,(r^{\cup}\{s_1,\ldots,s_m\}),\ldots,(r_n\,b_n)\},\ \{t_1,\ldots,t_n\} \cap \{s_1,\ldots,s_n\} \neq \varnothing}{P\{(r_1\,a_1)\ldots,(r^{\cup}(\{t_1,\ldots,t_n\} \cup \{s_1,\ldots,s_n\})),\ldots,(r_n\,b_n)\}}$$

Figure 2: Sample inference rules

maybe the price is worth paying after all, given the parsimonious predication device it admits with the associated "efficient" inference patterns. But more work remains to be done. A careful study of the proposed predication structure in the full context of a first-order language is perhaps the most pressing item on the agenda. In addition, a more careful study of the expressive power of the proposed language and the ontological questions the companion relational structure gives rise to are called for.

References

Davidson, D. (1967). The logical form of action sentences. In N. Recher (Ed.), *The Logic of Decision and Action* (pp. 81–95). Pittsburgh: University of Pittsburgh Press.

Diro, C. (2008). Non-symmetric relations. In D. Zimmerman (Ed.), *Oxford Studies in Metaphysics* (Vol. 1, pp. 155–192). Oxford: Oxford University Press.

Fine, K. (2000). Neutral Relations. *The Philosophical Review, 109*, 1–33.

Gilmore, C. (2013). Slots in universals. In K. Bennett & D. Zimmerman (Eds.), *Oxford Studies in Metaphysics* (Vol. 8, pp. 187–233). Oxford: Oxford University Press.

Johansson, I. (2011). Order, direction, logical priority and ontological categories. In J. Cumpa & E. Tegtmeier (Eds.), *Ontological Categories* (pp. 89–107). Heusenstamm: Ontos Verlag.

Johansson, I. (2014). On converse relations—what can we learn from Segelberg's controversies with Russell and Moore. In T. N. H. Malmgren & C. Svennerlind (Eds.), *Philosophy and Botany: Essays on Ivar Segelberg* (pp. 82–106). Stockholm: Bokförlaget Thales.

Kenny, A. (1963). *Action, Emotion and Will*. London: Routledge and Kegan Paul.

Leo, J. (2008a). The Identity of Argument Places. *The Review of Symbolic Logic, 1*, 335–354.

Leo, J. (2008b). Modeling Relations. *Journal of Philosophical Logic, 37*, 353–385.

Morton, A. (1975). Complex Individuals and Multigrade Relations. *Noûs, 9*, 309–318.

Oliver, J., & Smiley, T. (2004). Multigrade Predicates. *Mind, 113*, 609–681.

Parsons, T. (1990). *Events in the Semantics of English*. Cambridge: MIT Press.

Pickett, H. E. (1966). A Note on Generalized Equivalence Relations. *The American Mathematical Monthly, 73(8)*, 860–861.

Shapiro, S. (1986). Symmetric Relations, Intensional Individuals, and Variable Binding. *Proceedings of the IEEE, 74(10)*, 1354–1363.

Shapiro, S. (2000). SNePS: A logic for natural language understanding and commonsense reasoning. In Ł. Iwańska & S. Shapiro (Eds.), *Natural Language Processing and Knowledge Representation: Language for Knowledge and Knowledge for Language* (pp. 175–195). Menlo Park: AAAI Press/The MIT Press.

Shapiro, S., & Rapaport, W. (1987). SNePS considered as a fully intensional propositional semantic network. In N. Cercone & G. McCalla (Eds.), *The Knowledge Frontier* (pp. 263–315). New York: Springer-Verlag.

Taylor, B., & Hazen, A. P. (1992). Flexibly Structured Predication. *Logique et Analyse, 139–140*, 375–393.

Ušan, J., & Šešelja, B. (1981). Transitive n-ary Relations and Characterizations of Generalized Equivalences. *Review of Research Faculty of Science-University of Novi Sad, 11*, 231–245.

Ušan, J., & Šešelja, B. (2001). On the Family of $(n+1)$-ary Equivalnce

Relations. *Mathematica Moravica*, 5, 163–167.

Žižović, M. (1990). New Generalizations of Ordering Relations. *Zbornik Radova Filozofskog Fakulteta u Nišu, Serija Matematika*, 4, 109–111.

Haythem O. Ismail
Cairo University, Department of Engineering Mathematics
Egypt
E-mail: `haythem.ismail@guc.edu.eg`

Truthmaker Semantics: Fine Versus Martin-Löf

ANSTEN KLEV[1]

Abstract: Fine's truthmaker semantics for intuitionistic logic is here looked at through the lens of Martin-Löf's type theory.

Keywords: Intuitionistic logic, Truthmakers, Type theory

1 Introduction

The term 'truthmaker', or 'truth-maker', is a relatively recent addition to the philosophical vocabulary. It is said to have been employed by C. B. Martin in the late 1950's (Armstrong, 2004, p. 1), and it occurs in print in the 1970's in publications by some philosophers based in Australia, in particular in (Langtry, 1975, pp. 8–9) and (Tooley, 1977, *passim*). Quite independently, I believe, of this Australian tradition, the term was made the title of the well-known paper of Mulligan, Simons, and Smith (1984). These authors relied on the Central European, rather than the younger Australian, realist tradition, and when introducing the term they referred to Husserl, who in § 39 of his Sixth Logical Investigation (Husserl, 1901) speaks of *wahr-machender Sachverhalt*, true-making state of affairs. And indeed, the idea of a truthmaker is much older than the term 'truthmaker'. A truthmaker, as explained both by the Australians and by Mulligan, Simons, and Smith is something—'something in the world' as Langtry (1975) says—in virtue of which a truthbearer, provided it is true, is in fact true. The idea of a truthmaker could therefore be argued to be implicit in the ancient and venerable conception of truth as correspondence with something outside the realm of truthbearers.

Truthmaker semantics is semantics that makes use of the notion of a truthmaker. Most, if not all, systems of model-theoretic semantics could therefore be said to be systems of truthmaker semantics. Thus, the standard

---

[1] For discussion on topics related to this paper and for comments on its contents I am grateful to Adam Přenosil, Vít Punčochář, Stefan Roski and Göran Sundholm.

model ℕ of arithmetic could be taken to be a truthmaker of the sentence $\forall x \exists y (x < y \land \text{Prime}(y))$. More generally, a model $\mathfrak{M}$ in the usual, model-theoretic sense could be taken to be a truthmaker of a predicate-logical sentence A provided $\mathfrak{M} \models A$ holds. In the context of modal logic a possible world w could be taken to be a truthmaker, relative to a frame $\mathfrak{F}$ and valuation φ, of a formula A provided $\mathfrak{F}, \varphi, w \models A$ holds.

In Kit Fine's use of the term 'truthmaker semantics' the truthmakers employed are to be 'wholly relevant' to establishing the truth of the relevant truthbearers. If, following Mulligan, Simons, and Smith, we take so-called moments to be truthmakers, then the banana's yellowness alone could well be said to be wholly relevant to establishing the truth of 'the banana is yellow', whereas the fusion of the banana's yellowness and its curved shape would not be wholly relevant in this way, since also the yellowness of a deformed banana would make the sentence 'the banana is yellow' true. Fine calls a semantics employing truthmakers in this more restricted sense an *exact* semantics. It is clear that the standard systems of model-theoretic semantics for predicate and modal logic are not exact in this sense.

Fine notes that the proof-explanation of the logical connectives associated with the names of Brouwer, Heyting, and Kolmogorov can be regarded as providing an exact semantics for intuitionistic logic. The proof-explanation lays down, in the manner of an inductive definition, what is to count as a proof of a propositionally complex proposition. Several formulations can be found in the literature;[2] the following fits our purposes:

- A proof of $A \land B$ consists of a proof of A and a proof of B.

- A proof of $A \lor B$ consists of a proof of A or a proof of B.

- A proof of $A \supset B$ consists of a method for transforming any proof of A into a proof of B.

- There is no proof of $\bot$.

If, following intuitionists, one takes a mathematical proposition to be determined by what counts as a proof of it, then we have here indeed an explanation of the logical connectives. It should be noted, however, that these clauses by themselves do not constitute a precise semantics, since neither the notion of proof, nor the notion of a proof consisting of other proofs, nor the notion of a transformation of proofs have been explained. We should

---

[2]Cf. e.g. Heyting (1956, pp. 97–98), Troelstra (1969, p. 5), and Martin-Löf (1984, p. 12).

rather think of the above clauses as a scheme that can be made precise in various ways.

Fine's truthmaker semantics for intuitionistic logic (Fine, 2014) is one among several applications that Fine has given in recent years of the general idea of truthmaker semantics (for an overview, see Fine, 2016b). His truthmaker semantics for intuitionistic logic can be seen as making precise the proof-explanation of the logical connectives in the setting of order theory. Since order theory is also the setting of Kripke semantics, one may follow Fine in saying that his semantics is 'in some ways a cross between the construction-oriented semantics of Brouwer–Heyting–Kolmogorov and the condition-oriented semantics of Kripke' (Fine, 2014, p. 549). The relation between Fine's truthmaker semantics and Kripke semantics is relatively well understood owing to the main technical result of (Fine, 2014), which will be discussed below. But Fine (ibid. p. 559) records it as an open problem to understand better the relationship between his truthmaker semantics and the proof-explanation. This paper will be concerned with that problem. In particular, I shall be concerned with the relationship between Fine's semantics and Martin-Löf's type theory, a system of constructive logic that makes precise the proof-explanation in a very natural way. Along the way I shall also remark on similarities and differences between Fine's semantics and two other systems of formal semantics that make precise the proof-explanation, namely the realizability semantics of Läuchli (1970) and what is known as the Kreisel–Goodman theory of constructions (Kreisel 1962; Goodman 1970).

Although some of the discussion below will be technical, I will concentrate on conceptual issues in my comparison of the two systems. In particular, no theorem of the kind Fine has proved linking his semantics with Kripke semantics will be proved or postulated (it is not clear to me what such a theorem would look like). I begin by outlining Fine's semantics and thereafter explain the aspects of Martin-Löf's type theory that are relevant for present purposes. The comparison between the two systems will concentrate on four main differences.

2 Fine's truthmaker semantics

A complete lattice is a partial order $(S, \leq)$ on which there is defined a function $\sqcup : \wp(S) \to S$ such that for any $T \subseteq S$, $\sqcup T$ is the least upper bound of T; that is to say, for all $t \in T$, $t \leq \sqcup T$ and if $s \in S$ is such that for all $t \in T$,

$t \leq s$, then $\sqcup T \leq s$. We typically write $s \sqcup t$ for $\sqcup\{s,t\}$. The dual notion of greatest lower bound can be defined as $\sqcap T := \sqcup\{s \in S \mid (\forall t \in T) s \leq t\}$. A complete lattice $(S, \leq)$ has both a top element, namely $\sqcup S$, and a bottom element, namely $\sqcup \emptyset$, which we shall usually write as 0.

Let S be a complete lattice. We define a binary function $s \rightharpoonup t$ on S as follows:
$$s \rightharpoonup t := \sqcap\{u \in S \mid s \sqcup u \geq t\}$$
If for all $s, t \in S$ it holds that
$$s \sqcup (s \rightharpoonup t) \geq t,$$
then S is said to be *residuated*. Hence, if S is residuated, then for all $s, t \in S$, there is a least element u such that $s \sqcup u \geq t$, namely $u = s \rightharpoonup t$.[3] Examples of non-residuated complete lattices are provided by the non-distributive lattices M_3 and N_5. In fact, any residuated lattice is distributive.[4]

A truthmaker frame is a complete residuated lattice. The points in a truthmaker frame are to be thought of as exact truthmakers for formulae. Fine calls such points 'states', though he emphasizes (Fine, 2016b, p. 5) that this is a technical term and that any category of entities that 1) may be endowed with a mereological structure corresponding to a complete residuated lattice and 2) be regarded as truthmakers, may play the role of states.

The truthmaker relation between states and formulae is determined by an inductive definition. The base of the induction is provided by a valuation function φ from atomic formulae into sets of states, where an atomic formula is either a propositional letter or the constant $\bot$. That $\bot$ lies in the domain of the valuation function means, according to the intuitive interpretation, that there may be states making $\bot$ true. We follow Fine in using α as a variable ranging over atomic formulae in general and p as a variable

---

[3]Lattices that are residuated in Fine's sense are often called co-Heyting algebras in the literature. Fine's use of the term 'residuated' does not entirely coincide with its use in the literature, where 1) it has a more general meaning and 2) it includes Heyting algebras, but not co-Heyting algebras in general; in particular, the structure $(S, \leq, \rightharpoonup, 1)$ is *not* residuated in this sense, but $(S, \geq, \rightharpoonup, 0)$ is. I am grateful to Adam Přenosil for pointing this out to me and for providing me with the proof in the following footnote 4.

[4]Consider the following computation:

$$\begin{aligned} a \sqcup (b \sqcap c) \geq x \quad &\text{iff} \quad b \sqcap c \geq a \rightharpoonup x \\ &\text{iff} \quad b \geq a \rightharpoonup x \text{ and } c \geq a \rightharpoonup x \\ &\text{iff} \quad b \sqcup a \geq x \text{ and } c \sqcup a \geq x \\ &\text{iff} \quad (b \sqcup a) \sqcap (c \sqcup a) \geq x. \end{aligned}$$

ranging over propositional letters. Fine calls the following requirement the falsum condition: if $s \in \varphi(\bot)$, then for any propositional letter p, there is $s' \leq s$ such that $s' \in \varphi(p)$. If this condition is met, then a state that makes $\bot$ true contains a truthmaker of every propositional letter. A state s is called inconsistent if there is $t \leq s$ such that $t \in \varphi(\bot)$; otherwise s is consistent.

A truthmaker model $\mathfrak{T}$ is a couple $(\mathfrak{F}, \varphi)$ consisting of a truthmaker frame $\mathfrak{F}$ and a valuation function φ satisfying the falsum condition. The truthmaker relation $\mathfrak{T}, s \Vdash A$ is defined as follows.[5]

- $\mathfrak{F}, \varphi, s \Vdash \alpha$ iff $s \in \varphi(\alpha)$
- $\mathfrak{T}, s \Vdash A \vee B$ iff $\mathfrak{T}, s \Vdash A$ or $\mathfrak{T}, s \Vdash B$
- $\mathfrak{T}, s \Vdash A \wedge B$ iff $s = s_1 \sqcup s_2$ and $\mathfrak{T}, s_1 \Vdash A$ and $\mathfrak{T}, s_2 \Vdash B$
- $\mathfrak{T}, s \Vdash A \supset B$ iff there is a function $F : S \to S$ such that
  - $\mathfrak{T}, F(t) \Vdash B$ whenever $\mathfrak{T}, t \Vdash A$
  - $s = \sqcup \{t \rightharpoonup F(t) \in S \mid \mathfrak{T}, t \Vdash A\}$

Thus, a truthmaker of a disjunction $A \vee B$ is any truthmaker of one of the disjuncts. A truthmaker of a conjunction $A \wedge B$ is any fusion $s \sqcup t$ of a truthmaker s of A and a truthmaker t of B. And a truthmaker of an implication $A \supset B$ is any least upper bound of a set $\{t_1 \rightharpoonup F(t_1), t_2 \rightharpoonup F(t_2), t_3 \rightharpoonup F(t_3), \ldots\}$, where t_i runs through the truthmakers of A and each $F(t_i)$ is a truthmaker of B. Disregarding the treatment of $\bot$, it should be clear that these clauses are quite in accordance with those of the Brouwer–Heyting–Kolmogorov proof-explanation.

Recall that the forcing relation $\mathfrak{K}, s \Vdash A$ in Kripke semantics for intuitionistic logic is persistent: if $\mathfrak{K}, s \Vdash A$ and $s \leq s'$, then $\mathfrak{K}, s' \Vdash A$. The truthmaker relation $\mathfrak{T}, s \Vdash A$ is not persistent; indeed, even for atomic formulae α, we may have $\mathfrak{T}, s \Vdash \alpha$ and $s \leq s'$, but not $\mathfrak{T}, s' \Vdash \alpha$. This is in accordance with the intuitive understanding of the truthmaker relation as one of exact truthmaking: if s is the banana's colour and s' the fusion of its colour and its shape, then, intuitively, $s \leq s'$ but s' is not an exact truthmaker of 'the banana is yellow'.

A persistent truthmaker relation $\|\triangleright$ can be defined within a truthmaker model $\mathfrak{T}$ as follows:

$$\mathfrak{T}, s \|\triangleright A \text{ iff there is } s' \leq s \text{ such that } \mathfrak{T}, s' \Vdash A$$

---

[5]There are interesting similarities between these clauses and those of Punčochář (2016, p. 181).

The fusion of the banana's colour and shape will stand in this *inexact* truthmaker relation to 'the banana is yellow'. The main technical result of (Fine, 2014) establishes a close relationship between inexact truthmaking in truthmaker models and forcing in Kripke models.

Recall that a state s in a truthmaker model is inconsistent if for some $t \leq s$, $t \in \varphi(\bot)$, and consistent otherwise. Given a truthmaker model $\mathfrak{T} = (\mathfrak{F}, \varphi)$, let $\mathfrak{F}_\kappa$ be the frame got by restricting $\mathfrak{F}$ to consistent states and let φ_κ be the least persistent extension of φ restricted to $\mathfrak{F}_\kappa$, that is: $s \in \varphi_\kappa(p)$ if and only if s is a point in $\mathfrak{F}_\kappa$ and there is $s' \leq s$ such that $s' \in \varphi(p)$. Let $\kappa(\mathfrak{T})$ be $(\mathfrak{F}_\kappa, \varphi_\kappa)$. Thus, $\kappa(\mathfrak{T})$ is got from $\mathfrak{T}$ by excluding the inconsistent states and by imposing persistency on the propositional letters. It is readily seen that $\kappa(\mathfrak{T})$ is a Kripke model. Fine proves that for any consistent state s in $\mathfrak{T}$,

$$\mathfrak{T}, s \Vert\!\!\!> A \text{ iff } \kappa(\mathfrak{T}), s \Vdash A$$

Hence, for any truthmaker model $\mathfrak{T}$ there is a Kripke model $\kappa(\mathfrak{T})$ such that inexact truthmaking over consistent states in $\mathfrak{T}$ is equivalent to forcing in $\kappa(\mathfrak{T})$.

Two elements s, t in a partial order $(S, \leq)$ are incomparable if neither $s \leq t$ nor $t \leq s$ holds. Fine calls a partial order *tree-like* if no infinite ascending chain $s \leq s' \leq s'' \leq \ldots$ has an upper bound; no two incomparable elements have an upper bound; and S has a least element. A Kripke model $\mathfrak{K} = (S, \leq, \varphi)$ is called tree-like if $(S, \leq)$ is tree-like.

A subset $D \subset S$ is downward closed if $s \in D$ and $s' \leq s$ imply $s' \in D$. For any $s \in S$ there is a downward closed subset $[s] := \{t \in S \mid t \leq s\}$. A downward closed set is called principal if it is of the form $[s]$; otherwise it is called non-principal.

Given a tree-like Kripke model $\mathfrak{K} = (S, \leq, \varphi)$, let S_τ be the set of non-empty downward closed subsets of S; let $\leq_\tau$ be the subset relation on S_τ; and define φ_τ as follows:

$$\varphi_\tau(p) := \{[s] \in S_\tau \mid s \in \varphi(p)\} \cup \{D \in S_\tau \mid D \text{ is non-principal}\}$$
$$\varphi_\tau(\bot) := \{D \in S_\tau \mid D \text{ is non-principal}\}$$

Hence $[s]$ makes true precisely the atomic formulae that s makes true, while a non-principal set D makes true all atomic formulae. Fine shows that $\tau(\mathfrak{K}) = (S_\tau, \leq_\tau, \varphi_\tau)$ is a truthmaker model and that for any $s \in S$,

$$\mathfrak{K}, s \Vdash A \text{ iff } \tau(\mathfrak{K}), [s] \Vert\!\!\!> A$$

Truthmaker Semantics

In fact, the mapping $s \mapsto [s]$ is an isomorphism from $\mathfrak{K}$ to $\kappa(\tau(\mathfrak{K}))$. Fine proves a result to the effect that, in S_τ, $D \not\leq_\tau [s]$ for any principal $[s]$ and non-principal D. Hence we may think of $\tau(\mathfrak{K})$ as obtained from the tree-like Kripke model $\mathfrak{K}$ by extending it with a stock of inconsistent states.[6] More precisely, we may think of $\tau(\mathfrak{K})$ as obtained from $\mathfrak{K}$ as follows. To the domain S of $\mathfrak{K}$ we add the non-empty non-principal downward closed subsets of S ordered among themselves by $\subset$. An element $s \in S$ is to be comparable with such a non-principal set D iff $s \in D$, in which case $s \leq D$. Finally, we extend φ by stipulating, for any non-principal set D, that $D \in \varphi(\alpha)$ for any atomic formula α.

To understand in what sense truthmaker models are models of intuitionistic logic we must introduce a consequence relation. If Δ is a set of formulae, $s \mathrel{\Vert\!\!>} \Delta$ means that $s \mathrel{\Vert\!\!>} A$ for every $A \in \Delta$. Fine considers three consequence relations. In the definition of $\mathrel{\vdash_3}$ below, 0 is the bottom element of $\mathfrak{T}$.

$\Delta \mathrel{\vdash_1} A$ iff for any $\mathfrak{T}$ and any s: $\mathfrak{T}, s \mathrel{\Vert\!\!>} A$ whenever $\mathfrak{T}, s \mathrel{\Vert\!\!>} \Delta$

$\Delta \mathrel{\vdash_2} A$ iff for any $\mathfrak{T}$ and any *consistent* s: $\mathfrak{T}, s \mathrel{\Vert\!\!>} A$ whenever $\mathfrak{T}, s \mathrel{\Vert\!\!>} \Delta$

$\Delta \mathrel{\vdash_3} A$ iff for any $\mathfrak{T}$: $\mathfrak{T}, 0 \mathrel{\Vert\!\!>} A$ whenever $\mathfrak{T}, 0 \mathrel{\Vert\!\!>} \Delta$

The consequence relation for Kripke models $\mathfrak{K}$ is defined as

$\Delta \mathrel{\vdash_K} A$ iff for any $\mathfrak{K}$ and any s: $\mathfrak{K}, s \Vdash A$ whenever $\mathfrak{K}, s \Vdash \Delta$

Fine proves that all of these relations are equivalent:

$$\Delta \mathrel{\vdash_1} A \text{ iff } \Delta \mathrel{\vdash_2} A \text{ iff } \Delta \mathrel{\vdash_3} A \text{ iff } \Delta \mathrel{\vdash_K} A$$

From the soundness and completeness of Kripke semantics for intuitionistic logic it then follows in particular that A is a theorem of propositional intuitionistic logic iff $\mathfrak{T}, 0 \Vdash A$ for all truthmaker models $\mathfrak{T}$. Thus, in any truthmaker model, the bottom element, or 'null-state', makes true all theorems of propositional intuitionistic logic. On the other hand, let $\mathbb{B} = \{0, 1\}$ be the two-element lattice and let $\varphi(\alpha) = \emptyset$ for each atomic formula α. By induction on the complexity of formulae one can prove $\mathbb{B}, \varphi, 1 \not\Vdash A$ for every formula A.[7] Thus, the top element of this lattice makes true no formula whatsoever. Hence, no formula is made true by every state in every truthmaker model.

---

[6]Constructions of this kind are studied in more detail and generality in (Fine, 2016a).

[7]For $A = \alpha$, this follows from the definition of φ. Assume $A = B \wedge C$. If $\mathbb{B}, \varphi, 1 \Vdash A$,

Ansten Klev

3 Martin-Löf's type theory

It is not possible in this article to give a thorough introduction to Martin-Löf's type theory (cf. Martin-Löf 1975, 1982, 1984); a rough guide to those aspects of it that will be important for us will have to suffice. It was conceived as a system for reasoning constructively about mathematics, but for our purposes it may be useful to think of it simply as a system of what Hindley and Seldin (1986) call an applied typed λ-calculus. Terms are typed in the manner of Church (1940): the type of a term is intrinsic to it. By 'applied' is meant that the theory has primitive constants in its vocabulary. Hence, apart from abstraction and function application, common to every λ-calculus, the theory knows several other primitive notions. Functional abstraction in Martin-Löf's type theory is not indicated by means of 'λx', but rather by means of Curry's bracket notation '$[x]$'. The letter 'λ' serves another, albeit related, purpose. Hence I shall not speak of the λ-calculus, but simply of the *calculus*, of the terms of Martin-Löf's type theory.

The type system of Martin-Löf's theory is much richer than that of Church's simple type theory. We shall mainly be concerned with types of individuals,[8] although the theory also recognizes higher types (cf. Nordström, Petersson, & Smith, 1990, chs. 19–20).[9] In Church's simple type theory there are two types of individuals: the type o of truth-values and the type ι of all other individuals. In Martin-Löf's type theory there are, by contrast, an infinite range of types of individuals. These types are inductively generated by means of so-called formation rules. At the base of the induction we find for instance the type $\mathbb{N}$ of natural numbers and, for each natural number n (including 0), a type $\mathbb{N}_n$ of n members. In particular there is an empty type $\mathbb{N}_0$. One also finds so-called identity types: $\mathbf{Id}(A, s, t)$ is a type whenever s, t are terms of type A. There are many constructions for forming new

---

then, by the structure of $\mathbb{B}$, we must have $\mathbb{B}, \varphi, 1 \Vdash B$ or $\mathbb{B}, \varphi, 1 \Vdash C$; but this contradicts the induction hypothesis. One argues likewise that $\mathbb{B}, \varphi, 1 \nVdash A$ if $A = B \vee C$. Finally, assume $A = B \supset C$. By the induction hypothesis, $\mathbb{B}, \varphi, 1 \nVdash B$ and $\mathbb{B}, \varphi, 1 \nVdash C$. We distinguish two cases:

1. Both $\mathbb{B}, \varphi, 0 \Vdash B$ and $\mathbb{B}, \varphi, 0 \Vdash C$. Then $\mathbb{B}, \varphi, s \Vdash B \supset C$ iff $s = \sqcup\{0 \rightharpoonup 0\} = 0 \rightharpoonup 0 = 0$.

2. Either $\mathbb{B}, \varphi, 0 \nVdash B$ or $\mathbb{B}, \varphi, 0 \nVdash C$. Then $\mathbb{B}, \varphi, s \Vdash B \supset C$ iff $s = \sqcup\emptyset = 0$.

Hence, in either case, $\mathbb{B}, \varphi, 1 \nVdash A$.

[8] Types of individuals are sometimes called *sets* in the literature on Martin-Löf type theory; a set in this sense is not a set in the sense of axiomatic set theory; see (Klev, 2014, pp. 136–140) for some discussion.

[9] A description of the hierarchy of higher types can also be found in (Klev, 201x, sec. 3).

Truthmaker Semantics

types of individuals from already given types of individuals. The following are relevant for our purposes: if A and B are types of individuals, then so are $A \times B$, $A + B$ and $A \to B$.

With each formation rule apart from N_0-formation one or more so-called introduction rules are associated. Introduction rules lay down what the *canonical* terms of the type formed by the formation rule are. For instance, with N-formation there are associated the following two introduction rules:

$$0 : N \qquad \frac{n : N}{s(n) : N}$$

That a term t is of type N means that t is equal within the calculus of the theory to a canonical term of type N. For instance, assuming that we have made the suitable definitions, it follows that $2 + 2$ is of type N, since this term can be shown to be equal to $s(3)$, which is a canonical term of type N provided 3 is of type N. In general, that t is of type A, where A is a type of individuals, means that t is equal to a canonical term of type A.

A canonical term of type $A \times B$ is a pair $\langle a, b \rangle$, where a is of type A and b of type B. A canonical term of type $A+B$ is either $\mathbf{i}(a)$, where a is of type A, or $\mathbf{j}(b)$, where b is of type B. A canonical term of type $A \to B$ is $\lambda(F)$, where F is a function from A to B. Hence, F is an entity of higher type, namely of the type of functions from A to B. Identity types, $\mathbf{Id}(A, s, t)$, are less familiar and will play only a minor role in what follows; but let us note that provided one can show in the theory's calculus that two terms s, t are equal, then $\mathbf{refl}(s)$ is a canonical term of type $\mathbf{Id}(A, s, t)$. For instance, $\mathbf{refl}(\mathbf{s}(3))$ is a canonical term of type $\mathbf{Id}(N, \mathbf{s}(3), 2+2)$. Each introduction rule introduces a primitive constant of the theory. Thus, 0, s, $\langle -, - \rangle$, $\mathbf{i}$, $\mathbf{j}$, λ, and $\mathbf{refl}$ are primitive constants. Apart from 0, which is of type N, all of these constants are functions, thus entities of higher type.

With each formation rule there is associated an elimination rule. Also the elimination rules introduce primitive constants. We shall not need these primitive constants in their full generality; but we list here certain other constants definable from them. From the constant introduced by the elimination rule for $A \times B$ one can define two functions **fst** and **snd** satisfying:

$$\mathbf{fst}(\langle a, b \rangle) = a \qquad \mathbf{snd}(\langle a, b \rangle) = b$$

This justifies speaking of $\langle a, b \rangle$ as the pair of a and b. Likewise, from the constant introduced by the elimination rule for $A + B$ one can define two functions $\mathbf{i}^{-1}$ and $\mathbf{j}^{-1}$ satisfying:

$$\mathbf{i}^{-1}(\mathbf{i}(a)) = a \qquad \mathbf{j}^{-1}(\mathbf{j}(b)) = b$$

Finally, from the constant introduced by the elimination rule for $A \to B$ one can define a binary function **ap** satisfying:

$$\mathbf{ap}(\lambda(F), a) = F(a)$$

Thus we may think of $\lambda(F)$ as coding the function F, whose value for an argument a we may recover by means of **ap**. Readers of Frege's *Grundgesetze* (Frege, 1893) will here see the similarity with Frege's course-of-values operator $\acute{\alpha}f(\alpha)$ together with the associated application function $x \frown y$, satisfying the equation $d \frown \acute{\alpha}f(\alpha) = f(d)$. Note, however, that $\lambda(F)$ is of type $A \to B$, whereas A is the domain of F, hence we cannot apply F to $\lambda(F)$, whence Russell's contradiction is avoided. The difference between F and $\lambda(F)$ should be emphasized: whereas $\lambda(F)$ is an individual, F is a higher-order entity, an object of higher type.

A fundamental principle of Martin-Löf's type theory is the so-called Curry–Howard isomorphism, or propositions-as-types or formulae-as-types principle (cf. Howard, 1980). In Martin-Löf's type theory this principle is implemented by the identification of the category of types of individuals with the category of propositions: each proposition is a type of individuals, and each type of individuals a proposition. We are certainly not used to thinking about $\mathbb{N}$ as a proposition; from the point of view of the Curry-Howard isomorphism we may think of it as the proposition that $\mathbb{N}$ is inhabited. This remark generalizes: a type A of individuals determines the proposition that A is inhabited (cf. Martin-Löf, 1975, p. 77). When regarding A as a proposition, terms of type A are regarded as *proofs* of A. Thus a proposition is identified with the type of its proofs. It should be emphasized that 'proof' is here a technical term, meaning certain objects, more precisely, certain terms of a calculus. Proofs in this sense are sometimes called 'proof-objects' in order to distinguish them from proof-acts, as well as from traces of such acts (cf. Sundholm, 1998).

If propositions are identified with types, then operations on propositions may be identified with operations on types. We define $A \wedge B = A \times B$, $A \vee B = A + B$, $A \supset B = A \to B$ and $\bot = \mathbb{N}_0$. Instead of speaking of a canonical term of type A, we may, when regarding A as a proposition, speak of a canonical proof of A. Hence we may say that a canonical proof of $A \wedge B$ is a pair $\langle a, b \rangle$, where a is a proof of A and b a proof of B; a canonical proof of $A \vee B$ is $\mathbf{i}(a)$, where a is a proof of A, or $\mathbf{j}(b)$, where b is a proof of B; a canonical proof of $A \supset B$ is a code $\lambda(F)$, where F is a function from proofs of A to proofs of B; and that $\bot$ has no canonical proof. That a is a proof of a proposition A means that a is identical to a

Truthmaker Semantics

canonical proof of A. It should be obvious how all of this makes precise the Brouwer–Heyting–Kolmogorov proof-explanation.

That a proposition is true means that it has a proof, hence that it is inhabited as a type. Sundholm (1994, p. 117) remarks that we thereby have a truthmaker conception of truth, according to which the truth of a truthbearer is analyzed as the existence of a truthmaker.[10] The truthbearers here are propositions understood as types of individuals, and the truthmakers are terms typed by propositions.[11] Since a proposition is thus identified with the type of its truthmakers, it is natural also to regard the semantics of Martin-Löf's type theory as a truthmaker semantics. It must then at once be added, however, that semantics in this case does not mean model-theoretic semantics, but rather informal meaning explanations. The type theory as presented for instance by Martin-Löf himself is a language for reasoning *with*, not a formal system to reason *about*. As such, constructive type theory is closer in spirit to Frege's ideography and the language of Russell and Whitehead's *Principia Mathematica* than to the majority of logical systems studied by logicians nowadays. Its being a language for reasoning with does not exclude a perspective from which it is regarded as a language to reason about, a perspective that therefore also allows for model-theoretic semantics.[12] Here, however, we are interested in the intended interpretation of the language.

A note on the distinction between open and closed terms is in order. Open terms, such as variables, are typed, just as closed terms are. An open term is, however, not typed categorically, but only dependently.[13] For instance, the term $s(x)$ is not typed categorically by $\mathbb{N}$; rather it is typed by $\mathbb{N}$ given a context in which x is typed by $\mathbb{N}$. This means that whenever n is a closed term of type $\mathbb{N}$—hence, whenever n is typed categorically by $\mathbb{N}$—then $s(n)$ is also typed categorically by $\mathbb{N}$. Likewise, a variable x is typed by A, not categorically, but only given a context in which x is typed by A. We cannot assert that a proposition A is true simply by producing a variable of type A; only if we have a closed term of type A can we assert that A is true. A variable of type A should rather be thought of as a one-step open proof of A. Producing a variable of type A amounts to *assuming* that A is

---

[10] A similar observation is made by Ranta (1995, p. 54) and discussed, for the case of empirical propositions, in more detail by Hiipakka (1996).

[11] An account of the relevant notion of existence must also be given; see (Sundholm, 1994, pp. 122–123).

[12] For an example of such a semantics, see (Hofmann & Streicher, 1998).

[13] Thus Martin-Löf's type theory differs from Church's simple type theory in the typing of open terms. In the latter an individual variable, for instance, is typed categorically by ι.

true. An open term such as $\langle x, y \rangle$, where x is a variable of type A and y a variable of type B may, likewise, be regarded as a linearized form of the following open proof of $A \wedge B$:

$$\frac{A \qquad B}{A \wedge B}$$

We obtain a closed proof $\langle a, b \rangle$ of $A \wedge B$ from this open proof by providing a closed proof a of A and a closed proof b of B. These observations point to the close connections that obtain between proof-objects and natural deduction derivations.

4 Four differences

We are now ready to begin comparing Martin-Löf's type theory with Fine's truthmaker semantics. I shall concentrate on the differences. What seems to me the most important differences will be discussed under four headings: the structure of truthmakers, the treatment of absurdity, the truthmaker relation, and the universe of truthmakers.

4.1 The structure of truthmakers

A first difference is that, whereas λ-terms have internal structure, Finean truthmakers have only external structure. The points in a truthmaker frame have structure only by virtue of how they relate to each other by means of the $\leq$-relation. A term of type theory, by contrast, has internal structure in the sense that it is built up in a certain definite way. It is not just a point in an algebraic system, but it has a certain internal structure determined by the manner in which it is built up. The truthmaker $\lambda([x, y]\langle x, y \rangle)$ of $A \supset (B \supset A \wedge B)$, for instance, is built up as follows:[14]

$$\frac{\dfrac{\dfrac{x \qquad y}{\langle x, y \rangle}}{[x, y]\langle x, y \rangle}}{\lambda([x, y]\langle x, y \rangle)}$$

The two variables x and y form the pair $\langle x, y \rangle$ and are then abstracted on to yield a function $[x, y]\langle x, y \rangle$; to which the λ-operator is applied to yield the individual $\lambda([x, y]\langle x, y \rangle)$.

---

[14]Following a similar convention in ordinary λ-calculus, '$\lambda([x, y]\langle x, y \rangle)$' is shorthand for '$\lambda([x]\lambda([y]\langle x, y \rangle))$'.

Truthmaker Semantics

From the internal structure of a canonical term one can tell directly what the form is of the proposition it makes true. For instance, a term of the form $\langle a, b \rangle$ is a truthmaker of a conjunction, and a term of the form $\lambda(F)$ is a truthmaker of an implication. If a term is not in canonical form, then it can be evaluated within the calculus of the theory to canonical form. Whence one can tell what the form is of the proposition the term makes true by thus evaluating it. It is clear that simply by looking at a point in a truthmaker model one cannot say anything about the formulae it makes true.

The internal structure of terms also makes it possible in a systematic fashion to recover, for instance, a truthmaker of A and a truthmaker of B from a truthmaker of $A \wedge B$, and a truthmaker of B from a truthmaker of $A \supset B$ and a truthmaker of A. Thus, if c is a term of type $A \wedge B$, then it is equal to a term of the form $\langle a, b \rangle$, where a is of type A and b of type B. Hence $\mathbf{fst}(c) = \mathbf{fst}(\langle a, b \rangle) = a$ is of type A, that is, a truthmaker of A, and $\mathbf{snd}(c) = \mathbf{snd}(\langle a, b \rangle) = b$ is a truthmaker of B. Likewise, $\mathbf{ap}(c, a)$ will be a truthmaker of B provided c is a truthmaker of $A \supset B$ and a a truthmaker of A. In Fine's truthmaker semantics, by contrast, there is no general method for recovering a truthmaker of A or a truthmaker of B from a truthmaker of $A \wedge B$, or a truthmaker of B from a truthmaker of $A \supset B$ and a truthmaker of A. For instance, simply by knowing of the existence of two points s_1, s_2 such that $s = s_1 \sqcup s_2$ and $s_1 \Vdash A$ and $s_2 \Vdash B$, we do not know which points s_1 and s_2 are (for instance, in $\mathbb{B}$ we have both $1 = 1 \sqcup 0$ and $1 = 1 \sqcup 1$).

Recall that $\mathfrak{T}, s \;\|\!\!\!> A$ means that there is $t \leq s$ such that $\mathfrak{T}, t \Vdash A$. We have seen that this notion of inexact truthmaking plays an important role in the metatheory of Fine's truthmaker semantics, and it is natural to ask whether it can be made sense of in Martin-Löf's type theory. The answer to this question depends in part on the fact that terms have internal structure, whence the question can profitably be discussed in this subsection.

A definition that naturally suggests itself is the following:

> the term s is an inexact truthmaker of A if s contains a subterm t that is a truthmaker of A.

Thus, $\langle a, b \rangle$, where a is of type A and b of type B, would be an inexact truthmaker of both A and B. In terms of natural deduction derivations this is to say that

$$\frac{\begin{array}{cc} \mathcal{D}_1 & \mathcal{D}_2 \\ A & B \end{array}}{A \wedge B}$$

is an inexact truthmaker of both A and B, since it contains the two subproofs

$$\begin{array}{cc} \mathcal{D}_1 & \mathcal{D}_2 \\ A & B \end{array}$$

Thus described, inexact truthmaking is, however, not a well-defined notion: owing to the presence of elimination rules, there may be terms s and t such that $s = t$ and s, but not t, is an inexact truthmaker of A. For instance, $\mathbf{fst}(\langle a,b \rangle)$ is an inexact truthmaker of $A \wedge B$, since it contains the subterm $\langle a,b \rangle$; but $a = \mathbf{fst}(\langle a,b \rangle)$, and a need not be an inexact truthmaker of $A \wedge B$. In terms of natural deduction: the proof

$$\dfrac{\dfrac{\begin{array}{cc}\mathcal{D}_1 & \mathcal{D}_2 \\ A & B\end{array}}{A \wedge B}}{A}$$

is an inexact truthmaker of $A \wedge B$, but in Martin-Löf's calculus we have

$$\dfrac{\dfrac{\begin{array}{cc}\mathcal{D}_1 & \mathcal{D}_2 \\ A & B\end{array}}{A \wedge B}}{A} \quad = \quad \begin{array}{c}\mathcal{D}_1 \\ A\end{array}$$

and the proof on the right hand side need not contain any subproof that is a proof of $A \wedge B$.

To get a workable notion of inexact truthmaking we must rely on Martin-Löf's normalization theorem (Martin-Löf, 1975, Theorem 3.3). It follows from this theorem that every term has a unique normal form. Hence we may say

> the term s is an inexact truthmaker of A if the normal form of s contains a subterm t that is a truthmaker of A.

Since identical terms have the same normal form, this is a well-defined notion.

4.2 Absurdity

A second difference concerns the treatment of absurdity. We have already seen that $\bot$ is included in the domain of the valuation function in a Finean truthmaker model and that the only condition pertaining to the value of $\varphi(\bot)$ is that if $s \in \varphi(\bot)$, then for each propositional letter p there is $t \leq s$

such that $t \in \varphi(p)$. Therefore, given a truthmaker frame $\mathfrak{F} = (S, \leq)$ and provided $0 \in \varphi(p)$ for every atomic letter p, then we can let $\varphi(\bot) = S$. Hence, in the truthmaker model $(\mathfrak{F}, \varphi)$ every state will make $\bot$ true. In particular, therefore, if $|S| = \aleph_0$, then there are $\aleph_0$ many states in $\mathfrak{T}$ making $\bot$ true. In Martin-Löf's type theory, by contrast, the proposition $\bot$ has no truthmakers whatsoever. A truthmaker of $\bot$ would have to be a term that is identical to a canonical proof of $\bot$, but $\bot$ has no canonical proofs.

It may be remarked that also the realizability semantics of Läuchli (1970) allows, in the current language, that $\bot$ has truthmakers. In fact, Läuchli (ibid. p. 230) notes that if the set of truthmakers of $\bot$ is required to be empty, then the formulae valid according to his realizability semantics will not coincide with the set of theorems of intuitionistic logic; in particular, $\neg A \lor \neg\neg A$ will then be valid, though it is not an intuitionistic theorem.[15] A similar situation obtains in Fine's truthmaker semantics. If no state in a truthmaker model $\mathfrak{T}$ makes $\bot$ true, then only the bottom element 0 can make true a formula of the form $\neg A$.[16] Assume that $\mathfrak{T}$ is such a model and that $\mathfrak{T}, 0 \not\Vdash \neg A$. Then there is no t such that $\mathfrak{T}, t \Vdash \neg A$. Hence the identity function $\mathrm{id} : S \to S$ has the property that $\mathrm{id}(t) \in \varphi(\bot)$ whenever $\mathfrak{T}, t \Vdash \neg A$. Whence

$$\sqcup \{t \rightharpoonup \mathrm{id}(t) \mid \mathfrak{T}, t \Vdash \neg A\} = \sqcup \emptyset = 0$$

is a truthmaker of $\neg\neg A$. Therefore (by classical reasoning),

$$\mathfrak{T}, 0 \Vdash \neg A \lor \neg\neg A.$$

Fine (2014, p. 557) notes more generally that for no $n \in \mathbb{N}$ can we make the general requirement $|\varphi(\bot)| \leq n$ on valuation functions φ, provided we want the resulting logic to be intuitionistic.

Following standard practice we define $\neg A = A \supset \bot$. Although $\bot$ has no closed proofs, a proposition of the form $\neg A$ may have closed proofs.[17] For instance, the term $\lambda([x]\mathbf{ap}(x, \mathbf{refl}(0)))$ can be seen to be a closed proof of $\neg\neg \mathbf{Id}(\mathbb{N}, 0, 0)$,[18] the double negation of the proposition that 0 is identical

---

[15] For a detailed explanation of this point, see Lipton and O'Donnell (1996, p. 235); this whole article gives a very helpful explanation of Läuchli's semantics.

[16] Assume $\varphi(\bot) = \emptyset$ and $\mathfrak{T}, s \Vdash \neg A$. Then $s = \sqcup\{t \rightharpoonup F(t) \mid \mathfrak{T}, t \Vdash A\}$ where $F : S \to S$ is such that $F(t) \in \varphi(\bot)$ whenever $\mathfrak{T}, t \Vdash A$. Hence there is no t such that $\mathfrak{T}, t \Vdash A$, whence $s = \sqcup \emptyset = 0$.

[17] For a discussion of difficulties surrounding truthmakers for 'negative truths', see (Molnar, 2000).

[18] Let x be a variable of type $\neg \mathbf{Id}(\mathbb{N}, 0, 0)$. Then $\mathbf{ap}(x, \mathbf{refl}(0))$ is of type $\bot$, whence $\lambda([x]\mathbf{ap}(x, \mathbf{refl}(0)))$ is of type $\neg\neg \mathbf{Id}(\mathbb{N}, 0, 0)$.

to itself. Suppose one can construct a closed term s of type $\neg A$. Assuming the system to be consistent, one cannot then also construct a closed term t of type A, since in that case $\mathbf{ap}(s, t)$ would be a closed term of type $\bot$, contrary to the definition of $\bot$. Say that a proposition A is false if $\neg A$ is true. The foregoing reasoning shows that a false proposition has no truthmakers. This does not, however, mean that all false propositions are identical. For, if the types A and B are identical, then any term of type A—be it closed or open—is also a term of type B. The parenthetical remark here is important: although false propositions have no closed proofs, they may be differentiated by their open proofs. For instance, if x is a variable of type A and y a variable of type $\neg A$, then $\langle x, y \rangle$ is an open proof of $A \wedge \neg A$, but not of $\bot$, since the type of a term $\langle a, b \rangle$ must have the form $A \wedge B$. Hence, $\bot$ and $A \wedge \neg A$ are not identical propositions. It should here be added that Martin-Löf's type theory includes a theory of the identity of types, hence also of the identity of propositions. The theory is a higher-order λ-calculus and includes, for instance, principles to the effect that a type A and any α-, β-, or η-reduct of A are identical; a principle to the effect that a *definiendum* is identical with its *definiens*; and, of course, principles to the effect that identity between types is an equivalence relation and that substituting identicals for identicals preserves identity.

4.3 The truthmaker relation

A third difference concerns the truthmaker relation. Among the basic forms of statement in Martin-Löf's type theory are '$a : A$' and '$a = b : A$'. The first of these says that a is an individual of type A and the second that a and b are identical individuals of type A. Concrete examples are $2 : \mathbb{N}$ and $2 + 2 = \mathbf{s}(3) : \mathbb{N}$. If A is regarded as a proposition, then $a : A$ may be read as saying that a is a proof, or truthmaker, of A. Thus we see that truthmaker statements are here made in the language of the theory itself. The theory even allows for statements to the effect that two truthmakers s, t are equal, namely in the form '$s = t : A$'. In Fine's truthmaker semantics, by contrast, the truthmaker statement $\mathfrak{T}, s \Vdash A$ belongs only to the metatheory. The same holds for statements of equality between truthmakers. Such statements do occur in the definition of $\mathfrak{T}, s \Vdash A$, namely in the clauses for $\wedge$ and $\supset$, but they can be made only in the metatheory.

In Martin-Löf's type theory the relation between a truthmaker and the truthbearer it makes true is identified with the relation between an individual and its type. If there are such things as internal relations, then the relation between an individual and its type is a very good candidate for being one, for this is the relation between an individual and its *what-it-is*, its genus. What is 7? A natural number. What is $\langle 5, 7 \rangle$? A pair of natural numbers. However the notion of internal relation is understood (provided it is accepted at all), it would seem that the relation between an object and its genus, its kind, should count as one. Thus also the truthmaker relation would be regarded as an internal relation in Martin-Löf's type theory. What is **refl**(0)? A truthmaker of the proposition that 0 is identical with itself. That the truthmaker relation is internal has been emphasized, for instance, by Armstrong (2004, p. 8). Indeed, Rami (2009, p. 9) claims that 'most truth-maker theorists are internalists concerning the truth-maker relation'. While this intuition is thus easily accounted for in Martin-Löf's type theory, it seems more difficult to do so in Fine's semantics. The truthmaker relation there is just an ordinary relation between points in a mathematical structure and sentences in a formal language; given a truthmaker model $\mathfrak{T}$ and a suitable language L, the relation $\Vdash$ is, as it were, a further thing, not internal to $\mathfrak{T}$ and L.

Armstrong (ibid.) remarks that (on his conception of internal relations), if the truthmaker relation is internal, then a truthmaker statement does not itself require a truthmaker. Wittgenstein in the *Tractatus* claimed that the obtaining of an internal relation cannot be expressed in a proposition.[19] Hence, if propositions are the truthbearer-relata of the truthmaker relation, then Wittgenstein can be taken to hold that truthmaker statements *cannot*, on grounds of logical form, have truthmakers: only propositions have truthmakers, but truthmaker statements are not propositions. All of this conforms with how truthmaker theory is realized in Martin-Löf's type theory. A statement of the form '$a : A$' is called a *judgement*, and judgements are carefully distinguished form propositions (e.g. Martin-Löf, 1984, p. 3). In particular, only propositions are identified with types of individuals; and to say that an individual a is of type A can be done only in a judgement, namely $a : A$, not in a proposition. Hence, a judgement is not the kind of thing that can have a truthmaker.[20] It may be remarked that a regress is thereby blocked that would seem to ensue if a truthmaker statement itself required a truthmaker: A is true iff there exists a truthmaker of A iff there exists a truthmaker of 'there exists a truthmaker of A' etc. (cf. Sundholm, 1994, p. 119).

---

[19]TLP 4.122: "Das Bestehen solcher interner Eigenschaften und Relationen kann aber nicht durch Sätze behauptet werden,..."

[20]This is not to say that there is no notion of truth or correctness pertaining to judgements; see Martin-Löf (1987) and Sundholm (2004).

4.4 The universe of truthmakers

A fourth difference concerns how the universe of truthmakers is conceived. In Martin-Löf's type theory this universe is typed. Each truthmaker belongs to a specific type, indeed it is typed by the proposition it makes true. In Fine's semantics, by contrast, the universe of truth-makers is type-free, in the sense that all truthmakers belong to the same type. One may of course call $[\![A]\!]_{\mathfrak{T}} = \{s \in S \mid \mathfrak{T}, s \Vdash A\}$ a type, but the result of doing so is not a formulae-as-types interpretation, since also for a false formula A may the type $[\![A]\!]_{\mathfrak{T}}$ be inhabited. Nor are the types so called disjoint in general, since, for instance, $0 \in [\![A]\!]_{\mathfrak{T}}$ whenever A is a theorem of intuitionistic logic. And indeed, it seems to be more natural to call $[\![A]\!]_{\mathfrak{T}}$ a set in the ordinary sense of set theory rather than a type in the sense of type theory.

Fine's truthmaker semantics shares the property of being thus 'type-free' with the so-called theory of constructions of Kreisel and Goodman, which is presented by Goodman (1970, p. 101) as 'a type-free and logic-free theory directly about the rules and proofs which underlie constructive mathematics'.[21] It seems, however, that the similarity between these two interpretations of intuitionistic logic stops there, for the truthmakers of Kreisel and Goodman are λ-terms, hence they have internal structure, and (assuming the system to be consistent) no terms exist there that make $\bot$ true.

The realizability semantics of Läuchli (1970) is naturally viewed as developed within a type theory. It does, however, not constitute a formulae-as-types interpretation, since for any formula A, the type there associated with A is inhabited; it is inhabited, but it need not be inhabited by *realizers*. The distinction thus made between arbitrary elements of types and realizers has a conceptual parallel in Fine's distinction between truthmakers *simpliciter* and actual truthmakers. The manner in which this distinction is implemented mathematically differs, however, in the two systems: realizers are elements invariant under groups of permutations (cf. Lipton & O'Donnell, 1996, pp. 214–221), whereas actual truthmakers in a truthmaker model are elements of a non-empty ideal (in the lattice-theoretic sense) of consistent states (Fine, 2014, p. 571). But, again, we see that there are certain affinities between Läuchli's and Fine's systems.

An account of truthmaker theory in terms of type theory has been given by Jago (2011). This theory is, however, very different from Martin-Löf's, since Jago assigns all truthmakers to one and the same type (in one sense, therefore, Jago's theory may be called type-free), namely the type f of facts.

---

[21] For a discussion of this point, see (Sundholm, 1983, pp. 155–156).

Truthmaker Semantics

Important for Jago is a distinction between positive and negative facts; the distinction is to be so understood that if a positive fact t^+ makes A true, then the corresponding negative fact t^- would have made $\neg A$ true, and *vice versa*. Jago implements this distinction mathematically by means of what we may call signed function application: a function P of type $\tau \to f$ can be applied either positively or negatively to an element a of type τ, the result being written $[Pa]^+$ and $[Pa]^-$, respectively. Let me record here an argument that the rules of Jago's calculus lead to the identification of positive and negative facts. The calculus includes η-reduction

$$\lambda x.(tx) \twoheadrightarrow t \quad (x \notin \mathrm{FV}(t))$$

Jago (ibid. p. 41) glosses this rule by saying that 'the property you get by abstracting from a fact $[Fa]^+$ is none other than F itself'. For this to hold, η-reduction must also hold in the form

$$\lambda x.[Px]^+ \twoheadrightarrow P \quad (x \notin \mathrm{FV}(P))$$

and, lacking any statement to the contrary, I take it that Jago also assumes η-reduction to hold in the form

$$\lambda x.[Px]^- \twoheadrightarrow P \quad (x \notin \mathrm{FV}(P))$$

Jago regards two terms as identical if they have the same normal form (ibid. p. 42). But then $\lambda x.[Px]^+ = \lambda x.[Px]^-$, hence also $[Pa]^+ = [Pa]^-$ for any a of suitable type. It is not obvious how this conclusion can be avoided in a manner that is not ad hoc. (The moral, it seems to me, is that the idea of signed application introduces more complications than what would seem to be recognized in Jago's paper.)

4.5 Conclusion

Sundholm (1983, pp. 159–160) noted that there is an important difference between regarding the Brouwer–Heyting–Kolmogorov clauses as part of an explanation of the meaning of mathematical propositions and regarding them as a heuristic in setting up a formal semantics, such as realizability semantics or the theory of constructions. Martin-Löf's type theory, at least as presented by Martin-Löf himself, aims to offer a theory of meaning for the language of constructive mathematics. Naive set theory or the like is not assumed as an ambient theory; rather one starts from scratch and builds up a language by carefully explaining the meaning and syntax of the symbols one uses. Fine's truthmaker semantics, by contrast, is a formal semantics that presupposes a meaningful language for speaking about both formulae and certain algebraic structures. Here lies perhaps the most fundamental difference between these two systems. We have abstracted from this difference above, since only by doing so, it seems, can the two systems be compared in the first place. Thus, we have at times regarded type theory purely formally and at other times regarded truthmaker semantics as a fundamental form of semantics that takes no other mathematics for granted. We may thereby have been able to shed some light on both systems.

References

Armstrong, D. (2004). *Truth and Truthmakers*. Cambridge: Cambridge University Press.
Church, A. (1940). A formulation of the simple theory of types. *Journal of Symbolic Logic, 5,* 56–68.
Fine, K. (2014). Truth-maker semantics for intuitionistic logic. *Journal of Philosophical Logic, 43,* 549–577.
Fine, K. (2016a). *Constructing the impossible*. (Unpublished article available at www.academia.edu/11339241/Constructing_the_Impossible)
Fine, K. (2016b). *Truthmaker Semantics*. (Unpublished article available at https://www.academia.edu/10908756/survey_of_truthmaker_semantics)
Frege, G. (1893). *Grundgesetze der Arithmetik I*. Jena: Hermann Pohle.
Goodman, N. D. (1970). A theory of constructions equivalent to arithmetic. In A. Kino, J. Myhill, & R. E. Vesley (Eds.), *Intuitionism and Proof Theory* (pp. 101–120). Amsterdam: North-Holland.
Heyting, A. (1956). *Intuitionism. An Introduction*. Amsterdam: North-Holland.
Hiipakka, J. (1996). A philosophical problem in Martin-Löf's type theory. In T. Childers, P. Kolář, & V. Svoboda (Eds.), *LOGICA '95. Proceedings of the 9th International Symposium* (pp. 39–46). Prague: Filosofia.
Hindley, J. R., & Seldin, J. P. (1986). *Introduction to Combinators and λ-Calculus*. Cambridge: Cambridge University Press.
Hofmann, M., & Streicher, T. (1998). The groupoid interpretation of type theory. In G. Sambin & J. M. Smith (Eds.), *Twenty-five Years of*

Constructive Type Theory (pp. 83–111). Oxford: Oxford University Press.
Howard, W. A. (1980). The formulae-as-types notion of construction. In J. P. Seldin & J. R. Hindley (Eds.), *To H.B. Curry: Essays on Combinatory Logic, Lambda Calculus and Formalism* (pp. 479–490). London: Academic Press.
Husserl, E. (1901). *Logische Untersuchungen. Zweiter Band.* Max Niemeyer.
Jago, M. (2011). Setting the facts straight. *Journal of Philosophical Logic*, *40*, 33–54.
Klev, A. (2014). *Categories and Logical Syntax* (Unpublished doctoral dissertation). Leiden University.
Klev, A. (201x). Identity and sortals (and Caesar). *Erkenntnis*. (Forthcoming)
Kreisel, G. (1962). Foundations of intuitionistic logic. In E. Nagel, P. Suppes, & A. Tarski (Eds.), *Logic, Methodology and the Philosophy of Science* (pp. 198–210). Stanford: Stanford University Press.
Langtry, B. (1975). Similarity, continuity and survival. *Australasian Journal of Philosophy*, *53*, 3–18.
Läuchli, H. (1970). An abstract notion of realizability for which intuitionistic predicate calculus is complete. In A. Kino, J. Myhill, & R. E. Vesley (Eds.), *Intuitionism and Proof Theory* (pp. 227–234). Amsterdam: North-Holland.
Lipton, J., & O'Donnell, M. J. (1996). Some intuitions behind realizability semantics for constructive logic: Tableaux and Läuchli countermodels. *Annals of Pure and Applied Logic*, *81*, 197–239.
Martin-Löf, P. (1975). An intuitionistic theory of types: Predicative part. In H. E. Rose & J. C. Shepherdson (Eds.), *Logic Colloquium '73* (pp. 73–118). Amsterdam: North-Holland.
Martin-Löf, P. (1982). Constructive mathematics and computer programming. In J. L. Cohen, J. Łoś, et al. (Eds.), *Logic, Methodology and Philosophy of Science VI, 1979* (pp. 153–175). Amsterdam: North-Holland.
Martin-Löf, P. (1984). *Intuitionistic Type Theory*. Naples: Bibliopolis.
Martin-Löf, P. (1987). Truth of a proposition, evidence of a judgement, validity of a proof. *Synthese*, *73*, 407–420.
Molnar, G. (2000). Truthmakers for negative truths. *Australasian Journal of Philosophy*, *78*, 72–86.
Mulligan, K., Simons, P., & Smith, B. (1984). Truth-makers. *Philosophy*

and Phenomenological Research, *44*, 287–321.

Nordström, B., Petersson, K., & Smith, J. (1990). *Programming in Martin-Löf's Type Theory*. Oxford: Oxford University Press.

Punčochář, V. (2016). A nonstandard semantic framework for intuitionistic logic. In P. Arazim & M. Dančák (Eds.), *The LOGICA Yearbook 2015* (pp. 179–192). London: College Publications.

Rami, A. (2009). Introduction: truth and truthmaking. In E. J. Lowe & A. Rami (Eds.), *Truth and Truth-making* (pp. 1–36). Stocksfield: Acumen.

Ranta, A. (1995). *Type-Theoretical Grammar*. Oxford: Oxford University Press.

Sundholm, B. G. (1983). Constructions, proofs and the meaning of logical constants. *Journal of Philosophical Logic*, *12*, 151–172.

Sundholm, B. G. (1994). Existence, proof and truth-making: A perspective on the intuitionistic conception of truth. *Topoi*, *13*, 117–126.

Sundholm, B. G. (1998). Proofs as acts versus proofs as objects: Some questions for Dag Prawitz. *Theoria*, *64*, 187–216.

Sundholm, B. G. (2004). Antirealism and the roles of truth. In I. Niiniluoto, M. Sintonen, & J. Wolenski (Eds.), *Handbook of Epistemology* (pp. 437–466). Dordrecht: Kluwer.

Tooley, M. (1977). The nature of laws. *Canadian Journal of Philosophy*, *7*, 667–698.

Troelstra, A. S. (1969). *Principles of Intuitionism*. Berlin: Springer.

Wittgenstein, L. (1922). *Tractatus Logico-Philosophicus*. London: Routledge & Kegan Paul.

Ansten Klev
Czech Academy of Sciences
The Czech Republic
E-mail: anstenklev@gmail.com

Ignorance Without K(nowledge)

EKATERINA KUBYSHKINA

Abstract: We investigate the representation of such notions as knowledge and ignorance in a formal setting. In particular, we argue that these notions are not necessarily inter-definable. In order to defend this idea we compare two logics where ignorance is represented without any reference to K_a operator: the modal logic **Ig** introduced by van der Hoek and Lomuscio (2004) and the author's four-valued logic **RA+**. A connection between these two logics is established by defining a translation from **RA+** into **Ig**.

Keywords: Many-valued logics, Modal logics, Logic of rational agent, Rational agency, Translations, Knowledge representation

1 Introduction

Traditionally, the notion of knowledge is represented in epistemic logic by exploiting the K_a operator. Hintikka (1962) provided a semantic interpretation of this operator, which can be interpreted in terms of standard possible world semantics: "a knows ϕ" ($K_a\phi$) means that in all possible worlds compatible with what the agent a knows, it is the case that ϕ. The use of K_a operator together with the appropriate semantics provides a powerful tool to represent the notion of knowledge in a formal setting, to clarify its properties and formalize reasoning methods based on an agent's state of knowledge.

It is usual to use the K_a operator to represent not only the state of knowledge of an agent, but also his state of ignorance. Hintikka (1962) himself proposes two possible formalizations of an agent's state of ignorance: either as the negation of knowing (i) $\neg K_a \phi$; or, if one speaks about the ignorance of some true proposition, as (ii) $\phi \land \neg K_a \phi$. We add to these two definitions a third one, where ignorance is understood as not-knowing the truth value of a proposition: (iii) $\neg K_a \phi \land \neg K_a \neg \phi$. This corresponds to not-knowing neither this proposition, nor its negation. A sound representation of not-knowing formalized by (i) $\neg K_a \phi$ can not be really taken as ignorance representation, because it does not exclude the case when the agent does not know ϕ because he knows $\neg \phi$. In such a situation we can not say that the agent ignores ϕ. This leaves us with two possible definitions of ignorance in terms of K_a operator:

(ii) $\phi \wedge \neg K_a \phi$

and

(iii) $\neg K_a \phi \wedge \neg K_a \neg \phi$,

which represent two distinct characterizations of ignorance.

The use of (ii) $\phi \wedge \neg K_a \phi$ permits one to highlight the presuppositional nature of the verb "to know". As it is widely agreed in the literature, the verb "to know" is a factive presupposition trigger (see e.g. Kiparsky and Kiparsky (1970)). Hence its projection (in our case the negation of a proposition containing the presupposition trigger), should imply the same presuppositions. On this assumption we have that if the presupposition of $K_a \phi$ is ϕ, then the presupposition of ignoring ϕ is that ϕ is the case. The last is perfectly represented by (ii) $\phi \wedge \neg K_a \phi$. Nonetheless keeping the presupposition property for ignorance does not permit to represent ignorance from the agent's own point of view. Indeed, the agent is not capable to formulate his ignorance of a true proposition, because of his own ignorance about the truth value of the proposition under analysis. This formalization can be used only from some external, objective point of view, by someone who knows the state of affairs and reasons about the agent's incomplete epistemic state. In what follows we call this attitude the *objective approach* to ignorance characterization.

The other formalization, (iii) $\neg K_a \phi \wedge \neg K_a \neg \phi$, seems to be more suitable to represent ignorance from the agent's perspective. In this case ignorance is understood as not-knowing the truth value of a proposition (and thus not-knowing neither the proposition, nor its negation). This formalization can be exploited by the agent himself to reason about his own mental state, without any appeal to an objective state of affairs. We call this understanding of ignorance the *subjective approach* to ignorance characterization.

The use of K_a operator is not the only way to formalize the notion of ignorance. Van der Hoek and Lomuscio (2004) introduce a logic which formalizes ignorance by a special operator I taken as a primitive. Even though K_a operator is not used in this logic, the author's approach to ignorance characterization is the subjective one. Hence the subjective approach does not necessarily presuppose the representation of ignorance by the formula (iii). The objective approach can also be expressed via a logic that does not contain the K_a operator. The aim of the present paper is to analyze and compare these two representations of the notion of ignorance. In order to advance our analyses we describe a logic introduced by van der Hoek and

Lomuscio (2004) in section 2. Section 3 introduces a many-valued logic, where ignorance, as well as knowledge, is represented not by use of modal operators, but at the level of valuations. This logic is based on the objective approach to ignorance characterization. A connection between these two logics will be established by defining a translation procedure in section 4. We conclude in section 5.

2 Logic for ignorance

Logic for ignorance (**Ig**) constructed by van der Hoek and Lomuscio (2004) is a modal logic, used to formalize the notion of ignorance without the use of K_a operator. The authors explicitly state that by ignorance they mean a mental state in which the agent is unsure about the truth value of some proposition. This indicates that they adopt a subjective approach to ignorance interpretation.

Syntactically, **Ig** uses a mono-modal language $\mathcal{L}_I$ defined as follows (Backus-Naur Formalization):

$$\phi ::= p \mid \neg \phi \mid \phi \wedge \phi \mid I\phi$$

A formula $I\phi$ should be read as 'the agent is ignorant about ϕ'. The operators of disjunction, implication, and equivalence may be introduced by classical definitions.

Semantically, **Ig** uses standard possible world semantics with a special clause for I operator :

Definition 1 *A Kripke Frame $F = (W, R)$ is a tuple where W is a set of epistemic alternatives for the agent, and $R \subseteq W \times W$ is an accessibility relation. A Kripke Model $M = (F, \pi)$ is a tuple where F is a Kripke frame and $\pi : P \to 2^W$ is an interpretation for a set of propositional variables P.*

Given a model M and a formula ϕ, we say that ϕ is true in M at world w, written $M \models_{Ig}^{w} \phi$ if: [1]

- $M \models_{Ig}^{w} p$ *if* $w \in \pi(P)$;

- $M \models_{Ig}^{w} \neg \phi$ *if it is not the case that* $M, w \models_{Ig}^{w} \phi$;

- $M \models_{Ig}^{w} \phi \wedge \psi$ *if* $M \models_{Ig}^{w} \phi$ *and* $M \models_{Ig}^{w} \psi$;

---
[1] In what follows we call it an **Ig**-model.

- $M \models_{Ig}^{w} I\phi$ if there exist w', w'' such that Rww', Rww'', $M \models_{Ig}^{w'} \phi$ and $M \models_{Ig}^{w''} \neg\phi$.

A formula ϕ is valid, written $\models_{Ig} \phi$, if it is true in every world in every model.

The system **Ig** that is complete with respect to the semantics above is defined as follows:

Definition 2

I0 *All instances of propositional tautologies;*

I1 $I\phi \leftrightarrow I\neg\phi$;

I2 $I(\phi \wedge \psi) \to (I\phi \vee I\psi)$;

I3 $(\neg I\phi \wedge I(\alpha_1 \wedge \phi)) \wedge \neg I(\phi \to \psi) \wedge I(\alpha_2 \wedge (\phi \to \psi)) \to (\neg I\psi \wedge I(\alpha_1 \wedge \psi))$;

I4 $(\neg I\phi \wedge I\alpha) \to (I(\alpha \wedge \psi) \vee I(\alpha \wedge \neg\psi))$;

RI $\vdash_{Ig} \phi \Rightarrow \vdash_{Ig} \neg I\phi \wedge (I\alpha \to I(\alpha \wedge \phi))$;

MP *Modus Ponens;*

Sub *Substitution of equivalences.*

The completeness result for the system Ig is provided by van der Hoek and Lomuscio (2004). The authors introduced ignorance following the subjective approach to its characterization. Being ignorant about ϕ means that "the agent conceives two opposite alternatives for ϕ. In usual epistemic language this not just imples, but is semantically equivalent to saying that the agent does not know ϕ, and does not know $\neg\phi$ (...):

$$I\phi \equiv (\neg K\phi \wedge \neg K\neg\phi)."[2]$$

Interestingly enough, this equivalence does not allow the authors to prove the completeness of **Ig** by translating the epistemic operators used in usual modal systems for epistemic logic (e.g. $S5$). Moreover, they pointed out, that K_a operator cannot be defined in terms of I operator. The last two

---

[2] van der Hoek and Lomuscio (2004), p. 104.

observations make it clear that **Ig** should be considered as a way of formalizing the notion of ignorance independently of the notion of knowledge. In the next two sections, firstly we provide an alternative for representing the notion of ignorance and secondly we show the relationship between such an alternative and **Ig**.

3 Logic of rational agent

In this section we introduce a logic of generalized truth values. The semantical considerations underlying this logic were presented in (Kubyshkina & Zaitsev, 2016). The peculiarity of the semantics for the logic of rational agent lies in the fact that the truth values are considered to be structured entities each consisting of two components. The first component is the value *true* or *false*, T or F, respectively. The second one is a characterization of the epistemic state of the agent - *known* or *ignored*, 1 or 0, respectively. We multiply our two sets ($\{T, F\}$ and $\{1, 0\}$) and we have a set Q of four possible values $Q := \{T1, T0, F1, F0\}$. The value of a true proposition that the agent knows is '$T1$'; the value of a true proposition the truth of which is ignored by the agent is '$T0$'; the value of a false proposition the falsity of which is known by the agent is '$F1$'; and '$F0$' stands for a false proposition the value of which is ignored by the agent.

The logical system introduced in (Kubyshkina & Zaitsev, 2016) used the language $\mathcal{L}$ defined as follows:

$$\phi ::= p \mid \neg_4 \phi \mid \sim_4 \phi \mid \phi \wedge_4 \phi \mid \phi \vee_4 \phi$$

The operators are defined by the truth-tables 1, 2, 3, 4.

ϕ	$\neg_4 \phi$
T1	F1
T0	F0
F0	T0
F1	T1

Table 1: Ontological negation

We call operator $\neg_4$ *ontological negation*. This type of negation changes the ontological part of the truth value ('T' to 'F' and vice versa), without changing the epistemic state of an agent ('1' or '0'). We call operator $\sim_4$

ϕ	$\sim_4 \phi$
T1	T0
T0	T1
F0	F1
F1	F0

Table 2: Epistemic negation

$\phi \wedge_4 \psi$	**T1**	**T0**	**F0**	**F1**
T1	T1	T0	F0	F1
T0	T0	T0	F0	F1
F0	F0	F0	F0	F1
F1	F1	F1	F1	F1

Table 3: Conjunction

epistemic negation. It changes the epistemic state of an agent from known ('1') to ignored ('0'). The crucial point here is that this operator is not the analogue of "unknown" or "ignorant". We do not provide an abstract interpretation of the connective '$\sim_4$' in the natural language, but rather apply $\sim_4$ to some formula and give the interpretation. Suppose that a formula ϕ takes value $T1$ (i.e., ϕ is true and the agent knows it), then $\sim_4 \phi$ would take the value $T0$ (i.e., ϕ in this case is true, but $\sim_4 \phi$ indicates that the agent is ignorant about the truth of ϕ). In this case $\sim_4 \phi$ may be associated with "ϕ and ignored that ϕ", but we can not generalize this interpretation to any formula. Suppose that ϕ takes value $T1$, then $\sim_4\sim_4 \phi$ takes also the value $T1$. In this case $\sim_4\sim_4 \phi$ can not be interpreted as "ignorant about ignorance of the truth of ϕ", but as "ϕ is true and the agent knows it". This remark clarifies the difference between $\sim_4 \phi$ in $\mathcal{L}$ and the interpretation of epistemic operators ('K' or 'I').

In (Kubyshkina & Zaitsev, 2016) the system **LRA** based on the language $\mathcal{L}$ was introduced. At the semantical level **LRA** uses a consequence relation defined as preservation of the truth value '$T1$'.

The system **LRA** being quite limited, in this paper we present another system called **RA+**. Let us define the language $\mathcal{L}_{RA}$ as follows:

$$\phi ::= p \mid \neg_4\phi \mid \sim_4 \phi \mid \phi \wedge_4 \phi \mid \phi \vee_4 \phi \mid \phi \rightarrow_4 \phi$$

The language $\mathcal{L}_{RA}$ is the extension of the language $\mathcal{L}$ by implication,

Ignorance Without K(nowledge)

$\phi \vee_4 \psi$	$T1$	$T0$	$F0$	$F1$
T1	$T1$	$T1$	$T1$	$T1$
T0	$T1$	$T0$	$T0$	$T0$
F0	$T1$	$T0$	$F0$	$F0$
F1	$T1$	$T0$	$F0$	$F1$

Table 4: Disjunction

defined by truth-table 5.

$\phi \rightarrow_4 \psi$	$T1$	$T0$	$F0$	$F1$
$T1$	$T1$	$T0$	$F0$	$F1$
$T0$	$T1$	$T1$	$F0$	$F0$
$F0$	$T1$	$T1$	$T1$	$T0$
$F1$	$T1$	$T1$	$T1$	$T1$

Table 5: Implication

Taking conjunction and disjunction as the analogues of set theoretical operations of intersection and union respectively, we introduce a lattice **RA** of truth values and an ordering $\leq_{RA}$ (see figure 1).

We can now define the consequence relation for **RA+**.

Definition 3 *Let v^4 be a four-valued valuation function that maps all formulas of $\mathcal{L}_{RA}$ into the set $Q = \{T1, T0, F0, F1\}$. For all $\phi, \psi \in \mathcal{L}_{RA}$:*

$$\phi \models_{RA+} \psi \text{ iff } \forall v^4(v^4(\phi) \leq_{RA} v^4(\psi)).$$

The same consequence relation may be expressed as follows:
$\phi \models_{RA+}^{\mathfrak{X}\mathfrak{Y}} \psi$ iff $\forall v^4(v^4(\phi) \in \mathfrak{X} \Rightarrow v^4(\psi) \in \mathfrak{Y})$,
where $\mathfrak{X}, \mathfrak{Y}$ are the sets defined as follows: either $\mathfrak{X} = \{T1\}$ and $\mathfrak{Y} = \{T1\}$, or $\mathfrak{X} = \{T0\}$ and $\mathfrak{Y} = \{T1, T0\}$, or $\mathfrak{X} = \{F0\}$ and $\mathfrak{Y} = \{T1, T0, F0\}$, or $\mathfrak{X} = \{F1\}$ and $\mathfrak{Y} = \{T1, T0, F0, F1\}$.

We now present a system that is complete with respect to the semantics we described above. [3]

Definition 4 *The system **RA+** is defined as follows:* [4]

---

[3] The completeness result will be presented in full details in the author's Ph.D. thesis (in preparation).

[4] "$\phi \leftrightarrow_4 \psi$" is an abbreviation for $\phi \rightarrow_4 \psi$ and $\psi \rightarrow_4 \phi$.

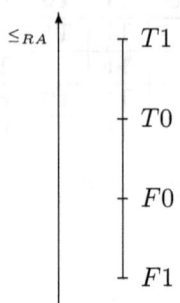

Figure 1: Lattice **RA**

1. $(\phi \wedge_4 \psi) \to_4 \phi$
2. $(\phi \wedge_4 \psi) \to_4 \psi$
3. $\phi \to_4 (\phi \vee_4 \psi)$
4. $\psi \to_4 (\phi \vee_4 \psi)$
5. $(\phi \wedge_4 (\psi \vee_4 \chi)) \leftrightarrow_4 ((\phi \wedge_4 \psi) \vee_4 (\phi \wedge_4 \chi))$
6. $\phi \leftrightarrow_4 \neg_4\neg_4\phi$
7. $\neg_4(\phi \wedge_4 \psi) \leftrightarrow_4 (\neg_4\phi \vee_4 \neg_4\psi)$
8. $\neg_4(\phi \vee_4 \psi) \leftrightarrow_4 (\neg_4\phi \wedge_4 \neg_4\psi)$
9. $(\phi \wedge_4 \neg_4\phi) \to_4 (\psi \vee_4 \neg_4\psi)$
10. $(\phi \to_4 \psi) \leftrightarrow_4 (\neg_4\psi \to_4 \neg_4\phi)$
11. $\phi \to_4 (\psi \to_4 (\phi \wedge_4 \psi))$
12. $(\phi \to_4 \psi) \to_4 ((\chi \to_4 \psi) \to_4 ((\phi \vee_4 \chi) \to_4 \psi))$
13. $(\phi \to_4 \psi) \to_4 ((\psi \to_4 \chi) \to_4 (\phi \to_4 \chi))$
14. $\phi \to_4 (\psi \to_4 \phi)$
15. $\phi \leftrightarrow_4 \sim_4 \sim_4 \phi$

16. $\phi \to_4 (\psi \vee_4 \neg_4\psi \vee_4 \sim_4 \psi \vee_4 \sim_4 \neg_4\psi)$

17. $\sim_4 (\phi \wedge_4 \psi) \to_4 (\sim_4 \phi \vee_4 \sim_4 \psi)$

18. $(\sim_4 \phi \wedge_4 \sim_4 \psi) \to_4 \sim_4 (\phi \vee_4 \psi)$

19. $\sim_4 (\phi \vee_4 \psi) \to_4 (\sim_4 \phi \vee_4 \sim_4 \psi)$

20. $(\sim_4 \phi \wedge_4 \sim_4 \psi) \to_4 \sim_4 (\phi \wedge_4 \psi)$

21. $\neg_4 \sim_4 \phi \leftrightarrow_4 \sim_4 \neg_4\phi$

22. $(\sim_4 \phi \wedge_4 \sim_4 \neg_4\phi) \to_4 (\phi \vee_4 \neg_4\phi)$

23. $(\neg_4\phi \wedge_4 \psi) \to_4 (\phi \to_4 \neg_4\psi)$

24. $\sim_4 (\phi \to_4 \psi) \to_4 (\sim_4 \phi \to_4 \sim_4 \psi)$

25. $((((\phi\vee_4 \sim_4 \phi) \to_4 \psi) \to_4 (((\neg_4\phi\vee_4 \sim_4 \neg_4\phi) \to_4 \psi) \to_4 (((\phi \vee_4 \neg_4\phi) \to_4 \psi) \to_4 (((\sim_4 \phi\vee_4 \sim_4 \neg_4\phi) \to_4 \psi) \to_4 \psi))))$

26. $((((\phi\vee_4 \sim_4 \phi) \to_4 ((\psi\vee_4 \sim_4 \psi) \wedge_4 (\psi \vee_4 \neg_4\psi))) \to_4 (((\neg_4\phi\vee_4 \sim_4 \neg_4\phi) \to_4 ((\psi\vee_4 \sim_4 \psi) \wedge_4 (\psi \vee_4 \neg_4\psi))) \to_4 (((\phi \vee_4 \neg_4\phi) \to_4 ((\psi\vee_4 \sim_4 \psi) \wedge_4 (\psi \vee_4 \neg_4\psi))) \to_4 (((\sim_4 \phi\vee_4 \sim_4 \neg_4\phi) \to_4 ((\psi\vee_4 \sim_4 \psi) \wedge_4 (\psi \vee_4 \neg_4\psi))) \to_4 \psi))))$

27. $((\phi\vee_4 \sim_4 \phi) \wedge_4 (\psi\vee_4 \sim_4 \psi)) \leftrightarrow_4 ((\phi \wedge_4 \psi)\vee_4 \sim_4 (\phi \wedge_4 \psi))$

28. $((\neg_4\phi\vee_4 \sim_4 \neg_4\phi) \vee_4 (\neg_4\psi\vee_4 \sim_4 \neg_4\psi)) \leftrightarrow_4 (\neg_4(\phi \wedge_4 \psi)\vee_4 \sim_4 \neg_4(\phi \wedge_4 \psi))$

29. $(((\phi \vee_4 \neg_4\phi) \wedge_4 (\psi \vee_4 \neg_4\psi)) \vee_4 ((\sim_4 \phi\vee_4 \sim_4 \neg_4\phi) \wedge_4 (\neg_4\psi\vee_4 \sim_4 \neg_4\psi) \wedge_4 (\psi \vee_4 \neg_4\psi)) \vee_4 ((\neg_4\phi\vee_4 \sim_4 \neg_4\phi) \wedge_4 (\phi \vee_4 \neg_4\phi) \wedge_4 (\sim_4 \psi\vee_4 \sim_4 \neg_4\psi))) \leftrightarrow_4 ((\phi \wedge_4 \psi) \vee_4 \neg_4(\phi \wedge_4 \psi))$

30. $(((\sim_4 \phi\vee_4 \sim_4 \neg_4\phi)\wedge_4(\sim_4 \psi\vee_4 \sim_4 \neg_4\psi))\vee_4((\sim_4 \phi\vee_4 \sim_4 \neg_4\phi)\wedge_4 (\psi\vee_4 \sim_4 \psi) \wedge_4 (\psi \vee_4 \neg_4\psi)) \vee_4 ((\phi\vee_4 \sim_4 \phi) \wedge_4 (\phi \vee_4 \neg_4\phi) \wedge_4 (\sim_4 \psi\vee_4 \sim_4 \neg_4\psi))) \leftrightarrow_4 (\sim_4 (\phi \wedge_4 \psi)\vee_4 \sim_4 \neg_4(\phi \wedge_4 \psi))$

31. $((\neg_4\psi\vee_4 \sim_4 \neg_4\psi) \vee_4 (\chi\vee_4 \sim_4 \chi)) \leftrightarrow_4 ((\psi \to_4 \phi)\vee_4 \sim_4 (\psi \to_4 \phi))$

32. $((\phi\vee_4 \sim_4 \phi) \wedge_4 (\neg_4\psi\vee_4 \sim_4 \neg_4\psi)) \leftrightarrow_4 (\neg_4(\phi \to_4 \psi)\vee_4 \sim_4 \neg_4(\phi \to_4 \psi))$

33. $(((\neg_4\phi \vee_4 \sim_4 \neg_4\phi) \wedge_4 (\phi \vee_4 \neg_4\phi)) \vee_4 ((\psi \vee_4 \sim_4 \psi) \wedge_4 (\psi \vee_4 \neg_4\psi)) \vee_4 ((\neg_4\phi \vee_4 \sim_4 \neg_4\phi) \wedge_4 (\sim_4 \psi \vee_4 \sim_4 \neg_4\psi)) \vee_4 ((\sim_4 \phi \vee_4 \sim_4 \neg_4\phi) \wedge_4 (\psi \vee_4 \sim_4 \psi)) \vee_4 ((\phi \vee_4 \neg_4\phi) \wedge_4 (\psi \vee_4 \neg_4\psi))) \leftrightarrow_4 ((\phi \rightarrow_4 \psi) \vee_4 \neg_4(\phi \rightarrow_4 \psi))$

34. $(((\phi \vee_4 \sim_4 \phi) \wedge_4 (\phi \vee_4 \neg_4\phi) \wedge_4 (\sim_4 \psi \vee_4 \sim_4 \neg_4\psi)) \vee_4 ((\phi \vee_4 \sim_4 \phi) \wedge_4 (\sim_4 \phi \vee_4 \sim_4 \neg_4\phi) \wedge_4 (\neg_4\psi \vee_4 \sim_4 \neg_4\psi)) \vee_4 ((\sim_4 \phi \vee_4 \sim_4 \neg_4\phi) \wedge_4 (\neg_4\psi \vee_4 \sim_4 \neg_4\psi) \wedge_4 (\psi \vee_4 \neg_4\psi))) \rightarrow_4 ((\sim_4 \phi \rightarrow_4 \psi) \vee_4 \sim_4 \neg_4(\phi \rightarrow_4 \psi))$

MP. *If* $\Gamma \vdash_{RA+} \phi$ *and* $\Gamma \vdash_{RA+} (\phi \rightarrow_4 \psi)$, *then* $\Gamma \vdash_{RA+} \psi$.

As already remarked, the system **RA+** provides an alternative way to represent an agent's ignorance in a formal setting. The semantical considerations about the truth values interpretation for this system represent the objective approach to ignorance characterization. Indeed, one is able to distinguish ignorance of a true proposition ($\vdash_{RA+} \sim_4 \phi$) from ignorance of a false proposition ($\vdash_{RA+} \sim_4 \neg_4\phi$). The subjective approach may be also expressed in terms of the system **RA+**: $\vdash_{RA+} \sim_4 \phi \vee \sim_4 \neg_4\phi$.

We have now two logical systems (**Ig** and **RA+**) where ignorance is represented independently from knowledge. The natural question we are interested in is whether there is a connection between these two logics. In order to answer this question we introduce a translation of **RA+** in terms of **Ig** in the next section.

4 Translation

To translate **RA+** into **Ig** we use the method proposed by Kooi and Tamminga (2013). The authors define a Translation Manual that converts any formula of any three-valued logic into a modal formula. Also they remark that the Translation Manual can be adapted to handle n-valued logics. In what follows we modify the translation proposed by Kooi and Tamminga in order to introduce a translation that preserves the interpretation of the four values of the logic of rational agent in terms of I operator. As our aim is not to introduce a general result, but to interpret **RA+** in terms of **Ig** in an intuitively clear way, we limit our construction to a translation that preserves valid arguments of **RA+** in an arbitrary world w in an **Ig**-model: $\Gamma \models_{RA+} \phi$ iff $T(\Gamma) \models_{Ig}^{w} T(\phi)$.

Ignorance Without K(nowledge)

The definition 5 introduces the translation procedure. In an **Ig**-model for each world there are four mutually exclusive and jointly exhaustive possibilities for each atomic formula: either the formula is true in a given world and its truth is not ignored by the agent, or it is true and the agent is ignorant about it, or the formula is false but the agent is ignorant about its falsity, or the formula is false and the agent does not ignore this fact. The translation we propose first maps the four possible truth-values of any atomic formula to these four possibilities, and then maps the four possible truth-values of any complex formula to truth-functional combinations of the mapped truth-values $\otimes_n(\phi_1, ..., \phi_n)$ (where $\otimes_n \in \{\neg_4, \sim_4, \wedge_4, \vee_4, \rightarrow_4\}$) of its constituent formulas according to the strictures of the operator's truth-table.

Definition 5 *Let p be an atomic formula, $\phi_1, ..., \phi_n \in \mathcal{L}_{RA}$, $f_{\otimes_n}$ is an n-placed function $\{T1, T0, F0, F1\}^n \rightarrow \{T1, T0, F0, F1\}$ that yields the truth value of a complex formula on the basis of the truth values of its constituent formulas. Then*

$$T1(p) = p \wedge \neg I p$$

$$T0(p) = p \wedge I p$$

$$F0(p) = \neg p \wedge I \neg p$$

$$F1(p) = \neg p \wedge \neg I \neg p$$

$$T1(\otimes_n(\phi_1, ..., \phi_n)) = \bigvee_{\langle \tau_1, ... \tau_n \rangle \in T1(f_{\otimes_n})} (\tau_1(\phi_1) \wedge ... \wedge \tau_n(\phi_n))$$

$$T0(\otimes_n(\phi_1, ..., \phi_n)) = \bigvee_{\langle \tau_1, ... \tau_n \rangle \in T0(f_{\otimes_n})} (\tau_1(\phi_1) \wedge ... \wedge \tau_n(\phi_n))$$

$$F0(\otimes_n(\phi_1, ..., \phi_n)) = \bigvee_{\langle \tau_1, ... \tau_n \rangle \in F0(f_{\otimes_n})} (\tau_1(\phi_1) \wedge ... \wedge \tau_n(\phi_n))$$

$$F1(\otimes_n(\phi_1, ..., \phi_n)) = \bigvee_{\langle \tau_1, ... \tau_n \rangle \in F1(f_{\otimes_n})} (\tau_1(\phi_1) \wedge ... \wedge \tau_n(\phi_n)),$$

where $\tau(f_{\otimes_n}) = \{\langle \tau_1, ..., \tau_n \rangle \in \{T1, T0, F0, F1\}^n : f_{\otimes_n}(\langle \tau_1, ..., \tau_n \rangle) = \tau\}$.

By using Definition 5 we can state the conditions under which a four-valued valuation is equivalent to its translation in a world w in a model M.

Definition 6 *Let v^4 be a four-valued valuation on $\mathcal{L}_{RA}$ and let w be a world in an **Ig**-model M. Then v^4 is 4-w-equivalent to its translation in a world w, if for all $\phi \in \mathcal{L}_{RA}$ it holds that*
$v^4(\phi) = T1$ *iff* $M \models^w_{Ig} T1(\phi)$
$v^4(\phi) = T0$ *iff* $M \models^w_{Ig} T0(\phi)$
$v^4(\phi) = F0$ *iff* $M \models^w_{Ig} F0(\phi)$
$v^4(\phi) = F1$ *iff* $M \models^w_{Ig} F1(\phi)$.

For each world in each **Ig**-model there is a 4-w-equivalent valuation and for each four-valued valuation there is a 4-w-equivalent translation in a world w in an **Ig**-model.

Lemma 1 *For each world in each **Ig**-model there is a 4-w-equivalent four-valued valuation v^4.*

Proof. Let M be an **Ig**-model, w is a world in M. We construct a four-valued valuation v_M by stipulating that for all atomic formulas p:

$v^w_M(p) = T1$ iff $p \in w$ and there does not exist w', w'' such that Rww', Rww'', $M \models^{w'}_{Ig} p$, and $M \models^{w''}_{Ig} \neg p$;

$v^w_M(p) = T0$ iff $p \in w$ and there exist w', w'' such that Rww', Rww'', $M \models^{w'}_{Ig} p$, and $M \models^{w''}_{Ig} \neg p$;

$v^w_M(p) = F0$ iff $p \notin w$ and there exist w', w'' such that Rww', Rww'', $M \models^{w'}_{Ig} p$, and $M \models^{w''}_{Ig} \neg p$;

$v^w_M(p) = F1$ iff $p \notin w$ and there does not exist w', w'' such that Rww', Rww'', $M \models^{w'}_{Ig} p$, and $M \models^{w''}_{Ig} \neg p$.

It is easy to see that v^w_M is a four-valued valuation of the logic of rational agent. It assigns to each formula exactly one of the truth-values $T1, T0, F0$ or $F1$ in a world w.

We show by structural induction on ϕ that v^w_M is 4-w-equivalent to the truth-conditions in each world in each **Ig**-model.

Ignorance Without K(nowledge)

Basis. The condition holds for each atomic formula due to the definition of v_M^w and the semantics of the **Ig**-model.

Induction Hypothesis. Suppose that the condition holds for all formulas ϕ with less operators than the formula $\otimes_n(\phi_1, ..., \phi_n)$.

Induction Step. Consider $\otimes_n(\phi_1, ...\phi_n)$. Suppose $v_M^w(\otimes_n(\phi_1, ..., \phi_n)) = T1$. Then $f_{\otimes_n}(\langle v_M^w(\phi_1), ..., v_M^w(\phi_n)\rangle) = T1$. Take an arbitrary ϕ_i and consider $v_M^w(\phi_i) = \tau_i$. By Induction Hypothesis, it holds that $M \models_{Ig}^w \tau_i(\phi_i)$ for all i with $1 \leq i \leq n$. Then, $M \models_{Ig}^w \tau_1(\phi_1) \wedge ... \wedge \tau_n(\phi_n)$. It is clear that $\langle \tau_1, ..., \tau_n \rangle \in T1(f_{\otimes_n})$. Therefore, $M \models_{Ig}^w T1(\otimes_n(\phi_1, ..., \phi_n))$.

Suppose $M \models_{Ig}^w T1(\otimes_n(\phi_1, ..., \phi_n))$. It follows there is an n-tuple $\langle \tau_1, ..., \tau_n \rangle \in T1(f_{\otimes_n})$ such that $M \models_{Ig}^w \tau_1(\phi_1) \wedge ... \wedge \tau_n(\phi_n)$. Hence, $M \models_{Ig}^w \tau_i(\phi_i)$ for all i with $1 \leq i \leq n$. Therefore, $v_M^w(\otimes_n(\phi_1, ..., \phi_n)) = f_{\otimes_n}(\langle v_M^w(\phi_1), ..., v_M^w(\phi_n)\rangle) = T1$.

The proof for other cases is similar. □

Lemma 2 *For each four-valued valuation v^4 there is an **Ig**-model where each world w is 4-w-equivalent to v^4.*

Proof. Let v^4 be a four-valued valuation. We construct an **Ig**-model $M_v = \langle W_v, R_v, \pi_v \rangle$ by stipulating that $W_v = \{w, w'\}$, Rww', Rww, $Rw'w$, $Rw'w'$ and for all atomic formulas p:

$\pi_v(p) = W$ iff $v^4(p) = T1$ in w;
$\pi_v(p) = \{w\}$ iff $v^4(p) = T0$ in w;
$\pi_v(p) = \{w'\}$ iff $v^4(p) = F0$ in w;
$\pi_v(p) = \{\emptyset\}$ iff $v^4(p) = F1$ in w.[5]

Obviously, M_v is an **Ig**-model. An adaptation of the inductive proof of lemma 1 shows that v^4 is 4-w-equivalent to its translation in a world w. □

We now prove the theorem stating that the logic **RA+** can be translated into the logic **Ig** in a given world w.

Theorem 1 *Let $\Gamma \subseteq \mathcal{L}_{RA}$, let $\phi \in \mathcal{L}_{RA}$, and let $\mathfrak{XY} \subseteq \{T1, T0, F0, F1\}$. Then*
$$\Gamma \models_{RA+}^{\mathfrak{XY}} \phi \text{ iff } \mathfrak{X}(\Gamma) \models_{Ig}^w \mathfrak{Y}(\phi).$$

Proof. Suppose that $\mathfrak{X}(\Gamma) \not\models_{Ig}^w \mathfrak{Y}(\phi)$. This means that $M \models_{Ig}^w \mathfrak{X}(\psi)$ for all ψ in Γ and $M \not\models_{Ig}^w \mathfrak{Y}(\phi)$. By lemma 1 there is a four-valued valuation v_M^4 such that $v_M^4(\psi) \in \mathfrak{X}$ for all ψ in Γ and $v_M^4(\phi) \notin \mathfrak{Y}$. Therefore, $\Gamma \not\models_{RA+}^{\mathfrak{XY}} \phi$.

---
[5] In case of the world w' we replace w by w' and vice versa in this definition.

Suppose that $\Gamma \not\models_{RA+}^{\mathfrak{X}\mathfrak{Y}} \phi$. Then there is a four-valued valuation v^4 such that $v^4(\psi) \in \mathfrak{X}$ for all ψ in Γ and $v^4(\phi) \notin \mathfrak{Y}$. By lemma 2 there is a world w in a **Ig**-model $M_v = \langle W_v, R_v, V_v \rangle$ such that $M_v \models_{Ig}^{w} \mathfrak{X}(\psi)$ for all ψ in Γ and $M_v \not\models_{Ig}^{w} \mathfrak{Y}(\phi)$, that is $\mathfrak{X}(\Gamma) \not\models_{Ig}^{w} \mathfrak{Y}(\phi)$. $\square$

As w is an arbitrary world in **Ig**-model, theorem 1 together with our translation gives us a method to express the four values of the logic of rational agent in terms of the logic for ignorance in each world. One may ask whether it is possible to introduce the equivalence result for the valuation v^4 and an arbitrary **Ig**-model without restriction to a concrete world. This can be done by introducing a generalization of the logic of rational agent, where a proposition can take more than one value. In such generalization we define the operators by the same truth-tables as for the system **RA+** (see the fig. 1) and add an additional condition. Suppose that the truth-value of a proposition consist of more than one element (for example it is $\{T0, F0\}$). To calculate the truth value for one-place operators first we take the biggest member of the truth-value with respect to the lattice **RA**, then we apply the corresponding four-valued connective to this value. For two-place operators we take the biggest members of both sub-formulas, then we apply the corresponding four-valued truth-table. It is obvious that this generalization together with the same definition of consequence relation (def. 3) will coincide with **RA+** and may be conservatively translated into the logic **Ig**. We leave this idea as future work.

5 Conclusion

To sum up, we have presented two logical systems where ignorance is formally represented without the use of K-operator. The system **Ig** represented ignorance understood according to the subjective approach to its characterization. The system **RA+** aims to represent ignorance according to the objective approach. Thus both approaches to ignorance characterization may be expressed in systems where the notion of ignorance is understood independently from the notion of knowledge. The translation of the logic **RA+** into **Ig** shows us the connection between these two logics. Both of them may be used to analyze the notion of ignorance and reasoning about someone's ignorance. The existence of at least two logics, where ignorance is expressed independently from knowledge, provides a formal basis to challenge the idea that ignorance should be defined in terms of knowledge. Indeed, by following an idea of Williamson (2000), one may argue that ignorance is a primitive epistemic notion in the same manner as knowledge is. This analysis is currently the subject of further investigations.

Acknowledgements

We would like to thank Paul Egré and Mattia Petrolo for the valuable remarks and comments on the current paper.

References

Hintikka, J. (1962). *Knowledge and Belief: An Introduction to the Logic of the Two Notions*. Cornell: Cornell University Press.

van der Hoek, W., & Lomuscio, A. (2004). A logic for ignorance. *Electronic Notes in Theoretical Computer Science, 85(2)*, 97–108.

Kiparsky, P., & Kiparsky, C. (1970). Fact. In M. Bierxisch & K. Heidolph (Eds.), *Progress in Linguistics* (pp. 143–173). The Hague: Mouton.

Kooi, B., & Tamminga, A. (2013). Three-valued logics in modal logic. *Studia Logica, 101*, 1061–1072.

Kubyshkina, E., & Zaitsev, D. V. (2016). Rational agency from a truth-functional perspective. *Logic and Logical Philosophy, 25*, 499–520.

Williamson, T. (2000). *Knowledge and its Limits*. Oxford University Press.

Ekaterina Kubyshkina
Université Paris 1 Panthéon - Sorbonne, Institut d'Histoire et de Philosophie des Sciences et des Techniques (IHPST)
France
E-mail: `ekaterina.kubyshkina@univ-paris1.fr`

Non-Normal Modal Logics: A Challenge to Proof Theory

SARA NEGRI

Abstract: A general procedure that follows the guidelines of inferentialism is presented for generating G3-style sequent calculi for non-normal modal logics on the basis of neighbourhood semantics.

Keywords: Neighbourhood semantics, Classical modal logics, Non-normal modal logics, Labelled sequent calculus

1 Introduction

The early decades of the modern study of modal logic were marked by the advent of possible worlds semantics. Earlier axiomatic studies of modal concepts were replaced by a solid and uniform semantic method that displayed the connections between modal axioms and conditions on the accessibility relation between possible worlds. The success of the semantic method, however, was not directly followed by equally powerful syntactic theories of modal and conditional concepts and reasoning and the literature until the 1990's shows a striking contrast between the generality of the semantic method and the scattered, goal-directed developments of the proof-theoretic method. Traditional Gentzen systems failed to establish basic properties such as normalization/cut-elimination and analyticity even for basic modal systems. Awareness of this gap was often expressed by defeatist statements among practitioners in the field.

The insufficiency of traditional Gentzen systems to meet the challenge of the development of a proof theory for modal and non-classical logic has led to the development of alternative formalisms which, in one way or other, extend the syntax of sequent calculus. In the proliferation of calculi beyond Gentzen systems, there have been two main lines of development, one that enriches the structure of sequents (display calculi, hypersequents, nested sequents, tree-hypersequents, deep inference), another that maintains their simple structure but adds *labels* and relations in the form of variables and atomic formulas.

Sara Negri

Through the conversion of what are known as geometric implications into rules that extend sequent calculus in a way that maintains the admissibility of structural rules, it has been possible to obtain a uniform presentation of a large family of modal logics, all those characterised by first-order frame conditions, and provability logics such as GL and Grz[1]. Parallel to the formal developments and contributions to widening the scope of such calculi, methodological reflections have been directed to the questions of how well they respond to the central issues of inferentialism (as analyzed in Negri & von Plato, 2015; Read, 2015), a discussion that goes under the wide umbrella of the extension of proof-theoretic semantics to non-classical logics.

Despite a wide range of applications, the powerful methods of possible worlds semantics are not a universal tool in the analysis of philosophical logics: they impose the straitjacket of *normality*, i.e. validity of the rule of necessitation, from $\vdash A$ to infer $\vdash \Box A$, and of the K axiom, $\Box(A \supset B) \supset (\Box A \supset \Box B)$. The limitative character of these imposed validities becomes clear in many of the logics that one encounters in the ever-expanding domains of applications of modal logic (in mathematics, philosophy, computer science, linguistics, cognitive science, social science). For instance, with the epistemic reading of the modality as a knowledge operator, an agent knows A if A holds in all the epistemic states available to her, and then the properties have the consequence that (1) *whatever has been proved is known* and (2) *an agent knows all the logical consequences of what she knows*. This leads to *logical omniscience*, clearly inadequate for cognitive agents with human capabilities, and thus to the rejection of both requirements. The same limitation is clear in the interpretation of the modality as a likelihood operator where one sees that the normal modal-logical validity of $\Box A \& \Box B \supset \Box(A \& B)$ should be avoided (Pacuit, 2007).

Another limitation in systems based on a standard Kripke-style semantics is that the propositional base is classical or intuitionistic logic. In both cases one is forced to material implication, shown since the analysis of C.I. Lewis to be an inadequate form of conditional if logical analysis is to be pursued in other venues than mathematics: the classical propositional base of modal logic is insufficient to treat conditionals beyond material or strict implication, as shown in (Lewis, 1973).

Among non-normal modal logics, *classical* modal logics are those obtained by requiring that the modality respects logical equivalence, that is

---

[1] Cf. Negri (2005); Dyckhoff and Negri (2015, 2016).

Non-Normal Modal Logics: A Challenge to Proof Theory

closure under the rule $\frac{A \supset B}{\Box A \supset \Box B}$. One can then obtain other systems below the normal modal logic **K** by removing the normality axiom and the necessitation rule and adding other axioms; combinations of the axiom schemes M, C, N give a lattice of eight different logics (cf. the diagram on p. 237 of Chellas, 1980). It is known that non-normal modal logics can be simulated through an appropriate translation by a normal modal logic with three modalities (Gasquet & Herzig, 1996; Kracht & Wolter, 1999), so that their proof theory can be approached indirectly through the translation: in the system proposed by Gilbert and Maffezioli (2015), the translation from non-normal to normal modal logics is used to define labelled sequent calculi with non-local *systems of rules* (in the sense of Negri, 2016) for basic systems of *classical* modal logics.

Rather than reducing non-normal modal logics to normal ones, we shall develop proof systems for them in a way that parallels the generation of labelled calculi for systems based on possible world semantics. To this end, we shall use the more general *neighbourhood semantics* which was introduced in the 1970's to provide a uniform semantic framework for philosophical logics that cannot be accommodated within the normal modal logic setting. Instead of an accessibility relation on a set of possible worlds one has for every possible world w a family of sets of possible worlds called neighbourhoods of w. The correspondence between relational frames and certain specific types of neighbourhood frames shows that neighbourhood semantics is a generalisation of the earlier possible world semantics. Further, it gives a way to transfer the intuition from one semantics to the other: roughly, worlds in a neighbourhood of w replace worlds accessible from w.

Our goal is to set the grounds for a proof theory of non-normal modal systems based on neighbourhood semantics, to achieve this directly, i.e. without the use of translations, with local rules, and in a way that makes possible extensions in various directions.

The goal will be accomplished by following the guidelines of inferentialism, that is, by starting from the meaning explanations of logical constants and converting them into well-behaved rules of a calculus through a five-stage procedure. In the systems obtained, all the logical rules are invertible and all structural rules admissible. On the one hand, these properties make the proof of metatheorems such as completeness a straightforward task; on the other, thay make the calculi obtained suitable for proof search.

Here we concentrate on the procedure of generation of calculi based on neighbourhood semantics and only mention the properties that these calculi enjoy. Detailed statements and proofs of the structural properties of such calculi as well as a more comprehensive bibliography will be given in (Negri, 2017). The development of calculi for conditional logics based on neighbourhood semantics and their application to the decision problem for such logics is presented in (Girlando, Negri, Olivetti, & Risch, 2016; Negri & Olivetti, 2015).

2 Neighbourhood Semantics

A *neighbourhood frame* is a pair $\mathcal{F} \equiv (W, I)$, where W is a set of worlds (states) and I is a neighbourhood function

$$I : W \longrightarrow \mathcal{P}(\mathcal{P}(W))$$

that assigns a collection of sets of worlds to each world in W. A *neighbourhood model* is then a pair $\mathcal{M} \equiv (\mathcal{F}, \mathcal{V})$, where $\mathcal{F}$ is a neighbourhood frame and $\mathcal{V}$ a propositional valuation, i.e. a map $\mathcal{V} : \text{Atm} \longrightarrow \mathcal{P}(W)$.

Worlds in a neighbourhood are the substitute, in this more general semantics, of accessible worlds. The inductive clauses for truth of a formula in a model are the usual ones for the propositional clauses; for the modal operator we have

$$\mathcal{M}, w \Vdash \Box A \equiv \text{ext}(A) \text{ is in } I(w),$$

where $\text{ext}(A) \equiv \{u \in W | \mathcal{M}, u \Vdash A\}$.

Given a relational frame (W, R), one can define a neighbourhood frame by taking as neighbourhoods of a world x the supersets of worlds accessible from x

$$I^R(x) \equiv \{a \mid R(x) \subseteq a\}$$

Conversely, given a neighbourhood frame (W, I) one can define a relational frame by

$$xR^I y \equiv y \in \bigcap I(x)$$

A neighbourhood frame is *augmented* if for all x, $\bigcap I(x) \in I(x)$ and is supplemented, i.e. closed under supersets. Relational frames correspond to augmented neighbourhood frames, in the sense that given a relational frame, there is an augmented neighbourhood frame that, as a model, validates the same formulas, and viceversa. [2]

---

[2] Details of the correspondence are found in Chellas (1980, p. 221). For an extensive survey on neighbourhood semantics see Pacuit (2007).

Non-Normal Modal Logics: A Challenge to Proof Theory

3 Five steps to good sequent calculi

In this section we shall present in detail the way to the determination of the rules of a G3-style sequent calculus that internalizes neighbourhood semantics. Much of the rationale is common to the determination of the rules of a G3-style sequent calculus based on possible worlds semantics, but there are new important elements that need to be taken into consideration when moving from possible worlds to the more general neighbourhood semantics.

The stages in the determination of the rules can be summarized as follows; each item will be further detailed together with the illustration of the procedure:

1. Turn the *semantic explanation* of logical constants into introduction rules of natural deduction.
2. Through *inversion principles*, find the corresponding elimination rules.
3. Translate the natural deduction system thus obtained into a sequent calculus. The resulting calculus is a sequent calculus with independent contexts.
4. Refine the calculus into a G3-style sequent calculus.

We observe that the above explanation is not really specific to the determination of labelled sequent calculi, but rather parametric and with an end-result that depends on the semantic explanation one starts with. With the BHK explanation of logical constants, the recipe gives the standard G3 sequent calculi, and in fact this is the route followed in (Negri & von Plato, 2001, Chapter 1). With possible worlds semantics, the meaning also depends on certain properties of an accessibility relation between worlds, so one has an additional step:

5. To obtain specific systems (e.g. intermediate logics) we add the rules for the accessibility relation following the method of "axioms as rules" (Negri & von Plato, 1998; Negri, 2003) for universal and geometric frame conditions.

The procedure for obtaining labelled calculi through this five-stage explanation is carried through both for intuitionistic logic and basic modal logics in (Negri & von Plato, 2015); the resulting calculi are those which have been investigated, respectively, in Dyckhoff and Negri (2012) and Negri (2005). We remark that one does not need to stop at geometric frame conditions, but one can expand the spectrum of frame conditions that can be dealt with: *generalised geometric implications* can be treated by the method of *systems of rules* as developed in Negri (2016) without extending the language but at the expense of locality. With the method of *geometrization of first-order logic* (Dyckhoff & Negri, 2015) and more specifically by the addition of new primitives together with a semidefinitional conservative extension, one obtains a splitting of the rules for frame conditions into possibly several geometric rules; the resulting sequent calculi allow to capture logics with *arbitrary* first-order conditions on their Kripke frames.

Next, we detail the method and the new step needed for neighbourhood semantics.

1. Convert semantic explanations into introduction rules.

We start with the truth condition for the necessity modality in terms of neighbourhood semantics

$$x \Vdash \Box A \equiv \text{for some } a \text{ in } I(x).a = \text{ext}(A)$$

i.e.

$$x \Vdash \Box A \equiv \exists a \in I(x).(\forall x(x \in a \to x \Vdash A) \,\&\, \forall x(x \Vdash A \to x \in a))$$

This cannot be converted into a local rule in a way similar to the condition in terms of relational semantics because of the nesting of quantifiers. To proceed we need a further step:

0. Add new primitives (definitional extension) and their rules.

The new primitives are the relation of "local" forcing, a forcing relation between neighbourhoods and formulas (here local is opposed to the pointwise forcing of possible worlds) and of "cover" between a formula and a neighbourhood, to express the mutual inclusions between the neighbourhood a and the extension of formula A.

$$a \Vdash^\forall A \equiv \forall x(x \in a \to x \Vdash A) \qquad A \triangleleft a \equiv \forall x(x \Vdash A \to x \in a)$$

Steps 0 and 1 give the following introduction rules:

$$\frac{[x \in a]}{\begin{array}{c} \vdots \\ x : A \\ \hline a \Vdash^\forall A \end{array}} \Vdash^\forall I, x\,\textit{fresh} \qquad \frac{[x : A]}{\begin{array}{c} \vdots \\ x \in a \\ \hline A \triangleleft a \end{array}} \triangleleft I, x\,\textit{fresh}$$

Non-Normal Modal Logics: A Challenge to Proof Theory

$$\frac{a \in I(x) \quad a \Vdash^\forall A \quad A \vartriangleleft a}{x : \Box A} \Box I$$

2. Obtain elimination rules through the *inversion principle*. The rule is determined in two stages: first, the elimination rule should be in accordance with the inversion principle; this states that whatever follows from the grounds for deriving a proposition must follows from that proposition (Negri & von Plato, 2001, p. 6). Application of this principle to $x : \Box A$ is unproblematic and gives the rule $\Box E$ below. For $a \Vdash^\forall A$ and $A \vartriangleleft a$ some further comments are in order. The grounds for deriving the proposition $a \Vdash^\forall A$ are given by a derivation of $x : A$ from $x \in a$ where x is arbitrary. Similarly for $A \vartriangleleft a$. In this case a direct application of the inversion principle would thus lead to higher-level elimination rules, where assumptions in derivations are given by other derivations. It is possible to avoid the recourse to higher-level rules by proceeding as for the elimination rule for implication in the system of natural deduction with general elimination rules and obtain the following elimination rules for $a \Vdash^\forall A$ and $A \vartriangleleft a$:[3]

$$\frac{a \Vdash^\forall A \quad x \in a \quad \begin{array}{c}[x:A]\\ \vdots \\ D\end{array}}{D} \Vdash^\forall E \qquad \frac{A \vartriangleleft a \quad x:A \quad \begin{array}{c}[x \in a]\\ \vdots \\ D\end{array}}{D} \vartriangleleft E$$

$$\frac{x : \Box A \quad \begin{array}{c}[a \in I(x), a \Vdash^\forall A, A \vartriangleleft a]\\ \vdots \\ D\end{array}}{D} \Box E$$

3. Translate ND rules to sequent calculus rules. This is a part of the general procedure for transforming a system of natural deduction into one of sequent calculus. First, the deducibility relation is internalized with an explicit notation for it, the sequent arrow in place of vertical dots; then the rules are made local by having the open assumptions listed in the left hand side of sequents at each step of a derivation. The introduction rules become

---

[3] The procedure that replaces a higher level rule that discharges derivations with an ordinary general elimination rule that discharges formulas has been called *flattening* and the limitation of its scope investigated by Olkhovikov and Schroeder-Heister (2014).

right rules and the elimination rules left rules. The format of rules that one obtains by this translation is that of a *sequent calculus with independent contexts* (von Plato, 2001):

$$\Vdash^\forall I \quad \leadsto \quad \frac{x \in a, \Gamma \Rightarrow x : A}{\Gamma \Rightarrow a \Vdash^\forall A} \, R \Vdash^\forall, \, x \, \textit{fresh}$$

$$\Vdash^\forall E \quad \leadsto \quad \frac{\Gamma \Rightarrow x \in a \quad \Gamma', x : A \Rightarrow D}{a \Vdash^\forall A, \Gamma, \Gamma' \Rightarrow D} \, L \Vdash^\forall$$

$$\lhd I \quad \leadsto \quad \frac{x : A, \Gamma \Rightarrow x \in a}{\Gamma \Rightarrow A \lhd a} \, R \lhd, \, x \, \textit{fresh}$$

$$\lhd E \quad \leadsto \quad \frac{\Gamma \Rightarrow x : A \quad \Gamma', x \in a \Rightarrow D}{A \lhd a, \Gamma, \Gamma' \Rightarrow D} \, L \lhd$$

$$\Box I \quad \leadsto \quad \frac{\Gamma \Rightarrow a \in I(x) \quad \Gamma' \Rightarrow a \Vdash^\forall A \quad \Gamma'' \Rightarrow A \lhd a}{\Gamma, \Gamma', \Gamma'' \Rightarrow x : \Box A} \, R \Box$$

$$\Box E \quad \leadsto \quad \frac{a \in I(x), a \Vdash^\forall A, A \lhd a, \Gamma \Rightarrow D}{x : \Box A, \Gamma \Rightarrow D} \, L \Box$$

4. Adapt the sequent calculus rules obtained in 3 to the G3 style. First, all rules are brought to the *shared context* form, that is, in two-premiss rules the same multisets of formulas appear in the contexts of both premisses. Second, the calculus has to be *multisuccedent*, so there are arbitrary multisets as contexts in the succedents of sequents. Third, rules that are not already invertible are made so by the repetition of the principal formulas in the premisses. Some optimization to reduce the number of premisses is also possible: rule $R\Box$ is rewritten as an equivalent two-premiss rule by having the formula $a \in I(x)$ in the antecedent of the conclusion so that one of the premisses becomes superfluous.

$$\frac{x \in a, \Gamma \Rightarrow \Delta, x : A}{\Gamma \Rightarrow \Delta, a \Vdash^\forall A} \, R \Vdash^\forall, \, x \, \textit{fresh} \qquad \frac{x \in a, x : A, a \Vdash^\forall A, \Gamma \Rightarrow \Delta}{x \in a, a \Vdash^\forall A, \Gamma \Rightarrow \Delta} \, L \Vdash^\forall$$

$$\frac{y \in a, A \lhd a, y : A, \Gamma \Rightarrow \Delta}{A \lhd a, y : A, \Gamma \Rightarrow \Delta} \, L \lhd \qquad \frac{y : A, \Gamma \Rightarrow \Delta, y \in a}{\Gamma \Rightarrow \Delta, A \lhd a} \, R \lhd, \, y \, \textit{fresh}$$

Non-Normal Modal Logics: A Challenge to Proof Theory

$$\frac{a \in I(x), a \Vdash^\forall A, A \triangleleft a, \Gamma \Rightarrow \Delta}{x : \Box A, \Gamma \Rightarrow \Delta} \; L\Box, \, a \, \textit{fresh}$$

$$\frac{a \in I(x), \Gamma \Rightarrow \Delta, x : \Box A, a \Vdash^\forall A \quad a \in I(x), \Gamma \Rightarrow \Delta, x : \Box A, A \triangleleft a}{a \in I(x), \Gamma \Rightarrow \Delta, x : \Box A} \; R\Box$$

Finally, initial sequents have only labelled atomic formulas (of the form $x : P$) and neighbourhood atoms (a priori, of the form $x \in a$ or $a \in I(x)$) as principal. We observe that formulas of the form $a \in I(x)$ are never active in the right-hand side of the rules we have listed above, and therefore the corresponding initial sequents can be dispensed with. So the only initial sequents actually needed in the calculus are

$$x : P, \Gamma \Rightarrow \Delta, x : P \qquad x \in a, \Gamma \Rightarrow \Delta, x \in a$$

By the procedure described and the addition of the propositional part of the labelled calculus G3K (cf. Negri, 2005) a G3 sequent calculus for the basic system **E** is obtained.

5. Extensions of system **E** (known as systems of *classical modal logics*) are obtained by adding axioms such as the following:

(M) $\Box(A \& B) \supset \Box A \& \Box B$

(C) $\Box A \& \Box B \supset \Box(A \& B)$

(N) $\Box \top$

To obtain complete sequent calculi for the classical systems defined by each of the above axioms, we incorporate in the basic calculus **G3E** the rules originated from the neighbourhood-semantic conditions that correspond to the axioms. Such rules are found through known correspondences in neighbourhood semantics and conversion into rules, or directly by a method of abduction from proof search in the basic calculus.

This process may involve the need for new primitives. The neighbourhood condition that corresponds to the first axiom states that supersets of a neighbourhood of x are themselves neighbourhoods of x; the condition that corresponds to the second axiom is closure under intersection: the intersection of two neighbourhoods of x is a neighbourhood of x. The third conditions requires that the set of all possible worlds (called the *unit*) is itself a neighbourhood of x. The table below summarizes for each system the modal axiom, the neighbourhood condition, and the corresponding rule. To complete the process, one needs rules for the new primitives of formal inclusion, intersection, and unit.

Axiom	NS property	Rule
(M) $\Box(A\&B) \supset \Box A\&\Box B$	$a \in I(x) \& a \subseteq b$ $\rightarrow b \in I(x)$	$\dfrac{a \in I(x), a \subseteq b, b \in I(x), \Gamma \Rightarrow \Delta}{a \in I(x), a \subseteq b, \Gamma \Rightarrow \Delta}$
(C) $\Box A\&\Box B \supset \Box(A\&B)$	$a \in I(x) \& b \in I(x)$ $\rightarrow a \cap b \in I(x)$	$\dfrac{a \in I(x), b \in I(x), a \cap b \in I(x), \Gamma \Rightarrow \Delta}{a \in I(x), b \in I(x), \Gamma \Rightarrow \Delta}$
(N) $\Box\top$	$W \in I(x)$	$\dfrac{W \in I(x), \Gamma \Rightarrow \Delta}{\Gamma \Rightarrow \Delta}$

Table 1: From modal axioms to NS rules

Formal inclusion between two neighbourhoods a, b is defined by[4]

$$a \subseteq b \equiv \forall x(x \in a \rightarrow x \in b)$$

and has the sequent calculus rules

$$\dfrac{x \in a, \Gamma \Rightarrow \Delta, x \in b}{\Gamma \Rightarrow \Delta, a \subseteq b} \; R\subseteq, x\,fresh \qquad \dfrac{x \in b, x \in a, a \subseteq b, \Gamma \Rightarrow \Delta}{x \in a, a \subseteq b, \Gamma \Rightarrow \Delta} \; L\subseteq$$

Formal intersection has the rules

$$\dfrac{x \in a, x \in b, x \in a \cap b, \Gamma \Rightarrow \Delta}{x \in a \cap b, \Gamma \Rightarrow \Delta} \; L\cap$$

$$\dfrac{\Gamma \Rightarrow \Delta, x \in a \cap b, x \in a \quad \Gamma \Rightarrow \Delta, x \in a \cap b, x \in b}{\Gamma \Rightarrow \Delta, x \in a \cap b} \; R\cap$$

Finally, the rule that defines the unit is simply

$$\dfrac{x \in W, \Gamma \Rightarrow \Delta}{\Gamma \Rightarrow \Delta} \; W$$

---

[4] Observe that to keep the notation simpler we use the same symbols ($\in, \subseteq, W$) both at the semantic and syntactic levels.

Non-Normal Modal Logics: A Challenge to Proof Theory

4 Streamlining

The method we have outlined in its main points produces complete calculi for the logical systems under examination, but can still be improved. In some cases, instead of adding extra neighbourhood properties, it is convenient to modify the forcing conditions in such a way that they become in-built. A similar move worked for the condition of Noetherianity for the provability logics GL and Grz (Negri, 2005, 2014; Dyckhoff & Negri, 2016). Unlike in that case, where the move was forced by the fact that Noetherianity is not expressible as a rule because it is not a first-order frame condition, here the move is optional, but it has a double advantage: it avoids the addition of some neighbourhood rules and simplifies the modal rules.

In the presence of monotonicity, the following forcing conditions give the same class of valid formulas (for a proof see Negri, 2017):

1. $x \Vdash^1 \Box A \equiv \exists a \in I(x)(a \Vdash^\forall A \,\&\, \forall y(y \Vdash A \to y \in a))$
2. $x \Vdash^2 \Box A \equiv \exists a \in I(x).a \Vdash^\forall A$

In logical systems closed under monotonicity, the rules for the necessity operator can thus be simplified into the following form with no added rule for monotonicity required:

$$\dfrac{a \in I(x), a \Vdash^\forall A, \Gamma \Rightarrow \Delta}{x : \Box A, \Gamma \Rightarrow \Delta}\; L\Box', a\, fresh$$

$$\dfrac{a \in I(x), \Gamma \Rightarrow \Delta, x : \Box A, a \Vdash^\forall A}{a \in I(x), \Gamma \Rightarrow \Delta, x : \Box A}\; R\Box'$$

5 Other modalities

It is often useful to have primitive rules also for modalities which can be defined through duality, such as the possibility modality. For this purpose, in addition to the universal forcing $\Vdash^\forall$ it is useful to consider another relation of local forcing, the existential one

$a \Vdash^\exists A$ is true iff there is some world x in a such that $x \Vdash A$

The corresponding rules are

$$\dfrac{x \in a, \Gamma \Rightarrow \Delta, x : A, a \Vdash^\exists A}{x \in a, \Gamma \Rightarrow \Delta, a \Vdash^\exists A}\; R\Vdash^\exists \qquad \dfrac{x \in a, x : A, \Gamma \Rightarrow \Delta}{a \Vdash^\exists A, \Gamma \Rightarrow \Delta}\; L\Vdash^\exists, x\,fresh$$

The rules that one obtains by unfolding the definition of forcing for the dual of necessity and by applying the process described above are

$$\frac{a \in I(x), x : \Diamond A, a \Vdash^\exists A, \Gamma \Rightarrow \Delta \quad a \in I(x), x : \Diamond A, \Gamma \Rightarrow \Delta, \neg A \triangleleft a}{a \in I(x), x : \Diamond A, \Gamma \Rightarrow \Delta} L\Diamond$$

$$\frac{a \in I(x), \neg A \triangleleft a, \Gamma \Rightarrow \Delta, a \Vdash^\exists A}{\Gamma \Rightarrow \Delta, x : \Diamond A} R\Diamond, a\,fresh$$

In the presence of monotonicity, the simplified rules are as follows:

$$\frac{a \in I(x), x : \Diamond A, a \Vdash^\exists A, \Gamma \Rightarrow \Delta}{a \in I(x), x : \Diamond A, \Gamma \Rightarrow \Delta} L\Diamond'$$

$$\frac{a \in I(x), \Gamma \Rightarrow \Delta, a \Vdash^\exists A}{\Gamma \Rightarrow \Delta, x : \Diamond A} R\Diamond'$$

The use of neighbourhood semantics in place of the relational semantics gives a splitting of the standard alethic modalities into four modalities, $[\,]$, $\langle\,]$, $[\,\rangle$, $\langle\,\rangle$, that correspond to the four different combinations of quantifiers in the semantic explanation:

$x \Vdash [\,] A$ iff *for every neighbourhood a of x, $a \Vdash^\forall A$*

$x \Vdash \langle\,] A$ iff *there is some neighbourhood a of x such that $a \Vdash^\forall A$*

$x \Vdash [\,\rangle A$ iff *for every neighbourhood a of x, $a \Vdash^\exists A$*

$x \Vdash \langle\,\rangle A$ iff *there is some neighbourhood a of x such that $a \Vdash^\exists A$*

It is then an easy exercise to convert the above semantic explanation into rules

Rules for NS-alethic modalities:

$$\frac{a \in I(x), \Gamma \Rightarrow \Delta, a \Vdash^\forall A}{\Gamma \Rightarrow \Delta, x : [\,]A} R[\,],\,a\,fresh \qquad \frac{a \in I(x), x : [\,]A, a \Vdash^\forall A, \Gamma \Rightarrow \Delta}{a \in I(x), x : [\,]A, \Gamma \Rightarrow \Delta} L[\,]$$

$$\frac{a \in I(x), \Gamma \Rightarrow \Delta, x : \langle\,]A, a \Vdash^\forall A}{a \in I(x), \Gamma \Rightarrow \Delta, x : \langle\,]A} R\langle\,] \qquad \frac{a \in I(x), a \Vdash^\forall A, \Gamma \Rightarrow \Delta}{x : \langle\,]A, \Gamma \Rightarrow \Delta} L\langle\,],\,a\,fresh$$

$$\frac{a \in I(x), \Gamma \Rightarrow \Delta, a \Vdash^\exists A}{\Gamma \Rightarrow \Delta, x : [\,\rangle A} R[\,\rangle,\,a\,fresh \qquad \frac{a \in I(x), x : [\,\rangle A, a \Vdash^\exists A, \Gamma \Rightarrow \Delta}{a \in I(x), x : [\,\rangle A, \Gamma \Rightarrow \Delta} L[\,\rangle$$

$$\dfrac{a \in I(x), \Gamma \Rightarrow \Delta, x : \langle\,\rangle A, a \Vdash^{\exists} A}{a \in I(x), \Gamma \Rightarrow \Delta, x : \langle\,\rangle A} \; R\langle\,\rangle \qquad \dfrac{a \in I(x), a \Vdash^{\exists} A, \Gamma \Rightarrow \Delta}{x : \langle\,\rangle A, \Gamma \Rightarrow \Delta} \; L\langle\,\rangle, \, a\,fresh$$

As a further example we consider the *ought* modality of deontic logic, perhaps the best known example of a logic for which the standard normal modal logic setting is inadequate.

We recall that the standard axiomatization of deontic logic is obtained by adding the axiom $\neg\mathcal{O}\bot$ to the axiomatisation of **K** and that the normal modal base leads to well known deontic paradoxes (e.g. the *gentle murder* of Forrester, 1984).

Non-normal systems of deontic logic have been proposed as a way out from paradoxes (Orlandelli, 2014). They are obtained as extensions of classical modal logics. System **ED**, **MD**, **RD**, and **KD** are obtained, respectively, as extensions of systems **E**, **M**, **N**, and **C** with the deontic axiom $\neg\,\mathcal{O}\bot$. The latter axiom corresponds to the rule

$$\dfrac{y \in a, a \in I(x), \Gamma \Rightarrow \Delta}{a \in I(x), \Gamma \Rightarrow \Delta} \; D, \, y\,fresh$$

G3MD has the modal rules:

$$\dfrac{a \in I(x), a \Vdash^{\forall} A, \Gamma \Rightarrow \Delta}{x : \mathcal{O}A, \Gamma \Rightarrow \Delta} \; L\mathcal{O}, \, a\,fresh$$

$$\dfrac{a \in I(x), \Gamma \Rightarrow \Delta, x : \mathcal{O}A, a \Vdash^{\forall} A}{a \in I(x), \Gamma \Rightarrow \Delta, x : \mathcal{O}A} \; R\mathcal{O}$$

Other systems of deontic logic are obtained by adding the rules that correspond to each neighbourhood property, as in Table 1. For the non-monotonic system, the rule for the deontic modality follows the general form of the rule for the alethic modality of system **E**.

6 Properties of NS-sequent calculi

The structural properties are established in a uniform way for any set of modalities and neighbourhood properties; the guidelines set for relational semantics require some additions and modifications. First, besides world labels, one has neighbourhood labels which are treated in the same way as world labels with respect to substitutions. Secondly, a suitable definition of formula weight is needed to reflect the nesting of the local forcing relations in the meaning explanation and the subsequent layering of the modal rules.

The following results are then established (for details see Negri, 2017):

1. Substitution is height-preserving admissible:

   (a) If $\vdash_n \Gamma \Rightarrow \Delta$, then $\vdash_n \Gamma(y/x) \Rightarrow \Delta(y/x)$;

   (b) If $\vdash_n \Gamma \Rightarrow \Delta$, then $\vdash_n \Gamma(b/a) \Rightarrow \Delta(b/a)$.

2. All the rules are height-preserving invertible.

3. The rules of left and right weakening and contraction are height-preserving admissible.

4. Cut is admissible.

5. The calculus is shown complete indirectly through equivalence to the axiomatic systems and known completeness results.

6. Direct completeness proof: All the rules are sound with respect to neighbourhood models and for every sequent, either there is a derivation or a countermodel in the intended class of neighbourhood models.

References

Chellas, B. (1980). *Modal Logic: An Introduction*. Cambridge University Press.

Dyckhoff, R., & Negri, S. (2012). Proof analysis in intermediate logics. *Archive for Mathematical Logic, 51*, 171–92.

Dyckhoff, R., & Negri, S. (2015). Geometrization of first-order logic. *The Bulletin of Symbolic Logic, 21*, 123–163.

Dyckhoff, R., & Negri, S. (2016). A cut-free sequent system for Grzegorczyk logic, with an application to the Gödel-McKinsey-Tarski embedding. *Journal of Logic and Computation, 26*, 169–187.

Forrester, J. (1984). Gentle murder, or the adverbial samaritan. *The Journal of Philosophy, 81*, 193–197.

Gasquet, O., & Herzig, A. (1996). From classical to normal modal logics. In H. Wansing (Ed.), *Proof Theory of Modal Logic* (pp. 293–311). Dordrecht: Kluwer.

Gilbert, D., & Maffezioli, P. (2015). Modular sequent calculi for classical modal logics. *Studia Logica, 103*, 175–217.

Girlando, M., Negri, S., Olivetti, N., & Risch, V. (2016). The logic of conditional beliefs: neighbourhood semantics and sequent calculus. In L. Beklemishev, S. Demri, & A. Máté (Eds.), *Advances in Modal Logic* (pp. 322–341). London: College Publications.

Kracht, M., & Wolter, F. (1999). Normal monomodal logics can simulate all others. *The Journal of Symbolic Logic, 64*, 99–138.

Lewis, D. (1973). *Counterfactuals*. Blackwell.

Negri, S. (2003). Contraction-free sequent calculi for geometric theories, with an application to Barr's theorem. *Archive for Mathematical Logic, 42*, 389–401.

Negri, S. (2005). Proof analysis in modal logic. *Journal of Philosophical Logic, 34*, 507–544.

Negri, S. (2014). Proofs and countermodels in non-classical logics. *Logica Universalis, 8*, 25–60.

Negri, S. (2016). Proof analysis beyond geometric theories: from rule systems to systems of rules. *Journal of Logic and Computation, 27*, 513–537.

Negri, S. (2017). Proof theory for non-normal modal logics: The neighbourhood formalism and basic results. To appear.

Negri, S., & Olivetti, N. (2015). A sequent calculus for preferential conditional logic based on neighbourhood semantics. In H. de Nivelle (Ed.), *Automated Reasoning with Analytic Tableaux and Related Methods* (Vol. 9323, pp. 115–134). Springer.

Negri, S., & von Plato, J. (1998). Cut elimination in the presence of axioms. *The Bulletin of Symbolic Logic, 4*, 418–435.

Negri, S., & von Plato, J. (2001). *Structural Proof Theory*. Cambridge University Press.

Negri, S., & von Plato, J. (2015). Meaning in use. In H. Wansing (Ed.), *Dag Prawitz on Proofs and Meaning* (pp. 239–257). Springer.

Olkhovikov, G., & Schroeder-Heister, P. (2014). On flattening elimination rules. *The Review of Symbolic Logic, 7*, 60–72.

Orlandelli, E. (2014). Proof analysis in deontic logics. In H. Wansing (Ed.), *Deontic Logic and Normative Systems* (Vol. 8554, pp. 139–148). Springer.

Pacuit, E. (2007). *Neighborhood Semantics for Modal Logic: An Introduction*. Lecture Notes of a course held at Essllis. (Available online at http://ai.stanford.edu/epacuit/classes/esslli/nbhdesslli.pdf)

von Plato, J. (2001). A proof of Gentzen's Hauptsatz without multicut.

Sara Negri

Archive for Mathematical Logic, *40*, 9–18.
Read, S. (2015). Semantic pollution and syntactic purity. *The Review of Symbolic Logic*, *8*, 649–661.

Sara Negri
University of Helsinki
Finland
E-mail: sara.negri@helsinki.fi

Revisiting Dummett's Proof-Theoretic Justification Procedures

HERMÓGENES OLIVEIRA[1]

Abstract: Dummett's justification procedures are revisited. They are used as background for the discussion of some conceptual and technical issues in proof-theoretic semantics, especially the role played by assumptions in proof-theoretic definitions of validity.

Keywords: Semantic inferentialism, Proof-theoretic semantics, Logical validity, Intuitionistic logic

1 The placeholder view of assumptions

In his contribution to the *Logica Yearbook 2007*, Schroeder-Heister (2008, § 3) pointed out some dogmas of proof-theoretic semantics. One of the dogmas was *the primacy of the categorical over the hypothetical*, or, as it was latter called, the *placeholder view of assumptions*.

According to this dogma, hypothetical arguments, or arguments with open assumptions, should be reduced to closed arguments, or closed proofs (proofs from no assumptions). In other words, assumptions are considered to be placeholders for closed proofs. The proof-theoretic definitions of validity for arguments proposed by Prawitz (1971, 1973, 2006, 2014) are prominent examples of the placeholder view of assumptions.

1.1 The problem with *reduction ad absurdum* arguments

In intuitionistic logic, *reductio ad absurdum* can be used to obtain negative sentences, or refutations. In such arguments, a contradiction (which in natural deduction systems is usually represented by an absurdity constant) is deduced from a collection of assumptions which are thereby shown to be jointly contradictory, or incompatible.

---

[1] I thank the organizers and participants of LOGICA 2016 for the stimulating environment provided for the discussion of themes related to my research. This work was supported by Deutscher Akademischer Austauschdienst (DAAD), grant number 91562976.

Hermógenes Oliveira

The task of explaining the validity of arguments that use *reductio ad absurdum* becomes problematic when assumptions are considered placeholders for closed proofs and validity is explained as a constructive function from closed proofs of the assumptions to closed proofs of the conclusion, because the explanation then needs to appeal to proofs of contradictions. These proofs do not need to be actual proofs, but must be at least possible or conceivable if the explanation is to be comprehensible at all. Whether proofs of contradictions can be conceived, or what does it mean to conceive such things, is one of the questions that the advocates of the placeholder view have to deal with.

In some sense, the conundrum with *reductio ad absurdum* is reminiscent of a problem that Prawitz (1971, § IV.1.1) already dealt with in his first attempt at defining a proof-theoretic notion of validity. There, the problem was the vacuous validation of implications with an unprovable antecedent. Prawitz's solution was to reformulate the semantic clause for implication so as to consider extensions of the underlying atomic system where the antecedent would be provable.[2] However, the problem becomes much more prominent when dealing with contradictions, because our intuition is that they are not supposed to be provable under any circumstances whatsoever.

1.2 The primacy of assertion

Walking side by side with the placeholder view of assumptions is what we can call *the primacy of assertion over other speech acts*. The rationale is that the speech act of assertion comes with a commitment on the part of the speaker to offer justifications for the asserted sentence and thus, in order to correctly assert the sentence, the speaker must be in possession of such justifications, or be able to produce them. In other words, in order to correctly assert a sentence, one needs to have a proof of the sentence.

From this picture emerges an approach to proof-theoretic semantics based on assertibility conditions, with proofs acting as justifications associated with assertions. Here, another dogma discussed by Schroeder-Heister (2008, § 3) comes into play: *the transmission view of consequence*. But, in contrast with semantics based on truth conditions, instead of truth, it is *correct assertibility* which is transmitted from premises to conclusion in valid arguments. Or, if one prefers to talk about what makes an assertion correct, or

---

[2]Unfortunately, the amendment was still insufficient to avoid validation of classical inferences in the implication fragment (Sanz, Piecha, & Schroeder-Heister, 2014, § 4).

justified, one can say that consequence transmits *proof* instead of *truth*. As a result, the approach assumes a distinctively epistemological character.

However, more complications related to hypothetical reasoning surface: it seems counterintuitive, to say the least, to hold that a speaker engaged in a hypothetical argument is committed to the *assertion* of either the assumptions or the conclusion of the argument. As a matter of fact, the speaker may even reject them and, given the argument is valid, her reasoning remains unassailable. In particular, the point reappears with renewed force when considered in the context of arguments that use *reductio ad absurdum*, since it would commit us to the possibility of correctly asserting absurdities.

One can appeal to a concept of *conditional assertion* to try and salvage the approach from such objections while preserving an unified explanatory model based primarily on assertion and proof. Thus, the conclusions of hypothetical arguments are taken not to be asserted outright but only under certain conditions. That is, the conclusions of hypothetical arguments are *conditionally asserted*. In terms of speech acts, however, it is not at all clear whether conditional assertion constitutes any assertion at all.

It seems to me that trying to explain deductive validity in terms of assertions and proofs is misguided. I am not trying to deny that deductive reasoning has epistemic importance or that deductive reasoning transmits evidence, or justification, from the premises to the conclusion. If there is a deductive relation between premises and conclusion, then, of course, the correct assertion of the premises would lead to the correct assertion of the conclusion and, similarly, if proof for the premises are provided then a proof for the conclusion is obtained. Rather, I contend that to *explain* deductive validity by reducing it to this transmission effect is to put the cart before the horse and confuse the cause with its effects, the disease with its symptoms.

2 BHK vs Gentzen

Proof-theoretic notions of validity have often been inspired by a mixture of ideas involving the BHK interpretation of the logical constants and Gentzen's well-known remarks on the rules of natural deduction. In particular, the conception of validity underlying the placeholder view of assumptions is largely informed by the BHK interpretation of implication: an argument from A to B is valid if, and only if, every proof of A can be transformed into a proof of B. Yet, with its unqualified reference to proofs, this view is not immediately amenable to the recursive treatment required of semantic clauses and definitions (Prawitz, 2007, § 2.1). In this context, Gentzen's ideas are often developed into a notion of canonical proof in order to achieve recursiveness for an approach primarily based on the BHK interpretation.

On the other hand, the core of Gentzen's ideas are independent of the BHK interpretation. They are best represented by what became known as *proof-theoretic harmony*. Harmony, as a fundamental principle of natural deduction systems, applies equally well to deductions from assumptions as to the particular case of proofs (deductions from no assumptions). By appealing to harmony while at the same time avoiding the BHK interpretation and the placeholder view of assumptions, we can develop a more appropriate proof-theoretic notion of validity.

2.1 A more Gentzenian approach to validity

The introduction rules for a logical constant γ can be seen as an explanation of the *canonical use* of a sentence *as a conclusion* in a deductive argument (where, of course, γ is the sentence's main connective). This is achieved by exhibiting the conditions for obtaining a sentence $A \gamma B$ as a conclusion of an argument (where γ is a binary connective). In the paradigmatic case, these conditions are expressed in terms of the component sentences A and B.[3]

In an analogous manner, the elimination rules for a logical constant can be seen as an explanation of the *canonical use* of a sentence *as an assumption* in a deductive argument. This is accomplished by exhibiting the consequences that can be extracted from the sentence (as a major premiss of an elimination rule, possibly in the context of some minor premisses).

Thus, introduction and elimination rules stand for two distinct aspects of the deductive use of the logical constants. Harmony arises as a requirement of balance between those two aspects such that there is an equilibrium between what is required by the introduction rules and what consequences are extracted by the elimination rules. As a result, among other things, harmony guarantees that there is nothing to be gained by performing roundabout derivations where sentences are obtained by an introduction rule to be immediately after analysed by the corresponding elimination rule. Therefore, for a proper understanding of the deductive practice, it suffices to look

---

[3]Notice that the conditions do not necessarily need to be expressed in terms of closed proofs of A and B, but can be expressed in terms of assumptions A and B or of arguments for A and B which may depend on other assumptions.

at the collection of direct derivations, also known as normal derivations.

The normal derivations have a very perspicuous form (Prawitz, 1965, Chapter IV, § 2, Theorem 2). They are composed of (can be divided into) two parts: an analytic part, where the assumptions are analysed (destructed), and a synthetic part, where the conclusion is synthesized (constructed) from the components resulted from the analysis.[4]

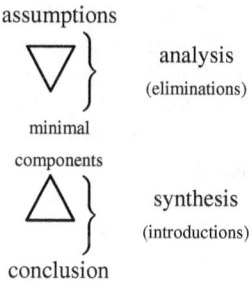

The equilibrium between introductions and eliminations suggest that, if we were to *supplement* the assumptions on top through a process of inversion by backward application of introductions, we would arrive at the minimal components required for the synthetic part. And, similarly, if we were to *complement* the conclusion by forward application of eliminations, we would arrive at the minimal components resulted from the analysis of the assumptions. Accordingly, the harmonious inferential behaviour of the logical constants has sometimes been expressed by pointing out that introductions and eliminations can be, in some sense, obtained from one another by *inversion principles*.

Gentzen's investigations into logical deduction can thus supply the basic pieces for a proof-theoretical notion of logical validity for arguments based on the inferential meaning conferred on the logical constants by either their introduction rules or their elimination rules. In particular with respect to the problems discussed in the previous section, the Gentzenian approach has the advantage of giving proper heed to assumptions and being fairly independent from specific speech acts.[5]

---

[4] In the general case, each of these parts can, of course, be empty.

[5] For instance, deductive arguments can be used to show someone who denies the conclusion that she has to deny at least one of the assumptions. They can also be used to explore the consequences of a conjecture. These applications of deductive arguments align very well with the Gentzenian approach, but none of them necessarily involves anyone making any assertions.

Gentzen's ideas suggest that, although a persistent dogma in much of the discussion around proof-theoretic semantics, the placeholder view of assumptions can be challenged from an authentic proof-theoretic perspective. In the next section, I revisit Dummett's justification procedures. I argue that, as a development of the Gentzenian approach just sketched, they afford a notion of proof-theoretic validity that incorporates assumptions in an essential way.[6]

I stay at the level of the core concepts, without going into rigorous definitions. Nonetheless, I hope that my explanations would be sufficient to give an overall idea of the relationship between the justification procedures (how they can be understood as emerging from a shared framework) and I also presume to have provided enough detail so that an interested and motivated reader would be able to attempt rigorous definitions of her own based on the approach outlined. In the last section, I discuss, to some extent, how Dummett's procedures perform with respect to adequacy to intuitionistic logic, especially in comparison with approaches that endorse the placeholder view of assumptions.

3 An overview of Dummett's approach

Dummett (1991, Chapter 11–13) proposed two proof-theoretic justification procedures for logical laws which amount to definitions of logical validity for arguments. The "verificationist" procedure defines validity of arguments on the basis of introduction rules for logical constants and the "pragmatist" procedure defines validity of arguments on the basis of elimination rules for logical constants.[7]

These proof-theoretic justification procedures play an important role in Dummett's philosophical anti-realist programme. They are central pieces of his very detailed and elaborate argument for rejection of classical logic in favor of intuitionistic logic.[8] In particular, Dummett (1975, 1991) has

---

[6]It is important to notice that, although the Dummettian approach that I advocate rejects the placeholder view of assumptions, other dogmatic characteristics, like the unidirectional and global character of the semantics, remain unchallenged (Schroeder-Heister, 2016, § 2.3 and 2.4).

[7]I adopt the characterizations "verificationist" and "pragmatist" from Dummett. However, without denying their existence, I do not imply with the adoption of the terminology any connections outside the domain of logical validity. Therefore, I refer to verificationism and pragmatism just as markers to distinguish between approaches to validity based on introduction rules and elimination rules, respectively.

[8]At the end of this argument, Dummett (1991, p. 300) writes: "We took notice of the

conjectured that proof-theoretic notions of validity would justify exactly intuitionistic logic.

Dummett's definitions of validity are based on canonical inference rules for the logical constants. These inference rules are thought to fix the meaning of the logical constants by displaying their canonical deductive use. They are, in Dummett's terminology, "self-justifying".

In contrast with some definitions found in the literature, Dummett's definitions are not based on semantic clauses for particular logical constants. Instead, he assumes that self-justifying rules are given. These self-justifying rules are introduction rules in the context of the verificationist procedure, and elimination rules in the context of the pragmatist procedure. In both procedures, the definitions are stated irrespective of the particular constants or rules provided. Therefore, Dummett's definitions can, at least in principle, be applied without modification to different logics by providing the appropriate self-justifying rules for the logical constants.

3.1 Core concepts

Both the verificationist and the pragmatist procedures can be seen as products of a basic common framework. The core notions of validity behind the justification procedures can be informally outlined as follows:

verificationism whenever the assumptions can be obtained in a canonical manner, the conclusion can also be obtained in a canonical manner

pragmatism any consequence that can be drawn in a canonical manner from the conclusion can also be drawn in a canonical manner from the assumptions.

The expression "canonical manner" is an allusion to *canonical arguments*. As usual in proof-theoretic notions of validity, canonical arguments are the main ingredients of the justification procedures. An important feature, however, is that Dummett's canonical arguments are *not closed proofs*,

---

problem what metalanguage is to be used in giving a semantic explanation of a logic to one whose logic is different. A metalanguage whose underlying logic is intuitionistic now appears a good candidate for the role, since its logical constants can be understood, and its logical laws acknowledged, without appeal to any semantic theory and with only a very general meaning-theoretical background. If that is not *the* right logic, at least it may serve as a medium by means of which to discuss other logics."

but instead *may depend on assumptions*. Consequently, when precisely formulated, the definitions of validity must take into account the assumptions on which the canonical arguments depend.

The canonical arguments are composed *primarily* of canonical inferences. However, they cannot be required to be *entirely* composed of canonical inferences. They must allow for the possibility of *subordinate subarguments*, that is, subarguments cultivated under the support of additional assumptions (Dummett, 1991, p. 260). These subordinate subarguments, when not already canonical arguments themselves, are *critical subarguments*. They are critical in the sense that the validity of the original canonical argument would recursively depend on their validity. This means, of course, that much care should be dispensed to guarantee that critical subarguments are of lower complexity than the original canonical argument.

In a verificationist context, critical subarguments are detected through the presence of assumption discharge. In a pragmatist context, they are detected through the presence of minor premises. These signs indicate, in their respective contexts, when assumptions are being added.

Now, returning to our informal notions of validity, in the verificationist procedure, the means to evaluate the conditions under which the assumptions may be obtained in a canonical manner are provided by *supplementations*. They result from substitution of the assumptions with canonical arguments. In the pragmatist procedure, the means to evaluate what consequences can be drawn from the conclusion are provided by *complementations*. They result from substitution of the conclusion with canonical arguments.

verificationism	pragmatism
canonical arguments	canonical arguments
(primarily introductions)	(primarily eliminations)
critical subarguments	critical subarguments
(revealed by assumption discharges)	(revealed by minor premises)
supplementation	complementation
(assumptions canonically unfolded)	(conclusion canonically unfolded)

Instead of as substitution operations, one can see the processes of supplementation and complementation more dynamically. The process of supplementation can be seen as the repeated backward application of introduction rules from the assumptions, thus growing the argument upwards, which is

Revisiting Dummett's Procedures

why Dummett also refers to the verificationist procedure as the *upwards justification procedure*. Similarly, the process of complementation can be seen as the repeated forward application of elimination rules to the conclusion of the argument as a major premiss, thus growing the argument downwards, which is why Dummett also refers to the pragmatist procedure as the *downwards justification procedure*.

In order to appraise the validity of an argument from Γ to C, the verificationist procedure examines its supplementations and investigates whether a canonical argument for C can be attained under the same conditions. Since supplementations result from canonical arguments *for* Γ, they may depend on assumptions Δ (remember that canonical arguments may depend on assumptions). Then, the canonical argument for C may not depend on other assumptions besides Δ.

In an analogous manner, in order to appraise the validity of an argument from Γ to C, the pragmatist procedure examines the complementations and investigates whether a canonical argument for the conclusion of the complementation, say Z, can be attained under the same conditions. Because complementations result from canonical arguments *from* C (as assumption and major premiss of elimination), they may depend on additional assumptions Δ required by minor premisses. Then, the canonical argument for Z may not depend on other assumptions besides Γ, Δ.

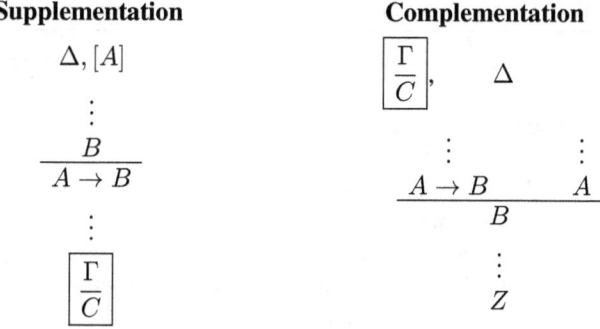

The canonical arguments used to supplement or complement may have critical subarguments. In the figures above, I indicate the form of possible supplementations and complementations of an illustrative argument from Γ to C. In the supplementation, the subargument from Δ, A to B may be critical. In the complementation, the subargument for the minor premiss A

of →E may be critical (for instance, if validly, but not canonically, obtained from assumptions in Δ).

3.2 An illustrative example

With the help of the concepts explained so far, let us try and show the validity of an argument according to the pragmatist procedure. Since I did not give any rigorous definitions, our validation of the argument must be carried out in an informal and intuitive manner. Still, we adhere to the overall pragmatist approach and appeal exclusively to the elimination rules for the logical constants.

$$\frac{A \to (B \land C)}{(A \to B) \land (A \to C)}$$

According to the pragmatist procedure, the argument depicted above would be valid if whatever consequences can be drawn from the conclusion in a canonical manner, can also be drawn from the assumptions in a canonical manner. In order to see what can be extracted canonically from the conclusion, we complement:

$$\cfrac{\cfrac{A \to (B \land C)}{(A \to B) \land (A \to C)} \quad }{\cfrac{A \to B \quad A}{B}} \qquad \cfrac{\cfrac{A \to (B \land C)}{(A \to B) \land (A \to C)} \quad }{\cfrac{A \to C \quad A}{C}}$$

The complementations were obtained by application of elimination rules until we arrived at a simple schematic letter (each consequence being the major premiss for the next application of an elimination rule). Let us concede that there is no loss of generality when we supply the minor premiss of →E with a simple assumption A. So, as we can see, we have two complementations, with conclusions B and C, respectively, and assumptions $A \to (B \land C)$ and A. In order to show validity, we must find canonical arguments from $A \to (B \land C)$ and A to B, and from $A \to (B \land C)$ and A to C.[9]

---

[9]The assumptions of the complementations happen to be the same in this case. In the general case, however, they have to be considered separately, e.g. each complementation has its own assumptions. In order to establish validity, we must show that the conclusion of the complementation can be obtained from the assumptions of the complementation, *for every complementation*.

$$\frac{A \to (B \wedge C) \quad A}{\frac{B \wedge C}{B}} \qquad \frac{A \to (B \wedge C) \quad A}{\frac{B \wedge C}{C}}$$

The example is a rather simple one (it does not involve, for instance, canonical arguments with critical subarguments). But it illustrates how a nontrivial argument (an argument whose *derivation* would need *both* eliminations and introductions) can be justified by appealing exclusively to the meaning conferred on the logical constants by their elimination rules.

4 Adequacy to intuitionistic logic

Sanz et al. (2014, § 4) showed that, in the fragment containing only implication, a strictly classical inference, Peirce's rule, is valid at the atomic level with respect to a proof-theoretic notion of validity proposed by Prawitz (1971, § IV.1.2). With implication already behaving classically at the atomic level, and assuming the validity of the usual rules for implication, conjunction and absurdity (but no disjunction), the result can be generalized to arbitrary complex sentences. As a consequence, we can have a set of logical constants powerful enough to account for all classically valid inferences in propositional logic (with the other constants being defined in terms of implication, conjunction, and absurdity).

In the context of Dummett's anti-realist philosophical programme, these kinds of results can have a very negative effect. They show that, contrary to what is intended, classical logic can be validated in a proof-theoretic setting, even when no classical principles are admitted in the semantics. Despite Dummett (1991, Chapter 15), there would be classical logic without bivalence (Sandqvist, 2009). Furthermore, the arguments used to establish these results can be expected to apply to any proof-theoretic notion based on the BHK view of implication and conservative extensions of production systems.

In contrast with Prawitz's early approach, Dummett's verificationist procedure does not adhere to the BHK view of implication as based on closed proofs, and is not conservative over extensions of production systems. Still, the verificationist procedure, as defined by Dummett (1991, p. 260), does not reject completely the placeholder view of assumptions, because the assumptions of canonical arguments are required to be atomic. As a result, from the point of view of a general approach to *logical* validity, too much emphasis is placed on the underlying atomic system and validity then ceases to be a schematic notion.[10]

One may be inclined to think that Dummett's approach based on atomic assumptions is basically equivalent to approaches based on production systems because atomic assumptions could be replaced by atomic axioms. However, the presupposition of monotonicity incorporated into production systems (by requiring that extensions of the production system be conservative) does not carry over to collections of assumptions. For instance, Goldfarb (2016, Counterexample 2) observed that, if the production system has no rules, the argument from $\varphi \to \psi$ to ψ is valid (where φ and ψ are atomic sentences). If the production system is extended with the rule φ/ψ, the argument ceases to be valid, which shows that verificationist validity is not monotonic over extensions of the production system.[11]

As a consequence, under the verificationist procedure with atomic assumptions, the atomic base cannot be correctly interpreted as expandable states of knowledge in the model of mathematics where, once a sentence is proved, it remains proved. Rather, conclusions cultivated under the support of some assumptions may not be available under other assumptions, and it is not intuitively compelling to restrict hypothetical arguments to conservatively extended collections of assumptions.

Since the arguments that show validation of Peirce's rule rely on the monotonicity of the atomic base, it is reasonable to expect that Dummett's verificationist procedure escapes validation of Peirce's rule, even in an atomic level. So, I would say that Dummett's procedure compares favourably in this respect with approaches based on closed proofs and conservative extensions of production systems.[12]

On the other hand, Dummett's pragmatist procedure incorporates as-

---

[10] I agree with Goldfarb's (2016, §4) assessment that Dummett's strategy, inspired by the BHK interpretation, of defining validity for actual sentences and then generalizing this notion in order to achieve logical validity is problematic.

[11] Disregarding other problematic features of the verificationist procedure, I agree with Michael Dummett and Ofra Magidor that Counterexample 2 is not really worrisome (Goldfarb, 2016, Postscript). In this context, the absence of atomic rules are more correctly interpreted, not as the absence of knowledge about the inferential relations between atomic sentences, but instead as the knowledge that there are no inferential relations between atomic sentences.

[12] Goldfarb (2016, Counterexample 3) pointed out another problem with the verificationist procedure: the validation of the intuitionistically underivable rule of distribution of implication over disjunction. Since proof-theoretic approaches based on closed proofs and conservative extensions are also affected (Piecha, Sanz, & Schroeder-Heister, 2015, §4), I suspect the roots of the problem here are more profound than the mere format of the atomic base, or the fact that assumptions are restricted to atomic sentences.

sumptions entirely (atomic and complex). As a matter of fact, in the pragmatist sense, the paradigmatic canonical argument is hypothetical. This is perfectly aligned with the view, formulated earlier, that the elimination rules express the deductive canonical use of sentences as assumptions. As a result, the pragmatist procedure is not subject to any of the objections raised in Section 1. It fits better with investigations grounded on other speech acts besides assertion and provides a more perceptive explanation of the meaning of conjectures and the validity of arguments that use *reductio ad absurdum*.

Furthermore, in contrast with the verificationist case, a little reflection shows that the pragmatist notion of validity is schematic. In particular, suppose some arbitrary atomic rules determining the inferential relationships between atomic sentences are supplied. These rules could be applied to the atomic conclusion of a canonical argument in order to extract further atomic consequences. Now, consider, for instance, our example in Section 3.2. It is easy to see that, even if B or C were complex, any further elimination or atomic rules applied in the complementations could be transferred to the canonical arguments without adding any assumptions with respect to the complementations.

Finally, I would like to suggest that the pragmatist procedure may offer a notion of validity more adequate with respect to intuitionistic logic. To this effect, I argue informally that atomic Peirce's rule is not valid according to the pragmatist procedure.

Claim *Let φ and ψ be atomic sentences. Then, Peirce's rule*

$$\frac{(\varphi \to \psi) \to \varphi}{\varphi}$$

is not valid with respect to the pragmatist procedure.

Informal argument. The conclusion φ is already an atomic sentence, therefore there is nothing to complement. Now, we need to ask ourselves whether it is possible to obtain a valid canonical argument for φ from $(\varphi \to \psi) \to \varphi$. Any such valid canonical argument will have $(\varphi \to \psi) \to \varphi$ as major premiss of $\to$E.

$$\frac{(\varphi \to \psi) \to \varphi \quad (\varphi \to \psi)}{\varphi}$$

At this point, since $(\varphi \to \psi)$ was not among the available assumptions, Peirce's rule would be valid only if we could validly deduce $(\varphi \to \psi)$ from no assumptions (a critical subargument). But this is not the case, because complementations of $(\varphi \rightarrow \psi)$ will have ψ as conclusion with only φ as assumption

$$\frac{(\varphi \rightarrow \psi) \quad \varphi}{\psi}$$

and there is no general way to obtain a canonical argument for ψ from φ unless, of course, in the particular cases where $\psi = \varphi$ or ψ can be extracted from φ by accepted atomic inferences. □

Since I favored a more conceptual discussion in detriment of technical developments, I am not able to produce a more rigorous proof of this claim here. But I still hope that my informal discussion and explanations deliver some evidence for the adequacy of the pragmatist procedure to intuitionistic logic.

Perhaps I should not end without a warning. While revisiting Dummett's procedures, my agenda was to advance what I think is a more appropriate approach to proof-theoretic validity. As a consequence, I may have presented the justification procedures under a perspective at odds with Dummett's own.[13] Albeit exegetical correctness was obviously not among my primary concerns, I do believe that Dummett's procedures supply the essential elements for an interesting proof-theoretic notion of validity, one that is free from the influence of the BHK interpretation and rejects completely the placeholder view of assumptions. I attempted, so to say, to separate the wheat from the chaff.

References

Dummett, M. (1975). The philosophical basis of intuitionistic logic. In H. Rose & J. Shepherdson (Eds.), *Logic Colloquium '73* (Vol. 80, pp. 5–40). Amsterdam: North Holland.

Dummett, M. (1991). *The Logical Basis of Metaphysics*. Cambridge, Massachusetts: Harvard University Press.

Goldfarb, W. (2016). On Dummett's proof-theoretic justification of logical laws. In T. Piecha & P. Schroeder-Heister (Eds.), *Advances in Proof-Theoretic Semantics* (Vol. 43, pp. 195–210). Springer.

---

[13]For instance, Dummett seems to be committed to the primacy of assertion and to endorse the BHK interpretation as the authoritative source on the meaning of the intuitionistic logical constants.

Piecha, T., Sanz, W. d., & Schroeder-Heister, P. (2015). Failure of completeness in proof-theoretic semantics. *Journal of Philosophical Logic*, *44*(3), 321–335.

Prawitz, D. (1965). *Natural Deduction: A Proof-Theoretical Study*. Stockholm: Almqvist & Wiksell.

Prawitz, D. (1971). Ideas and results in proof theory. In *Proceedings of the Second Scandinavian Logic Symposium* (Vol. 63, pp. 235–307).

Prawitz, D. (1973). Towards a foundation of a general proof theory. In P. Suppes, L. Henkin, A. Joja, & G. C. Moisil (Eds.), *Logic, Methodology and Philosophy of Science IV* (Vol. 74, pp. 225–250). Amsterdam: North Holland.

Prawitz, D. (2006). Meaning approached via proofs. *Synthese*, *148*(3), 507–524.

Prawitz, D. (2007). Pragmatist and verificationist theories of meaning. In R. E. Auxier & L. E. Hahn (Eds.), *The Philosophy of Michael Dummett* (Vol. XXXI, pp. 455–481). Chicago and La Salle, Illinois: Open Court Publishing Company.

Prawitz, D. (2014). An approach to general proof theory and a conjecture of a kind of completeness of intuitionistic logic revisited. In L. C. Pereira, E. H. Haeusler, & V. de Paiva (Eds.), *Advances in Natural Deduction* (Vol. 39, pp. 269–279). Dordrecht: Springer.

Sandqvist, T. (2009, April). Classical logic without bivalence. *Analysis*, *69*(2), 211–218.

Sanz, W. d., Piecha, T., & Schroeder-Heister, P. (2014). Constructive semantics, admissibility of rules and the validity of Peirce's law. *Logic Journal of the IGPL*, *22*(2), 297–308.

Schroeder-Heister, P. (2008). Proof-theoretic versus model-theoretic consequence. In M. Peliš (Ed.), *The Logica Yearbook 2007* (pp. 187–200). Prague: Filosofia.

Schroeder-Heister, P. (2016). Open problems in proof-theoretic semantics. In T. Piecha & P. Schroeder-Heister (Eds.), *Advances in Proof-Theoretic Semantics* (Vol. 43, pp. 253–283). Springer.

Hermógenes Oliveira
Eberhard Karls Universität Tübingen
Germany
E-mail: oliveira@informatik.uni-tuebingen.de

Double-Line Harmony in a Sequent Setting

NORBERT GRATZL AND EUGENIO ORLANDELLI[1]

Abstract: This paper concentrates on how to capture harmony in sequent calculi. It starts by considering a proposal made by Tennant and some objections to it which have been presented by Steinberger. Then it proposes a different analysis which makes use of a double-line presentation of sequent calculi in the style of Došen and it shows that this proposal is able to dismiss disharmonious operators without thereby adopting any global criterion.

Keywords: Proof-theoretic semantics, Harmony, Sequent calculi, Double-line rules, Inversion principles.

1 Introduction

Logical inferentialism maintains that the meaning of a logical operator $ should be explained solely in terms of the rules of inference governing its behaviour. But, as shown by Prior's (1960) infamous operator $tonk$, not any set of rules introduces a meaningful operator. Thus, many different provisos have been provided to rule out tonkish operators, both in natural deduction (ND) and in sequent calculi (SC). One idea that has received much attention is the claim that the rules of inference for $ must be in *harmony*, i.e. that there must be 'a certain consonance between the two aspects [i.e. introduction and elimination rules in ND and right and left rules in SC] of the use of a given form of expression' (Dummett, 1973, p. 397). There has been little agreement on how to make precise this intuitive notion of harmony. In the literature there have been four kinds of explication. We may talk of (i) *global harmony* when harmony is explicated in terms of conservativity and uniqueness (Belnap, 1962); (ii) *intrinsic harmony* when it is explicated in terms of reduction procedures (Dummett, 1991; Prawitz, 1965); (iii) *general elimination harmony* when the elimination rules can be read off the introduction rules (Negri & von Plato, 2015; Read, 2000); and

---
[1] Many thanks to the audience of LOGICA 2016, in particular to Hermógenes Oliveira and to Mattia Petrolo, for helpful discussions.

(iv) *harmony as deductive equilibrium* when harmony is explicated directly in terms of an equilibrium between the two kinds of rules (Tennant, forth.).

One point of agreement between most proposals is that the rules governing an operator can be disharmonious in two different ways. The lack of consonance between the right and left rules in SC—or, equivalently, between the introduction and elimination rules in ND—may depend either (i) on the left rules being too strong with respect to the right ones, or (ii) on their being too weak. Following (Steinberger, 2011b) we talk of *S-disharmony* in the first case and of *W-disharmony* in the second one. A paradigmatic case of S-disharmony is Prior's *tonk* (▲), which is an operator having the left rules of conjunction and the right ones of disjunction.[2] A paradigmatic case of W-disharmony is its dual *knot* (▼), which is an operator having the left rule of disjunction and the right one of conjunction. A satisfactory analysis of harmony must rule out both *tonk* and *knot*.

Even though harmony is usually considered in the context of ND, when considering logics other than intuitionistic logic SC behave better. One advantage of SC is that they allow for multiple conclusions which are useful for some logics—e.g. for classical logic—and, possibly, essential for some others—e.g. for linear logic and for dual-intuitionistic logic. Another advantage is that SC allow to capture many modal logics which have no (known) formulation in ND. Furthermore, SC represent deductions more appropriately for those who take the concept of consequence as more fundamental than truth, (Schroeder-Heister, 2012). Thus, it is important to discuss the notion of harmony in SC.

In the context of SC, the analysis of harmony which is more widespread is the one in terms of global harmony. Belnap's seminal paper (1962) was precisely devoted to the introduction of global harmony in SC. Similar proposals have been given by Hacking (1979) for classical logic and, more recently, by Wansing (1998) for modal logics. Nevertheless, the global harmony of an operator $ may depend on whether some other operator is present, and we prefer to analyse the harmony of $ in terms of the rules governing its behaviour and independently of other operators—i.e. we opt for a so-called *local* analysis of harmony.

The local analyses in terms of intrinsic harmony and of general elimination harmony seems to be tailored for ND. Thus, this paper concentrates on how to capture harmony as (local) deductive equilibrium in SC (as will

---

[2]Usually *tonk* is defined by taking only rules $L\blacktriangle_2$ and $R\blacktriangle_1$ from Table 2. We prefer this equivalent and more symmetric version.

Double-Line Harmony in a Sequent Setting

Table 1: $\{\vee, \exists\}$-fragment of LK

$$\dfrac{A_1, \Gamma \Longrightarrow \Delta \qquad A_2, \Gamma \Longrightarrow \Delta}{A_1 \vee A_2, \Gamma \Longrightarrow \Delta} \, L\vee \qquad \dfrac{\Gamma \Longrightarrow \Delta, A_i}{\Gamma \Longrightarrow \Delta, A_1 \vee A_2} \, R\vee_i, i \in \{1,2\}$$

$$\dfrac{A[y/x], \Gamma \Longrightarrow \Delta}{\exists x A, \Gamma \Longrightarrow \Delta} \, L\exists,\ y \text{ fresh} \qquad \dfrac{\Gamma \Longrightarrow \Delta, A[t/x]}{\Gamma \Longrightarrow \Delta, \exists x A} \, R\exists$$

$$\dfrac{}{A \Longrightarrow A} \, \textit{Ref} \qquad \dfrac{\Gamma \Longrightarrow \Delta, A \qquad A, \Gamma' \Longrightarrow \Delta'}{\Gamma, \Gamma' \Longrightarrow \Delta, \Delta'} \, \textit{Cut} \qquad \dfrac{\Gamma \Longrightarrow \Delta}{\Gamma', \Gamma \Longrightarrow \Delta, \Delta'} \, W$$

Table 2: Rules for ▲ and ▼

$$\dfrac{A_i, \Gamma \Longrightarrow \Delta}{A_1 \blacktriangle A_2, \Gamma \Longrightarrow \Delta} \, L\blacktriangle_i, i \in \{1,2\} \qquad \dfrac{\Gamma \Longrightarrow \Delta, A_i}{\Gamma \Longrightarrow \Delta, A_1 \blacktriangle A_2} \, R\blacktriangle_i, i \in \{1,2\}$$

$$\dfrac{A_1, \Gamma \Longrightarrow \Delta \qquad A_2, \Gamma \Longrightarrow \Delta}{A_1 \blacktriangledown A_2, \Gamma \Longrightarrow \Delta} \, L\blacktriangledown \qquad \dfrac{\Gamma \Longrightarrow \Delta, A_1 \qquad \Gamma \Longrightarrow \Delta, A_2}{\Gamma \Longrightarrow \Delta, A_1 \blacktriangledown A_2} \, R\blacktriangledown$$

be presented in Definition 1). It starts by considering a proposal made by Tennant (2010, forth.) and some objections to it which have been presented in (Steinberger, 2009, 2011a). Then it proposes a different analysis which makes use of a double-line presentation of SC in the style of (Došen, 1989) and it shows that this proposal, which is purely local, avoids Steinberger's objections to Tennant's proposal. Without loss of generality, we will consider only the $\{\vee, \exists\}$-fragment of Gentzen's LK with context-as-sets and we consider a standard first-order language, see (Negri & von Plato, 2001). Table 1 gives the rules of this calculus and Table 2 gives the rules of the operators *tonk* and *knot* which will be used here as test for disharmony.[3]

---

[3] All we will say works equally good for the single conclusion intuitionistic calculus LJ. The addition of negation to LK is not a problem to our approach.

Norbert Gratzl and Eugenio Orlandelli

2 Harmony in sequent calculus

Tennant analyses harmony in SC as a deductive equilibrium between left and right introduction rules for an operator.[4] Harmony is 'a kind of Nash equilibrium between introduction and elimination rules' (Tennant, forth., p. 22). Roughly, for a monadic $, the idea is that $A has to be the strongest formula derivable by R$ and the weakest one derivable by L$ (and vice versa), or equivalently:

Definition 1 (H-DE)   *If we take L$ as primitive, then R$ has to allow us to derive* no more *(no less) sequents than those already derivable from a possible premiss of R$ (conclusion of L$).*

Notice that H-DE contains a *no-gain* condition, which is given by its 'no more...' clause and which is meant to rule out S-disharmonious operators, and it contains a *no-loss* condition, which is given by its 'no less...' clause and which is meant to rule out W-disharmonious operators. This will be extremely important in section 3 of this paper.

Tennant (2010) makes this precise by requiring that the rules for $ satisfy three constraints: *harmony*, *maximality* and *admissibility*. The constraint of harmony is based on the following Fregean definition of (logical) strength:

> a proposition ϕ is at least as strong as ψ iff the derivability of the sequent $\psi, \Gamma \Longrightarrow \Delta$ entails the derivability of $\phi, \Gamma \Longrightarrow \Delta$.

This allows Tennant to define harmony of $, $h(R\$, L\$)$, as follows:

(S) If ψ satisfy the conditions on $A in rule R$, then by making full use of L$ and no use of R$ we can show that $A is at least as strong as ψ.

(W) If ψ satisfy the conditions on $A in rule L$, then by making full use of R$ and no use of L$ we can prove that $A is at least as weak as ψ.

It is immediate to notice that harmony holds for $\vee$ and $\exists$. To illustrate, if we take a ψ with the following right introduction rule: $\dfrac{\Gamma \Longrightarrow \Delta, A[t/x]}{\Gamma \Longrightarrow \Delta, \psi} R\psi$, then we can prove that the rules for $\exists$ comply with (S) as follows:

$$\dfrac{\dfrac{\dfrac{\overline{A[y/x] \Longrightarrow A[y/x]}\ Ref}{A[y/x] \Longrightarrow \psi}\ R\psi \quad \psi, \Gamma \Longrightarrow \Delta}{\dfrac{A[y/x], \Gamma \Longrightarrow \Delta}{\exists x A, \Gamma \Longrightarrow \Delta}\ L\exists}}{}\ Cut \quad (1)$$

---

[4] Tennant's (forth.) analysis is in terms of ND and he introduces a SC-version in his (2010) to show that Steinberger's (2009) objection goes wrong.

Double-Line Harmony in a Sequent Setting

and, by taking a ψ with the following left rule: $\dfrac{A[y/x], \Gamma \Longrightarrow \Delta}{\psi, \Gamma \Longrightarrow \Delta} L\psi$, y fresh, we can prove (W) as follows:

$$\dfrac{\dfrac{\dfrac{\overline{A[y/x] \Longrightarrow A[y/x]}\ Ref}{A[y/x] \Longrightarrow \exists x A}\ R\exists \qquad \exists x A, \Gamma \Longrightarrow \Delta}{A[y/x], \Gamma \Longrightarrow \Delta}\ Cut}{\psi, \Gamma \Longrightarrow \Delta}\ L\psi \qquad (2)$$

The notion of harmony eliminates some S-disharmonious operators, as it is witnessed by the fact that in trying to prove that (S) and (W) hold for *tonk* we cannot make full use of the rules for ▲—e.g. in a proof of (S) we can use only one of the rules $L▲_1$ and $L▲_2$:

$$\dfrac{\dfrac{\dfrac{\overline{A_i \Longrightarrow A_i}\ Ref}{A_i \Longrightarrow \psi}\ R\psi_i \qquad \psi, \Gamma \Longrightarrow \Delta}{A_i, \Gamma \Longrightarrow \Delta}\ Cut}{A_1 ▲ A_2, \Gamma \Longrightarrow \Delta}\ L▲_i$$

Moreover, harmony eliminates some W-disharmonious operators, as it is witnessed by the fact that there is no way of proving either (S) or (W) for *knot*— e.g. if ψ satisfies the conditions on $A ▼ B$ in rule $L▼$, we have the following failed proof-attempt of (W):

$$\dfrac{\dfrac{\dfrac{\overline{A \Longrightarrow A}\ ref}{A, B \Longrightarrow A}\ W \qquad \dfrac{\overline{B \Longrightarrow B}\ ref}{A, B \Longrightarrow B}\ W}{A, B \Longrightarrow A ▼ B}\ R▼ \qquad A ▼ B, \Gamma \Longrightarrow \Delta}{\dfrac{A, B, \Gamma \Longrightarrow \Delta}{?}}\ Cut$$

Thus, harmony helps in determining harmonious operators, which is much in line with Tennant's aim. Nevertheless, he argues that this constraint is necessary but not sufficient for H-DE because it doesn't determine uniquely the rules of the operators. To wit, harmony is satisfied not only by the rules for ordinary disjunction (∨) but also by the ones for quantum disjunction (Υ):

$$\dfrac{A_1 \Longrightarrow \Delta \qquad A_2 \Longrightarrow \Delta}{A_1 \curlyvee A_2 \Longrightarrow \Delta}\ L\curlyvee \qquad \dfrac{\Gamma \Longrightarrow \Delta, A_i}{\Gamma \Longrightarrow \Delta, A_1 \curlyvee A_2}\ R\curlyvee_i \qquad (3)$$

In fact a proof that ∨ satisfies (S) and (W) is, at the same time, a proof that Υ satisfies them. Given that it is well known that it is problematic to

Norbert Gratzl and Eugenio Orlandelli

have a calculus with both standard and quantum disjunction, Tennant wants to rule out one of the two operators, and, therefore, he ensures the unique determination of the operators of LK by requiring *maximality*:

> $R\$$ has to be the strongest right rule that is in harmony with $L\$$; and $L\$$ the strongest left rule that is in harmony with $R\$$.

This condition forces the adoption of standard disjunction and the dismissal of quantum disjunction since (i) these two operators have the same right rules and (ii) the rule $L\lor$ is stronger than $L\curlyvee$ in that any sequent derivable with the latter is derivable with the former, but not vice versa.

The joint requirement of harmony and maximality eliminates all forms of W-disharmony. But, as shown in (Steinberger, 2009), they introduce some form of S-disharmony for the quantifiers because, instead of the existential quantifier $\exists$, they pick up the rogue quantifier $\mathfrak{Z}$, whose rules are

$$\frac{A[t/x], \Gamma \Longrightarrow \Delta}{\mathfrak{Z}xA, \Gamma \Longrightarrow \Delta} \; L\mathfrak{Z} \qquad \frac{\Gamma \Longrightarrow \Delta, A[t/x]}{\Gamma \Longrightarrow \Delta, \mathfrak{Z}xA} \; R\mathfrak{Z} \qquad (4)$$

with no variable condition on the left rule. Given that the variable restriction on $L\exists$ played no major role in the proofs given in (1) and (2), these proofs show also that $\mathfrak{Z}$ satisfies (S) and (W). Moreover, maximality forces us to adopt $\mathfrak{Z}$ in place of $\exists$ given that (i) they have the same right rule and (ii) $\mathfrak{Z}$ has a stronger left rule which allows to derive the schematic sequent $A[s/x] \Longrightarrow A[t/x]$ for any pair of terms s, t. This last fact means that $\mathfrak{Z}$ is S-disharmonious in that it has a left rule that is unduly too permissive with respect to its right rule.

Tennant (2010) argues that the S-disharmonious operator $\mathfrak{Z}$ is ruled out by the *admissibility* constraint: the existence of a syntactic proof of cut-elimination.[5] This constraint rules out $\mathfrak{Z}$ since the rules of this operator do not satisfy even the weaker requisite of *cut-inductiveness*—i.e. eliminability of a cut with cut formula principal in both premises—as it is witnessed by the fact that the following cut

$$\frac{\dfrac{\Gamma \Longrightarrow \Delta, A[s/x]}{\Gamma \Longrightarrow \Delta, \mathfrak{Z}xA} \; R\mathfrak{Z} \qquad \dfrac{A[t/x], \Pi \Longrightarrow \Sigma}{\mathfrak{Z}xA, \Pi \Longrightarrow \Sigma} \; L\mathfrak{Z}}{\Gamma, \Pi \Longrightarrow \Delta, \Sigma} \; Cut \qquad (5)$$

---

[5] We have transformed Tennant's request for a proof of cut-admissibility into the request for a proof of cut-elimination. The two requisites are equivalent in the present setting and we have chosen to follow Gentzen in taking cut as a primitive rule of inference of LK.

Double-Line Harmony in a Sequent Setting

is not eliminable whenever $s \neq t$ and t occurs free in Π, Σ.

Steinberger (2011a) argues that the problems with admissibility are that (i) it makes the notion of harmony redundant and that (ii) it is a global criterion. It is known that cut-inductiveness disposes of S-disharmonious operators—such as ▲—and maximality disposes of W-disharmonious ones—such as ▼. Thus, we can replace the two constraints of harmony and admissibility with the local one of cut-inductiveness. Steinberger claims that an analysis of H-DE in terms of cut-inductiveness and maximality behaves better than Tennant's one in that (i) it eliminates the same disharmonious operators, but (ii) it is not redundant and (iii) it is does not contain the global requisite of admissibility.

Even though we may agree that cut-inductiveness behaves better than harmony+admissiblity, we do not think that Steinberger's analysis is purely local because maximality is not a local criterion. If we consider again the case of $\vee$ and Υ, we see immediately that maximality rules out Υ precisely because of the existence of the stronger $\vee$—i.e. it goes global in that it sanctions an operator as disharmonious because of the existence of another one. Dicher (2016) argues that no local (extensional) criterion is able to rule out Υ. If this is true, we have to look for an analysis which rules out both S- and W-disharmonious operators without thereby ruling out also Υ.

To sum up, we are looking for a purely local analysis of harmony as H-DE in SC. Tennant (2010) analyses H-DE in terms of harmony, maximality and admissibility, and Steinberger (2011a) in terms of maximality and cut-inductiveness. H-DE has the following advantages w.r.t. other well known explications of harmony: (i) in banning both rules that are too strong and rules that are too weak it allows to determine one rule from the other in a unique way—this doesn't hold for intrinsic harmony; and (ii) it is in principle a local analysis since it refers only to the rules of the operator under consideration—this doesn't hold for global harmony.

We believe that H-DE is on the right track. However, we also believe that a purely local analysis of harmony as deductive equilibrium has to differ from Tennant's and from Steinberger's proposals.

3 Double-line harmony

We want to capture H-DE in a way that (i) ensures the unique determination of one rule from the other without thereby (ii) having to adopt the global requirement of maximality. In ND this is feasible thanks to the *generalized inversion principle* (GIP): 'whatever follows from the direct grounds for deriving a proposition must follow from that proposition' (Negri & von Plato, 2001, p. 6). Already in (Prawitz, 1965) the inversion principle has been the essential element for defining harmony in ND. Even though Prawitz's formulation is not strong enough to ensure the unique determination of the elimination rules, GIP allows us to determine the (generalized) elimination rules as unique functions of the introduction rules (Negri & von Plato, 2015, p. 243); see also (Moriconi & Tesconi, 2008) on inversion principles. This fact lies at the core of general elimination harmony. We are now going to show how to give an explication of H-DE in SC which exploits something like GIP.

Došen (1989) introduces double-line rules to show the logicality of operators, where a double-line rule $\frac{S_1}{S_2}$ ↑↓$\mathcal{R}$ is a rule that can be applied both in downward (↓ $\mathcal{R}$) and in upward (↑ $\mathcal{R}$) direction. As emphasised in (Došen, 2015, p. 151), the possibility of formulating SC via double-line rules is intimately related with the inversion principle. Thus, it should be possible to express H-DE by means of double-line rules. Nevertheless, as shown in (Bonnay & Simmenauer, 2005), formulating SC in terms of double-line rules does not in itself capture any notion of harmony. Roughly, the problem is that, unless we uses $G3$-style calculi, each operator has only either the left or the right rule that is invertible. The source of this problem points directly to its solution: we have to exploit the possibility of formulating either $L\$$ or $R\$$ as a double-line rule as a kind of GIP that allows to determine the unique set of rules $R\$$ or $L\$$ that are in deductive equilibrium with it. This idea can be captured formally as follows:

Definition 2 (dl-harmony) *The rules for an operator $\$$ are in dl-harmony whenever either $L\$$ or $R\$$ is the unique set of rules that has the same deductive strength of the ↑ direction of the (respectively right or left) double-line rule for $\$$. More precisely*

(dl-S) we have to derive an arbitrary instance of (some) $L\$$ (resp. $R\$$) from any instance of $\$$ ↑; and

(dl-W) we have to derive an arbitrary instance of $\$$ ↑ from an instance of (some) $L\$$ (resp. $R\$$).

Notice that dl-harmony contains both a no-gain condition—given by dl-S—and a no-loss condition—given by dl-W—and that these two conditions are precisely the constituents of H-DE. Thus, we propose that the rules for $\$$ are in H-DE whenever they satisfy dl-harmony.

Double-Line Harmony in a Sequent Setting

In order to substantiate our claim, we will show (i) that the operators of the $\{\vee, \exists\}$-fragment of LK satisfy dl-harmony, and (ii) that none of the disharmonious operators ▲, ▼, and 3 satisfies it. In order to prove (i), let us consider the following double-line rules for $\vee$ and for $\exists$:

$$\frac{A_1, \Gamma \Longrightarrow \Delta \quad A_2, \Gamma \Longrightarrow \Delta}{A_1 \vee A_2, \Gamma \Longrightarrow \Delta} \uparrow\downarrow\vee \qquad \frac{A[y/x]\Gamma \Longrightarrow \Delta}{\exists x A, \Gamma \Longrightarrow \Delta} \uparrow\downarrow\exists, \ y \text{ fresh} \quad (6)$$

For $\vee$ we have to prove that the bottom-up rule $\uparrow \vee$ is interderivable with the rules $R\vee_i$.[6] We have the following proof of the no-gain condition dl-S:

$$\frac{\Gamma \Longrightarrow \Delta, A_1 \quad \dfrac{\dfrac{}{A_1 \vee A_2 \Longrightarrow A_1 \vee A_2} \ ref}{A_1 \Longrightarrow A_1 \vee A_2} \uparrow\vee}{\Gamma \Longrightarrow \Delta, A_1 \vee A_2} \ Cut \quad (7)$$

and we have the following proof of the no-loss condition dl-W:

$$\frac{\dfrac{\dfrac{}{A_1 \Longrightarrow A_1} \ ref}{A_1 \Longrightarrow A_1 \vee A_2} R\vee_1 \quad A_1 \vee A_2, \Gamma \Longrightarrow \Delta}{A_1, \Gamma \Longrightarrow \Delta} \ Cut \quad (8)$$

When we come to the proof of the no-loss condition dl-S for the existential quantifier, we have to make essential use of the facts that (i) the double-line rule has a variable condition and that (ii) the rule of substitution of terms for variables[7] is (height-preserving) admissible in LK. The proof goes as follows:

$$\frac{\Gamma \Longrightarrow \Delta, A[t/x] \quad \dfrac{\dfrac{\dfrac{}{\exists x A \Longrightarrow \exists x A} \ Ref}{A[y/x] \Longrightarrow \exists x A} \uparrow\exists}{A[t/x] \Longrightarrow \exists x A} [t/y]}{\Gamma \Longrightarrow \Delta, \exists x A} \ Cut \quad (9)$$

Finally, we have the following proof of the no-loss condition dl-W for $\exists$:

$$\frac{\dfrac{\dfrac{}{A[y/x] \Longrightarrow A[y/x]} \ Ref}{A[y/x] \Longrightarrow \exists x A} R\exists \quad \exists x A, \Gamma \Longrightarrow \Delta}{A[y/x], \Gamma \Longrightarrow \Delta} \ Cut \quad (10)$$

In order to prove (ii), we consider the disharmonious operators ▲, ▼, and 3. For the S-disharmonious ▲, it is immediate to notice that it does

---

[6] Without loss of generality, we consider only the 'left half' of $\uparrow \vee$ and $R\vee_1$.

[7] With the usual proviso to avoid capture of free variables.

not satisfy dl-harmony because there is no single double-line rule that determines it. Suppose that we take the double-line version of either $R\blacktriangle_1$ or of $R\blacktriangle_2$, that is we take either

$$\frac{\Gamma \Longrightarrow \Delta, A_1}{\Gamma \Longrightarrow \Delta, A_1\blacktriangle_1 A_2} \uparrow\downarrow\blacktriangle_1 \quad \text{or} \quad \frac{\Gamma \Longrightarrow \Delta, A_2}{\Gamma \Longrightarrow \Delta, A_1\blacktriangle_2 A_2} \uparrow\downarrow\blacktriangle_2 \qquad (11)$$

then with either rule, even if we can prove the no-loss condition, we cannot prove the no-gain condition. If we start with $\uparrow\downarrow \blacktriangle_1$ we can prove the no-gain condition only for $L\blacktriangle_1$ and, vice versa, if we start with $\uparrow\downarrow \blacktriangle_2$ we can prove the no-gain condition only for $L\blacktriangle_2$. This shows that the only left rules that are in dl-harmony—i.e. interderivable—with the rules in (11) are, respectively

$$\frac{A_1, \Gamma \Longrightarrow \Delta}{A_1\blacktriangle_1 A_2, \Gamma \Longrightarrow \Delta} L\blacktriangle_1 \quad \text{and} \quad \frac{A_2, \Gamma \Longrightarrow \Delta}{A_1\blacktriangle_2 A_2, \Gamma \Longrightarrow \Delta} L\blacktriangle_2$$

and in neither case we determine the disharmonious operator $\blacktriangle$. These sets of rules determine, respectively, the harmonious first-projection operator $\blacktriangle_1$ and second-projection operator $\blacktriangle_2$. Roughly, the paradoxicality of *tonk* is explained as an equivocation between two distinct projection operators.

For the W-disharmonious operator $\blacktriangledown$, the following problem would arise with the double-line version of any of its rules. Suppose that we start with the double-line version of its right rule

$$\frac{\Gamma \Longrightarrow \Delta, A_1 \quad \Gamma \Longrightarrow \Delta, A_2}{\Gamma \Longrightarrow \Delta, A_1\blacktriangledown A_2} \uparrow\downarrow\blacktriangledown \qquad (12)$$

then we can prove that $L\blacktriangledown$ satisfies the no-gain condition but not that it satisfy the no-loss condition since we are at best able to prove

$$\cfrac{\Gamma \Longrightarrow \Delta, A_1\blacktriangledown A_2 \quad \cfrac{\cfrac{A_1 \Longrightarrow A_1}{A_1 \Longrightarrow A_1, A_2} W \quad \cfrac{A_2 \Longrightarrow A_2}{A_2 \Longrightarrow A_1, A_2} W}{A_1\blacktriangle A_2 \Longrightarrow A_1, A_2} L\blacktriangledown}{\Gamma \Longrightarrow \Delta, A_1, A_2} Cut$$

whose conclusion is not, and cannot be transformed into, the desired conclusion of $\blacktriangledown$. We conclude that also $\blacktriangledown$ is sanctioned as disharmonious by dl-harmony.[8]

---

[8] Without using the rule of weakening, the rule $\uparrow\downarrow \blacktriangledown$ given in (12) would determine the left rules of conjunction of LK, thus transforming $\blacktriangledown$ into (a notational variant of) conjunction.

Double-Line Harmony
in a Sequent Setting

Finally, we consider the S-disharmonious rogue quantifier 3. The double-line rule $\uparrow\downarrow$ 3 is like the one given in (6) for the existential quantifier except that it does not have the variable restriction.[9] On the one hand, this operator satisfies the no-loss condition of dl-harmony. In fact the proof of dl-W given in (10) for $\exists$ made no use of the variable condition and, therefore, it works also for 3. On the other hand, 3 does not satisfy the no-gain condition and, therefore, is not in dl-harmony. This can be shown as follows. First, we reduce the search-space by noticing that if dl-S holds for 3, it has to be provable as we did in (9) for $\exists$. That is, we have to transform

$$\cfrac{\Gamma \Longrightarrow \Delta, A[t/x] \qquad \cfrac{\cfrac{\cfrac{}{3xA \Longrightarrow 3xA}\;Ref}{A[y/x] \Longrightarrow 3xA}\;\uparrow 3}{(A[y/x])[t/y] \Longrightarrow (3xA)[t/y]}\;[t/y]}{\Gamma \Longrightarrow \Delta, (3xA)[t/y]}\;Cut$$

into a proof showing that 3 satisfies dl S. But this would be possible only if either y is t or if y does not occur free in $3xA$.[10] The latter assumption is feasible (without loss of generality) just in case 3 satisfies the same variable condition as $\exists$. The former assumption—i.e. the idea of exemplifying $\uparrow 3$ directly with the term t and not with an arbitrary term y—does not work because in proving the no-gain condition we cannot rely on a specific instance of $\uparrow 3$. It might be objected that the instance of $\uparrow 3$ which concludes $A[t/x] \Longrightarrow 3xA$ is a legal one. Nevertheless, we do not accept it because the double-line rule for 3 talks of all terms and not of a particular one and, therefore, in proving dl-harmony we have to apply it in all its generality. In a sense the quantifiers without variable condition that can be shown to be in dl-harmony are, for any given t, the innocuous t-specific quantifier 3^t which is like 3 save that its rules are applicable only for the given term t. These rules give us an harmonious and meaningful operator, whose meaning implies that $A[t/x]$ is equivalent to $3^t xA$.

We have thus shown that dl-harmony gives the expected results—i.e. it is satisfied by $\vee$ and $\exists$ and it is not satisfied by ▲, ▼, and 3. One nice aspect of dl-harmony is that it has a no-gain condition—i.e. dl-S—that rules out S-disharmonious operators such as ▲ and 3 and it has a no-loss condition—i.e. dl-W—that rules out W-disharmonious operators such as ▼. Moreover

---

[9] Nothing essential relies on taking the double-line version of the left rule; by taking the right one we obtain the rogue universal quantifier.

[10] Notice that these two are the very same moves that would transform (5) into an effective proof of cut-inductiveness for 3. Notice also that the admissibility of the rule of substitution depends essentially on the variable restriction of the rules for the quantifiers, see (Negri & von Plato, 2001, p. 69).

dl-harmony is a purely local requisite in that it relies only on the rules governing the operator under examination and on the structural rules of Cut and of Reflexivity. Notice that these two structural rules are necessary to prove its no-gain condition. For the no-loss condition it would be enough to have these rules as admissible. For example we could have proved dl-W also in the calculus LK without the rule of Cut since this rule is admissible in it. On the other hand, these two structural rules are not eliminable from the proofs of dl-S. These considerations show that dl-harmony could not hold for the operators of non-reflexive or non-transitive substructural logics. We don't take this as a severe limitation.

4 Conclusions

It has been shown that it is possible to give a purely local analysis of harmony as H-DE in SC by imposing the requisite of dl-harmony. As opposed to both Tennant's and Steinberger's proposals, the requisite of dl-harmony guarantees that the quantum disjunction operator Υ is harmonious. This happens because the proofs given in (7) and (8) still work if we replace the rule $\uparrow\downarrow \vee$ by the following one

$$\frac{A_1 \Longrightarrow \Delta \qquad A_2 \Longrightarrow \Delta}{A_1 \Upsilon A_2 \Longrightarrow \Delta} \uparrow\downarrow\Upsilon$$

where the left-context is empty. This will be taken as a limitation of our approach by anyone who is inclined to consider Υ as (intuitively) disharmonious. We are not disturbed by this since we take Υ as a perfectly harmonious operator. Its only problem might be that it interacts badly with $\vee$, but for anyone who takes harmony to be a purely local matter, as we did, this should not be relevant for assessing whether it is harmonious or not.[11]

All in all, the analysis of harmony as H-DE in terms of dl-harmony differs from both Tennant's and Steinberger's in that it is a purely local analysis which is based on the inversion principle. Neither the requisite of maximality nor that of admissibility are acceptable in a purely local analysis of harmony. Notice that even if Tennant and Steinberger were to agree that Υ is harmonious, they cannot drop the global requisites from their analyses since (i) Tennant needs admissibility to rule out $\mathfrak{Z}$ and (ii) Steinberger needs maximality to rule out ▼. This is not intended by us as a criticism of their

---

[11] See Dicher (2016) for a critical presentation of many proposals that disagree with us on this point.

proposals inasmuch as they have not asked for a purely local analysis and, therefore, can use global requisites.

Like other approaches based on GIP, the present proposal satisfies a notion of unique determination which differs from the one given by maximality. With maximality unique determination holds in the sense that there cannot be two harmonious operators sharing some rule, as it happens for $\vee$ and Υ. With dl-harmony unique determination holds in the sense that starting from the double-line rule governing an operator we determine one unique set of rules that are in harmony with it. A more comprehensive analysis of the relationship between dl-harmony and other concepts of harmony which are based on GIP—such as the ND-based ones in terms of general elimination harmony (Negri & von Plato, 2015; Read, 2000) and the SC-based ones in terms of so-called reflection principles (Sambin, Battilotti, & Faggian, 2000; Schroeder-Heister, 2007)—goes beyond the scope of this paper and we leave it to future research.

The present approach to harmony can be extended beyond the $\{\vee, \exists\}$-fragment of LK. At the present stage it is already clear that it can be extended to logics containing any additive or multiplicative operator of linear logic and to their extensions with the structural rules of weakening and contraction. It is also clear that it cannot be extended to Tennant's core logic ($\mathbb{C}$); because, from the present perspective, the operators of $\mathbb{C}$ are not harmonious in that they are not determinable according to GIP. For example, the rules for conjunction in $\mathbb{C}$ are the right rule for multiplicative conjunction and (a version of) the left rules for additive conjunction, neither of which is invertible. One natural future line of research is to give dl-harmonious rules for the operators of many modal logics by means of generalizations of Gentzen's SC such as display logics (Wansing, 1998) and labelled calculi (Negri & von Plato, 2015).

References

Belnap, N. D. (1962). Tonk, plonk and plink. *Analysis*, *22*, 130–134.
Bonnay, D., & Simmenauer, B. (2005). Tonk strikes back. *Australasian Journal of Logic*, *3*, 33–44.
Dicher, B. (2016). Weak disharmony: Some lessons for proof-theoretic semantics. *The Review of Symbolic Logic*, *9*, 583–602.
Došen, K. (1989). Logical constants as punctuation marks. *Notre Dame Journal of Formal Logic*, *30*, 362–381.

Došen, K. (2015). Inferential semantics. In H. Wansing (Ed.), *Dag Prawitz on Proofs and Meaning* (pp. 147–162). Springer.

Dummett, M. (1973). *Frege: Philosophy of Language*. Harvard University Press.

Dummett, M. (1991). *The Logical Basis of Metaphysics*. Harvard University Press.

Hacking, I. (1979). What is logic? *The Journal of Philosophy, 76*, 285–319.

Moriconi, E., & Tesconi, L. (2008). On inversion principles. *History and Philosophy of Logic, 29*, 103–113.

Negri, S., & von Plato, J. (2001). *Structural Proof Theory*. Cambridge University Press.

Negri, S., & von Plato, J. (2015). Meaning in use. In H. Wansing (Ed.), *Dag Prawitz on Proofs and Meaning* (pp. 239–257). Springer.

Prawitz, D. (1965). *Natural Deduction: A Proof-Theoretical Study*. Almqvist & Wiskell.

Prior, A. (1960). The runabout inference ticket. *Analysis, 21*, 38–39.

Read, S. (2000). Harmony and autonomy in classical logic. *Journal of Philosophical Logic, 29*, 123–154.

Sambin, G., Battilotti, G., & Faggian, C. (2000). Basic logic. *Journal of Symbolic Logic, 65*, 979–1013.

Schroeder-Heister, P. (2007). Generalized definitional reflection and the inversion principle. *Logica Universalis, 1*, 355–376.

Schroeder-Heister, P. (2012). Proof-theoretic semantics, self-contradiction, and the format of deductive reasoning. *Topoi, 31*, 77–85.

Steinberger, F. (2009). Not so stable. *Analysis, 69*, 655–661.

Steinberger, F. (2011a). Harmony in a sequent setting: A reply to Tennant. *Analysis, 71*, 273–280.

Steinberger, F. (2011b). What harmony could and could not be. *Australasian Journal of Philosophy, 89*, 617–639.

Tennant, N. (2010). Harmony in a sequent setting. *Analysis, 70*, 462–468.

Tennant, N. (forth.). Inferentialism, logicism, harmony, and a counterpoint. In A. Miller (Ed.), *Essays for Crispin Wright: Logic, Language and Mathematics*. Oxford University Press.

Wansing, H. (1998). *Displaying Modal Logic*. Kluwer.

Double-Line Harmony
in a Sequent Setting

Norbert Gratzl
MCMP, Ludwig-Maximilians-Universität München
Germany
E-mail: `Norbert.Gratzl@lrz.uni-muenchen.de`

Eugenio Orlandelli
University of Bologna
Italy
E-mail: `eugenio.orlandelli@unibo.it`

A Normal Paradox

MATTIA PETROLO[1] AND PAOLO PISTONE

Abstract: We challenge the idea that the lack of a normalization procedure can be taken as a distinctive feature of paradoxes. In particular, we present a counterexample to Tennant's proof-theoretic characterization of paradoxes in intuitionistic natural deduction. Our counterexample follows a simple technique for eliminating cuts introduced by Kreisel and Takeuti.

Keywords: Paradoxes, Russell's paradox, Intuitionistic natural deduction, Normalization, Computational consistency.

1 Proof-theoretic analysis of paradoxes

Prawitz (1965) considered a system for naive set theory obtained by extending minimal logic with an introduction and elimination rule for formulas of the form $t \in \{x : A\}$, expressing set-theoretical comprehension:

$$\frac{A[t/x]}{t \in \{x : A\}} \in I \qquad \frac{t \in \{x : A\}}{A[t/x]} \in E$$

Following Prawitz's analysis, an application of $\in I$ immediately followed by $\in E$ constitutes a redundancy which can be eliminated by one $\in$-reduction step:

$$\frac{\begin{array}{c} \mathcal{D} \\ A[t/x] \end{array}}{\dfrac{t \in \{x : A\}}{A[t/x]}} \quad \leadsto_\in \quad \begin{array}{c} \mathcal{D} \\ A[t/x] \end{array}$$

By exploiting these rules and their associated reduction, Prawitz showed that Russell's paradox in naive set theory yields a derivation with a non-terminating reduction[2]. The investigation of paradoxical phenomena in natural deduction was further developed and refined in (Tennant, 1982). Tennant considered a wide range of examples and showed that all prominent

---

[1]This work has been partially funded by the French-German ANR-DFG project *Beyond-Logic* (ANR-14-FRAL-0002).

[2]More precisely, a derivation in which all possible reduction sequences loop.

mathematical and logical paradoxes follow a looping reduction pattern. The steps playing the role of paradoxical rules are called *id est* inferences, as they result from extra-logical principles. The outcome of his analysis is that the fact that derivations of absurdity loop, and thus do not normalize, is a distinctive feature of paradoxes[3]. Tennant (1995) (p.202*f.*) summarizes this idea as follows:

> [...] enumerate proofs of absurdity; start normalizing those that are not in normal form; and check to see whether the reduction sequences ever enter loops, or manifest any other conclusive evidence that they will not terminate. As soon as a reduction sequence does enter a loop, or manifest such evidence, one can check off the proof concerned as a 'paradoxical' proof.

Tennant develops his analysis further by making a sharp distinction between paradoxes and inconsistencies. He distinguishes between the derivation of $\bot$ generated via the paradoxical rules for $\in$, and open derivations of $\bot$, such as the one below

$$\cfrac{\cfrac{A \land \neg A}{A} \quad \cfrac{A \land \neg A}{\neg A}}{\bot}$$

which is already in normal form. The difference between a paradox and an inconsistency can be spelled out as follows: Whereas a derivation of $\bot$ with a looping reduction sequence shows that the sentences involved in its *id est* inferences are paradoxical, a normalizable open derivation of $\bot$ shows the inconsistency of (at least one of) its undischarged assumptions[4].

Recently, several authors challenged Tennant's analysis. The objections raised can be grouped in two classes: 1) those claiming that looping is not a *necessary* condition (von Plato, 2000; Rogerson, 2007); 2) those claiming that looping is not a *sufficient* condition (Schroeder-Heister and Tranchini (2016)). Rogerson found a way to normalize a formal counterpart of Curry's paradox in classical logic, but her counterexample only works in the $\lor, \exists$-free fragment of classical natural deduction (see Schroeder-Heister and Tranchini (2016)). Von Plato shows that by using general elimination

---

[3]Tennant (1995) broadens the test to non-terminating reduction sequences in order to account for paradoxes without self-reference, such as Yablo's.

[4]Tennant calls *pure* paradox a paradox whose non-normalizing derivations of $\bot$ has no undischarged assumptions.

rules it is possible to obtain a derivation for the so-called Ekman's paradox in normal form, but nothing is said about other more standard paradoxes[5].

Following Schroeder-Heister and Tranchini (2016), looping is not a sufficient condition for being a paradox. Additionally, one must require that: (1) reduction does not trivialize identity of proof, i.e. that computational consistency holds; (2) *id est* inferences provide an *isomorphism* between the conclusions of paradoxical rules. We will come back to these conditions in §4.

In what follows, we present a different counterexample to Tennant's proof-theoretic characterization of paradoxes in intuitionistic logic.

2 Normalization for free

Our counterexample exploits a technique introduced by Kreisel and Takeuti (1974) (KT for short) in the investigation of cut-free systems for second order logic. The KT technique allows to transform certain proofs with cuts into cut-free ones in a simple and direct way, without passing through a Gentzen style inductive argument. The easiest way to illustrate this trick is by considering a sequent calculus derivation in second order logic like the one below

$$\frac{\Gamma, \forall X(X \to X) \vdash A, \Delta \quad \Gamma', A \vdash \Delta'}{\Gamma, \Gamma', \forall X(X \to X) \vdash \Delta, \Delta'} \; cut$$

the cut on A can be made "disappear" by replacing it with a sequence of rules (a left $\to$-rule, a left $\forall$-rule and a contraction rule), by exploiting the second order hypothesis $\forall X(X \to X)$:

$$\frac{\dfrac{\Gamma, \forall X(X \to X) \vdash A, \Delta \quad \Gamma', A \vdash \Delta'}{\dfrac{\Gamma, \Gamma', \forall X(X \to X), A \to A \vdash \Delta, \Delta'}{\dfrac{\Gamma, \Gamma', \forall X(X \to X), \forall X(X \to X) \vdash \Delta, \Delta'}{\Gamma, \Gamma', \forall X(X \to X) \vdash \Delta, \Delta'} C} \forall L}}{} \to L$$

A consequence of this technique is that a "trivial" cut-elimination result for derivations of sequents of the form $\Gamma, \forall X(X \to X) \vdash \Delta$ holds: every derivation can be reduced into a cut-free one in a primitively recursive way (as the reduction is direct and does not require complex inductive

---
[5]Remark that the KT technique introduced in §2 can be straightforwardly adapted to handle the case of general elimination rules.

measures). As one might expect, from this "trivial" cut-elimination theorem (see Kreisel and Takeuti (1974) or Girard (1987), p.192) one cannot deduce all consequences of usual cut-elimination theorems. In particular, consistency cannot be inferred: one argues from cut-elimination to consistency by supposing that a derivation of $\vdash \bot$ exists and concluding that then a cut-free derivation of $\vdash \bot$ exists, which is absurd. However, "trivial" cut-elimination cannot be applied to derivations of sequents of the form $\vdash \bot$. This fact suggests that "trivial" cut-elimination can be performed also in systems containing "paradoxical" rules.

In natural deduction, the KT technique corresponds to normalizing derivations by introducing some *ad hoc* undischarged assumptions (or, equivalently, by enriching the system with some axioms): a redundancy like the one below

$$\cfrac{\cfrac{[A]^n \\ \vdots \\ B}{A \to B} \to I,n \quad \cfrac{\vdots \\ A}{} }{B} \to E$$
$$\vdots$$

is reduced to

$$\vdots$$
$$\cfrac{[A \to A]^m \quad \cfrac{\vdots}{A}}{A} \to E$$
$$\vdots$$
$$B$$
$$\vdots$$

by introducing a fresh undischarged assumption $[A \to A]^m$, which can also be seen as an axiom (expressing an obviously valid formula).

More precisely, in the case of Gentzen's system of intuitionistic natural deduction ($\mathcal{NJ}$) the following result holds: let $\mathcal{NJ}_{A \to A}$ be the system obtained by adding to $\mathcal{NJ}$ all axioms of the form $A \to A$, with A a formula. Let $\longrightarrow$ denote the usual natural deduction reductions (as defined for instance in (Prawitz, 1965)) and let, moreover, $\rightsquigarrow$ be the transitive and context closure of the reduction below:

A Normal Paradox

$$
\begin{array}{c}
[A]^n \\
\vdots \\
\dfrac{\dfrac{B}{A \to B} \to I, n \quad \vdots}{B} \\
\vdots
\end{array}
\quad \rightsquigarrow \quad
\begin{array}{c}
\dfrac{\dfrac{A \to A \quad A}{A} \quad A}{\vdots} \to E \\
\vdots \\
B \\
\vdots
\end{array}
$$

Given two reduction relations r_1 and r_2 we indicate by $r_1 \cup r_2$ the reduction relation given by $t \; r_1 \cup r_2 \; u$ if either $t \; r_1 \; u$ or $t \; r_2 \; u$.

Proposition 1 *The reduction $\longrightarrow \cup \rightsquigarrow$ on the system $\mathcal{NJ}_{A \to A}$ is strongly normalizing.*

Proof. The reduction $\rightsquigarrow$ is obviously terminating as it eliminates cuts in one step (see also (Girard, 1987), theorem 3.B.11 p.192). To any reduction path π of $\longrightarrow \cup \rightsquigarrow$ one can associate a reduction path π_0 of $\longrightarrow$ obtained by reducing by $\longrightarrow$ all redexes which are reduced by $\rightsquigarrow$ in π. Clearly, if π is infinite, π_0 is infinite too. The result follows then from the fact that the reduction $\longrightarrow$ is strongly normalizing. □

It is well-known that the so-called Curry-Howard correspondence (see (Sørensen & Urzyczyn, 2006)) establishes an isomorphism between the system $\mathcal{NJ}$ and the simply-typed λ-calculus $\lambda_\to$, with the reduction $\longrightarrow$ corresponding to β-reduction.

The KT technique can be then reproduced in $\lambda_\to$: it corresponds to adding new *constants* $\mathrm{k}_A : A \to A$ with the following reduction rule

$$(\lambda x^A.t^B)u^A \rightsquigarrow t^B[(\mathrm{k}_A)u^A/x].$$

Hence, a result similar to Proposition 1 holds for the simply typed λ-calculus $\lambda_{\to,\mathrm{k}_A}$ enriched with the new constants k_A.

The modification of $\mathcal{NJ}$ obtained by adding formulas of the form $A \to A$ is admittedly *ad hoc*, but harmless. First, from a logical point of view, one adds some valid formulas as axioms, hence consistency is preserved. Secondly, from a computational point of view, the reduction does not violate normalization. Moreover, as we show below in $\lambda_{\to,\mathrm{k}_A}$, this reduction preserves an important property of β-reduction, called *computational consistency*. Such a property guarantees that the identity of proofs provided by β-reduction is not trivial: it is always possible to exhibit (at least) two derivations of the same conclusion belonging to two distinct equivalence classes. Roughly speaking, computational consistency prevents the identification of too many terms by the computation. Indeed, this would preclude the possibility of a proof-theoretic analysis of derivations.

Let $\to_\beta^1$ and $\leadsto^1$ indicate, respectively, one-step β-reduction and one-step $\leadsto$-reduction.

Lemma 1 *The reductions $\to_\beta$ and $\leadsto$ commute in the following sense: if $t \leadsto^1 u \to_\beta^1 v$, then there exists a term u' such that $t \to_\beta^1 u' \leadsto v$.*

Let us call *pure* a term without constants. Given a well-typed term t, let us call $|t|$ the term obtained by replacing every subterm of the form $k_A u$, for some formula A and term u, with the term u. The reader can easily verify that $|t|$ is well-typed. If t is pure and $t \leadsto u$, then $|u|$ is pure and moreover $t =_\beta |u|$. In the following we indicate by $=$ syntactic equality[6] between λ-terms, by $=_\beta$ the equivalence induced by β-reduction, and by $\equiv$ the equivalence induced by the reduction $\to_\beta \cup \leadsto$.

Definition 1 (strict subterm) *For t, u λ-terms, we say that t is a* strict subterm *of u (noted $t < u$) if one of the conditions below holds:*

- $u = x$ and $t = u$;

- $u = \lambda x.u'$, for some term u' and $t < u'$;

- $u = u_1 u_2$, for some terms u_1, u_2 and $t < u_1$ or $t < u_2$.

We say that t is a subterm *of u (noted $t \leq u$) when $t < u$ or $t = u$. Given terms t, u, v with $t < v$ and $u < v$, we call t and u* incomparable *(with respect to v) if neither $t \leq u$ nor $u \leq t$ holds.*

Proposition 2 *If t, u are pure terms and, for some v, $t \to_\beta \cup \leadsto v$ and $u \to_\beta \cup \leadsto v$, then $t =_\beta u$.*

Proof sketch. Let t, u, v be terms such that $t \to_\beta \cup \leadsto v$ and $u \to_\beta \cup \leadsto v$, with both reductions of length at least 1. By lemma (1) we can suppose w.l.o.g. that $t \leadsto v$ and $u \leadsto v$. We show, by induction on the length $l + l'$ of the reductions leading, respectively, from t to v and from u to v, that $|t| =_\beta |u|$.

---
[6]Subject to usual α-equivalence.

A Normal Paradox

If $l + l' = 2$, then t contains a subterm $t' \leq t$ of the form $(\lambda x.u_1)v_1$ (with v_1 of a certain type A) and u contains a subterm $u' \leq u$ of the form $(\lambda y.u_2)v_2$ (with v_2 of a certain type B) and v is the term obtained either by replacing, in t, the subterm t' with $t'' = u_1[\mathrm{k}_A v_1/x]$, or by replacing, in u, the subterm u' with $u'' = u_2[\mathrm{k}_B v_2/y]$. One must distinguish three cases, depending on whether $u'' \leq t'$, $t'' \leq u'$ or none of the two holds. If $u'' \leq t'$, then either $u'' = t'$, either $u'' = \lambda x.u_1$, either $u'' \leq u_1$, or $u'' \leq v_1$. We only consider the first case and leave the others to the reader: if $u'' = t'$, then $(\lambda x.u_1)v_1 = u_2[\mathrm{k}_B v_2/y]$, hence one has $u' = (\lambda y.u_1^0[\mathrm{k}_A v_1^0/x])v_2$, for some terms u_1^0, v_1^0 such that $u_1^0[\mathrm{k}_B v_2/y] = u_1$ and $v_1^0[\mathrm{k}_B v_2/y] = v_1$, and we can suppose that v_2 contains no free occurrence of the variable y (indeed, if it does, we can replace $\lambda y.u_2$ with the equivalent term $\lambda y'.u_2[y'/y]$ for some fresh y'). Then $|t'| =_\beta |u_1[v_1/x]| = |u_1^0[v_2/y][v_1^0[v_2/y]/x]| = |u_1^0[v_1^0/x][v_2/y]| =_\beta |u'|$, from which the thesis follows.

If $t'' < u'$ one can argue symmetrically to the case above. Finally, if the two terms are incomparable, then $u'' < t$ and $t'' < u$, with u'' and t'' incomparable, from which one can conclude $|t| =_\beta |u|$.

If $l + l' = n + 3$, then one of the reductions $t \rightsquigarrow v$ $u \rightsquigarrow v$, say the first one, factors as $t \rightsquigarrow^1 t' \rightsquigarrow v$. By induction hypothesis we have that $|t'| =_\beta |u|$. Moreover $|t| =_\beta |t'|$, from which we deduce $|t| =_\beta |u|$.

If t, u are pure, then $|t| =_\beta |u|$ implies $t =_\beta u$. □

Corollary 1 *The equivalence $\equiv$ is computationally consistent.*

Proof. From proposition (2) it follows that the equivalence $\equiv$ induced by the reduction $\rightarrow_\beta \cup \rightsquigarrow$ is conservative over β-equivalence in the following sense: if t, u are pure and $t \equiv u$, then $t =_\beta u$. Hence if t and u are pure terms such that $t \neq_\beta u$ (which exist as $=_\beta$ is computationally consistent), then $t \not\equiv u$. □

3 Normalizing Russell's paradox

Let us now apply the KT technique to the case of paradoxes. We take the derivation of Russell's paradox as a case study, but the technique can be easily applied to the other paradoxes analyzed in (Tennant, 1982). The leading idea here is that by adding an *ad hoc* assumption[7], redundancies in Russell's paradox can be normalized. This can be done by adding to $\mathcal{NJ}$ a new formula ρ and an axiom $\overline{\neg \rho \rightarrow \neg \rho}$.

---
[7] Such assumption will be considered as an axiom.

By taking ρ to be $r \in r$, where r is the Russell term $\{x : x \notin x\}$, one obtains the following paradoxical rules for ρ (see (Prawitz, 1965) and (Tennant, 1982)).

$$\frac{\neg \rho}{\rho} \rho I \qquad \frac{\rho}{\neg \rho} \rho E$$

By exploiting the KT technique, it is now possible to define an *ad hoc* reduction designed to prevent the appearance of redundancies on the formula $\neg \rho$:

$$\frac{\vdots}{\frac{\neg \rho}{\rho} \rho I} \rho E \quad \rightsquigarrow_\rho \quad \frac{\neg \rho \rightarrow \neg \rho \qquad \neg \rho}{\neg \rho} \rightarrow E$$
$$\vdots \qquad\qquad\qquad \vdots$$

We call $\mathcal{NJ}_\rho$ the system obtained by adding to $\mathcal{NJ}$ the formula ρ, its axiom $\overline{\neg \rho \rightarrow \neg \rho}$, as well as the rules ρI and ρE and the reduction $\rightsquigarrow_\rho$. The reduction rule translating Prawitz's example would be, on the contrary, the reduction generated by

$$\frac{\vdots}{\frac{\neg \rho}{\rho} \rho I} \rho E \quad \rightsquigarrow_p \quad \neg \rho$$
$$\vdots$$

From the Curry-Howard perspective, the system $\mathcal{NJ}_\rho$ corresponds to a system $\lambda_{\rightarrow, k_\rho}$ obtained by adding to the simply typed λ-calculus a ground type $\bot$, a new type ρ, a new constant k_ρ and two new constructors in $: (\rho \rightarrow \bot) \rightarrow \rho$ and out $: \rho \rightarrow \rho \rightarrow \bot$ along with a reduction rule out(inu) $\rightsquigarrow_\rho k_\rho u$. The reduction rule which translates Prawitz's example would be, on the contrary, out(inu) $\rightsquigarrow_p u^8$.

---

[8] Similar issues are well-investigated in the literature on recursive types. There one usually considers types ρ satisfying *recursive equations* of the form $\rho \simeq F(\rho)$, where $\simeq$ denotes isomorphism and F is a *positive* operator over types, along with constructors in $: F(\rho) \rightarrow \rho$ and out $: \rho \rightarrow F(\rho)$, along with a reduction out(inu) $\rightarrow_\rho u$. When the operator F is not positive (as is the case of $F(X) = X \rightarrow \bot$), the reduction $\rightarrow_\rho$ is not normalizing (indeed $R \rightarrow_\rho R$). For an introduction to recursive types, see (Mendler, 1987) and (Cardone & Coppo, 1991).

A Normal Paradox

As a consequence of proposition (1) the system $\mathcal{NJ}_\rho$ as well as the system $\lambda_{\to,k_\rho}$ are normalizing. In $\mathcal{NJ}_\rho$, Russell's paradox has the form of the following derivation of absurdity (see (Prawitz, 1965) and (Tennant, 1982)).

$$\cfrac{\cfrac{\cfrac{[\rho]^n}{\neg\rho}\,\rho E \quad [\rho]^n}{\to E}}{\cfrac{\bot}{\neg\rho}\to I, n} \quad \cfrac{\cfrac{\cfrac{[\rho]^m}{\neg\rho}\,\rho E \quad [\rho]^m}{\to E}}{\cfrac{\cfrac{\bot}{\neg\rho}\to I, m}{\rho}\,\rho I}\to E}{\bot}$$

which is represented in $\lambda_{\to,k_\rho}$ by the term $R = (\lambda x.\mathrm{out}\,xx)\mathrm{in}\lambda x.\mathrm{out}\,xx$.

The redundancy obtained by the successive applications of the $\to I$ and $\to E$ rules produces a second redundancy on the occurrences of an application of a ρI rule immediately followed by a ρE rule:

$$\cfrac{\cfrac{\cfrac{[\rho]^m}{\neg\rho}\,\rho E \quad [\rho]^m}{\to E}}{\cfrac{\cfrac{\bot}{\neg\rho}\to I, m}{\cfrac{\rho}{\neg\rho}\,\rho I}\,\rho E} \quad \cfrac{\cfrac{\cfrac{[\rho]^m}{\neg\rho}\,\rho E \quad [\rho]^m}{\to E}}{\cfrac{\cfrac{\bot}{\neg\rho}\to I, m}{\rho}\,\rho I}\to E}{\bot}$$

Reducing this redundancy with Prawitz's reduction $\leadsto_p$ leads back to the original derivation, thus entering a looping sequence in the reduction procedure. Hence, with the reduction $\longrightarrow \cup \leadsto_p$, the derivation above is not normalizing (as well as the term R with the reduction $\to_\beta \cup \leadsto_p$). On the contrary, from proposition (1) it follows that, with the reduction $\longrightarrow \cup \leadsto_\rho$ a normal form can obtained. The normal form is illustrated below:

$$\cfrac{\neg\rho \to \neg\rho \quad \cfrac{\cfrac{\cfrac{[\rho]^n}{\neg\rho}\,\rho E \quad [\rho]^n}{\to E}}{\cfrac{\bot}{\neg\rho}\to I, n} \to E}{\neg\rho} \quad \cfrac{\cfrac{\cfrac{[\rho]^m}{\neg\rho}\,\rho E \quad [\rho]^m}{\to E}}{\cfrac{\cfrac{\bot}{\neg\rho}\to I, m}{\rho}\,\rho I}\to E}{\bot}$$

(1)

corresponding to the normal term $k_\rho(\lambda x.\mathrm{out}\,xx)\lambda x.\mathrm{out}\,xx$.

The derivation above suggests a way to construct a normal closed derivation, without axioms, whose conclusion is false. Indeed, if the axiom

$\overline{\neg\rho \to \neg\rho}$ is replaced by an assumption $[\neg\rho \to \neg\rho]^q$, by discharging it we obtain a closed derivation of the formula $(\neg\rho \to \neg\rho) \to \bot$:

$$\cfrac{[\neg\rho \to \neg\rho]^q \qquad \cfrac{\cfrac{\cfrac{[\rho]^n}{\neg\rho}\rho E \quad [\rho]^n}{\bot} \to E}{\neg\rho} \to I, n}{\cfrac{\bot}{(\neg\rho \to \neg\rho) \to \bot} \to I, q} \qquad \cfrac{\cfrac{\cfrac{[\rho]^m}{\neg\rho}\rho E \quad [\rho]^m}{\bot} \to E}{\cfrac{\neg\rho}{\rho}\rho I} \to I, m}{\to E}$$

We presented two normal derivations which are both closed and have a false conclusion. The first one uses an axiom, and has conclusion $\bot$, while the second one does not contain axioms and has conclusion $(\neg\rho \to \neg\rho) \to \bot$, i.e. a false formula of the form "something true $\to \bot$". Moreover, these derivations do not use any potentially invalid rule but the standard rules ρI and ρE for ρ.

Hence, if one accepts the (admittedly *ad hoc*) examples just sketched as cases of a genuine paradox, then the lack of a normalization procedure cannot be taken as a distinctive feature of paradoxes.

4 Conclusion

We can sum up the situation described above as follows: reductions making paradoxical rules normalizing can be devised as soon as a natural deduction system is enriched with axioms corresponding to valid formulas. It must be stressed that the technique presented here for Russell's paradox can be applied in full generality: it can be straightforwardly adapted to any paradox described by means of *id-est* inferences.

Obviously, by adding valid axioms, derivability is in no way altered. However, the proof-theoretic properties of the system are altered: it is well-known that, in presence of axioms, the introduction property (namely, the fact that a closed normal derivation ends with an introduction rule) must be weakened, as the derivation might end with an elimination rule with an axiom as major premiss. Nonetheless, in the case considered here, in which axioms are formulas of the form $A \to A$, one can respond that from a closed normal proof violating the introduction property one can extract in a straightforward way a subproof, with the same conclusion, ending with an introduction.

A Normal Paradox

The derivations presented, obtained by applying the KT technique, have some puzzling features, and one may be skeptical whether they can be considered as genuine paradoxes.

In line with what is argued in (Schroeder-Heister & Tranchini, 2016), one might also object that, with the equality of proofs induced by the reduction $\leadsto_\rho$, while computational consistency still holds, the isomorphism criterion is not satisfied: the derivation $\dfrac{\dfrac{\neg\rho}{\rho}\,\rho I}{\neg\rho}\,\rho E$ is not equal to the assumption $\neg\rho$.

If one endorses the criterion put forward in (Schroeder-Heister & Tranchini, 2016), then it is possible to conclude that our "paradoxes in normal form" are not genuine paradoxes. However, in that case, our examples could be used to construct an argument of the form: as soon as one accepts that looping is all that matters in order to characterize a paradox, a pathological example (i.e. an *ad hoc* normalizing derivation with false conclusion) can be produced. One concludes then that some other property (for instance, isomorphism) must be invoked in order to reject our example and characterize genuine paradoxes.

Incidentally, we observe that the isomorphism condition presupposes that the *id-est* inferences have a very specific form: the (unique) premiss of the introduction rule must be the conclusion of the elimination rule. While the inferences giving rise to the paradoxes (like Russell's or Curry's or the Liar's paradox) analyzed in (Tennant, 1982) are of this form, investigations on recursive types (see footnote 8) show that one can easily construct looping derivations starting from inference rules of a different form.

Finally, one might raise the objection that Kreisel-Takeuti's normalization technique is a mere syntactical device which masks the real semantical nature of paradoxical reasoning. First, from the discussion above it should be clear that one could hardly raise objections from a model-theoretic viewpoint, as the introduction of valid axioms does not alter any property related to derivability and satisfiability. On the other hand, is it not clear how to settle the question whether the normal derivation here presented is valid, or not, from a proof-theoretic viewpoint. We leave the task of a semantical analysis of this phenomenon for future investigations.

References

Cardone, F., & Coppo, M. (1991). Type inference with recursive types:

Syntax and Semantics. *Information and Computation*, 92, 48–80.
Girard, J.-Y. (1987). *Proof Theory and Logical Complexity. Vol. 1*. Naples: Bibliopolis.
Kreisel, G., & Takeuti, G. (1974). Formally self-referential propositions for cut-free classical analysis and related systems. *Dissertationes Mathematicae*, 118, 1–55.
Mendler, P. (1987). *Inductive Definitions in Type Theory* (Unpublished doctoral dissertation). Cornell University.
von Plato, J. (2000). A problem of normal form in natural deduction. *Mathematical Logic Quarterly*, 46, 121–124.
Prawitz, D. (1965). *Natural Deduction. A Proof-Theoretic Study*. Stockholm: Almqvist & Wiksell.
Rogerson, S. (2007). Natural deduction and Curry's paradox. *Journal of Philosophical Logic*, 36, 155–179.
Schroeder-Heister, P., & Tranchini, L. (2016). Ekman's paradox. *Notre Dame Journal of Formal Logic*. (To appear)
Sørensen, M., & Urzyczyn, P. (2006). *Lectures on the Curry-Howard Isomorphism*. Amsterdam: Elsevier.
Tennant, N. (1982). Proof and paradox. *Dialectica*, 36, 265–296.
Tennant, N. (1995). On paradox without self-reference. *Analysis*, 55, 199–207.

Mattia Petrolo
IHPST, CNRS, ENS, Université Paris 1 Panthéon-Sorbonne
France
E-mail: `mattia.petrolo@univ-paris1.fr`

Paolo Pistone
Dipartimento di Matematica e Fisica, Università Roma Tre
Italy
E-mail: `paolo.pistone@uniroma3.it`

The Definitional View of Atomic Systems in Proof-Theoretic Semantics

THOMAS PIECHA AND PETER SCHROEDER-HEISTER[1]

Abstract: Atomic systems, that is, sets of rules containing only atomic formulas, play an important role in proof-theoretic notions of logical validity. We consider a view of atomic systems as definitions that allows us to discuss a proposal made by Prawitz (2016). The implementation of this view in the base case of an inductive definition of validity leads to the problem that derivability of atomic formulas in an atomic system does not coincide with the validity of these formulas. This is due to the fact that, on the definitional view of atomic systems, there are not just production rules, but both introduction and elimination rules for atoms, which may even generate non-normalizable atomic derivations. This shows that the way atomic systems are handled is a fundamental issue of proof-theoretic semantics.

Keywords: Proof-theoretic semantics, Validity of atomic formulas, Definitions, Definitional reflection, Dummett's fundamental assumption

1 Introduction

The proof-theoretic semantics of logical constants can be given by an inductive definition of a notion of validity for logical formulas. Such an inductive definition contains a semantic clause for each logical constant under consideration. In its base case the validity of atomic formulas (in short: atoms) is defined in terms of derivability of these formulas in atomic systems. Atomic systems can be sets of atoms or sets of atomic rules, that is, sets of rules that contain only atoms. Atomic rules can be production rules, but one may also consider atomic rules that can discharge atomic assumptions, or even generalize further and consider higher-level atomic rules that can discharge assumed atomic rules (see Section 2).

---

[1]This work was supported by the French-German ANR-DFG project "Beyond Logic: Hypothetical Reasoning in Philosophy of Science, Informatics, and Law", DFG grant Schr 275/17-1.

Atomic systems can be given different interpretations. One possibility is to view atomic systems as *knowledge bases* that are invariant with respect to extensions with further knowledge. Propositions that are valid for a given knowledge base should remain valid if this knowledge base is extended. Under this view, consequence with respect to atomic systems S can be explained by making reference to such extensions of atomic systems: A proposition B is a consequence of a proposition A with respect to an atomic system S (in short: B is an S-consequence of A) if, and only if, for all extensions S' of S, it holds that whenever A is valid w.r.t. S', then B is valid w.r.t. S'. Thus S-consequence is monotone with respect to extensions of atomic systems, if they are understood as knowledge bases.

Another interpretation of atomic systems is given by the *definitional view*, where atomic systems are understood as definitions of certain atoms. Under this view we do not expect monotonicity of S-consequence with respect to extensions, since extending a definition will in general change the meaning of the defined atoms. An explanation of S-consequence should then no longer refer to extensions of atomic systems.

The definitional view of atomic systems is preferred by Prawitz (2016), who sees an atomic system (which he calls a base) "as determining the meanings of the atomic sentences" (ibid., p. 15). According to him, extensions should not be considered in the definitional view: "To consider extensions of the given base [...] is natural when a base is seen as representing a state of knowledge, but is in conflict with the view adopted here that a base is to be understood as giving the meanings of the atomic sentences." (ibid., fn. 12, p. 18).

We implement the definitional view by using a theory of definitions based on the principle of definitional reflection[2] (see Section 3). What we consider to be essential for the definitional interpretation of atomic systems is that the atomic rules for a specific atom completely determine its meaning. This does not exclude partial or non-wellfounded definitions. Complete determination of meaning consists rather in the fact that once we write down an atomic system as a definition of certain atoms it is assumed that the atomic system is complete in the sense that nothing else defines these atoms. Under this assumption (which corresponds to the extremality condition in standard inductive definitions) the principle of definitional reflection is justified, and a certain kind of derivability relation for atoms is induced.

Based on this kind of derivability we then inductively define a proof-

---

[2] See Hallnäs (1991) and Schroeder-Heister (1993).

theoretic notion of validity (see Section 4). The base case in this inductive definition defines the validity of an atom with respect to an atomic system by its derivability in that atomic system; inductive clauses for logical constants are given and logical validity is defined. We then show that for atoms validity does not coincide with derivability. This indicates that the definitional view of atomic systems might not be the proper foundation for proof-theoretic validity.

That the non-definitional view of atomic systems as knowledge bases, where arbitrary extensions of atomic systems are considered, is problematic, was already shown in Piecha, de Campos Sanz, and Schroeder-Heister (2015). Together these results show that the role played by atomic systems in a definition of validity for complex formulas is far from trivial. This problem has been widely neglected in proof-theoretic semantics.

2 Atomic systems

We use letters $a, b, c, \ldots, a_1, a_2, \ldots$ for atoms. The most simple kind of atomic systems (besides sets of atoms) are sets of production rules for atoms. Such a *(first-level) atomic system* S is a (possibly empty) set of atomic rules of the form

$$\frac{a_1 \quad \ldots \quad a_n}{b}$$

Such rules are also called *first-level rules*. The set of premises $\{a_1, \ldots, a_n\}$ in a rule can be empty, in which case the rule is called an *atomic axiom* and is of *level 0*.

In the context of proof-theoretic semantics, atomic systems of this kind were considered, for example, by Prawitz (1971) and Dummett (1991). But one does not have to stop at first-level atomic systems. We also consider second-level and arbitrary higher-level atomic systems.

A *second-level atomic system* S is a (possibly empty) set of atomic rules of the form

$$\frac{[\Gamma_1] \qquad [\Gamma_n]}{a_1 \quad \ldots \quad a_n}{b}$$

where the a_i and b are atoms, and the Γ_i are finite sets of atoms. Such a *second-level rule* expresses that from the premises $a_1, \ldots, a_n$ one may conclude b, where b need no longer depend on assumptions belonging to Γ_i on which the premises a_i might still depend, for each i; that is, assumptions in Γ_i may be discharged. If the sets $\Gamma_1, \ldots, \Gamma_n$ are empty, then the rule is a first-level rule. If the set of premisses $\{a_1, \ldots, a_n\}$ is empty in addition, then the rule is an axiom.

We now further generalize second-level atomic systems to the higher-level case by allowing for atomic rules that can discharge not only atoms but also atomic rules as assumptions.[3] We use the following linear notation to introduce *higher-level rules*:

1. Every atom a is a rule of level 0.

2. If $R_1, \ldots, R_n$ are rules ($n \geq 1$), whose maximal level is ℓ, and a is an atom, then $(R_1, \ldots, R_n \triangleright a)$ is a rule of level $\ell + 1$.

In tree notation, higher-level rules have the form

$$\frac{[\Gamma_1] \quad \quad [\Gamma_n]}{b} \quad \frac{a_1 \quad \cdots \quad a_n}{b}$$

where the a_i and b are atoms, and the Γ_i are finite sets $\{R_1^i, \ldots, R_k^i\}$ of rules, which may be empty. The set of premisses $\{a_1, \ldots, a_n\}$ of such a rule can again be empty, in which case the rule is an axiom. A *higher-level atomic system* S is a (possibly empty) set of higher-level rules.

We now define the notion of *derivation* for higher-level atomic systems:

1. For a level-0 rule a,

$$\frac{}{a}\, a$$

is a *derivation* of a from $\{a\}$.

2. Now consider a level-$(\ell+1)$ rule $(\Gamma_1 \triangleright a_1), \ldots, (\Gamma_n \triangleright a_n) \triangleright b$. Suppose that for each i ($1 \leq i \leq n$) a derivation

$$\begin{array}{c} \Sigma_i \cup \Gamma_i \\ \mathscr{D}_i \\ a_i \end{array}$$

of a_i from $\Sigma_i \cup \Gamma_i$ is given. Then

$$\frac{\begin{array}{ccc} \Sigma_1 & & \Sigma_n \\ \mathscr{D}_1 & & \mathscr{D}_n \\ a_1 & \cdots & a_n \end{array}}{b}\, (\Gamma_1 \triangleright a_1), \ldots, (\Gamma_n \triangleright a_n) \triangleright b$$

is a *derivation* of b from $\Sigma_1 \cup \ldots \cup \Sigma_n \cup \{(\Gamma_1 \triangleright a_1), \ldots, (\Gamma_n \triangleright a_n) \triangleright b\}$.

---

[3] Atomic rules of this kind are thus a special case of the higher-level rules in Schroeder-Heister (1984), which are not restricted to atomic formulas.

The Definitional View of Atomic Systems in Proof-Theoretic Semantics

An atom b is *derivable* from Σ in a higher-level atomic system S, if there is a derivation of b from $\Sigma \cup S$. This is written as follows: $\Sigma \vdash_S b$.

We give an example derivation for the higher-level atomic system

$$S^\star \begin{cases} a \\ ((a \triangleright b) \triangleright c) \triangleright d \\ ((b \triangleright c) \triangleright f) \triangleright g \end{cases}$$

and the set of assumptions $\Sigma = \{e, ((d, e) \triangleright f)\}$:

$$\cfrac{\cfrac{1\ \cfrac{\cfrac{\cfrac{\overline{a}\ \langle a \rangle}{b}\ [a \triangleright b]^1}{c}\ [b \triangleright c]^2}{d}\ \langle((a \triangleright b) \triangleright c) \triangleright d\rangle \qquad \cfrac{\overline{e}\ e}{f}\ (d,e) \triangleright f}{2\ \cfrac{f}{g}\ \langle((b \triangleright c) \triangleright f) \triangleright g\rangle}}{}$$

We use angle brackets $\langle\ \rangle$ to indicate that a rule from the atomic system $S^\star$ is applied, and we use square brackets $[\ ]$ together with numerals to indicate the discharge of assumed rules. In the left branch we start by introducing the premiss a with the 0-level rule $a \in S^\star$. Assuming the rule $a \triangleright b$ we conclude b, and assuming the rule $b \triangleright c$ we obtain c from b. We have thus derived c under the assumption of rules $a \triangleright b$ and $b \triangleright c$. An application of $((a \triangleright b) \triangleright c) \triangleright d \in S^\star$ yields d and discharges the first assumption $a \triangleright b$ (as indicated by the numeral 1). In the right branch we use the assumption $e \in \Sigma$ to get e. Now the rule $(d, e) \triangleright f \in \Sigma$ can be applied to obtain f, which still depends on the assumed rule $b \triangleright c \notin \Sigma$. This assumption is discharged in the last step (as indicated by the numeral 2) by applying $((b \triangleright c) \triangleright f) \triangleright g \in S^\star$ to conclude g. The derivation thus shows $\Sigma \vdash_{S^\star} g$.

3 The definitional view of atomic systems

Atomic systems can be understood as inductive definitions[4] of atomic formulas. Consider an atomic system

$$S \begin{cases} \Gamma_1 \triangleright a \\ \vdots \\ \Gamma_k \triangleright a \end{cases}$$

---

[4] See Aczel (1977).

of k higher-level atomic rules. The atomic rules $\Gamma_i \triangleright a$ can be read as *definitional clauses* for the atom a with *defining conditions* Γ_i, for $1 \leq i \leq k$, and the whole set S of atomic rules can thus be read as a *definition* of the atom a.

The defining conditions in definitional clauses can be empty. Clauses of this form, which we write as $\emptyset \triangleright a$, or simply as a, are read as the base clauses in an inductive definition of a. Clauses of the form $\Gamma_i \triangleright a$, for non-empty Γ_i, are the inductive clauses of such a definition.

A direct application of a definitional clause $\Gamma_i \triangleright a$ consists in passing from the defining condition Γ_i of a to the defined atom a:

$$\frac{\Gamma_i}{a} \; \langle \Gamma_i \triangleright a \rangle$$

We refer to rules of this kind as (steps of) *definitional closure*. They figure as *introduction rules for atoms*.

Definitional closure alone is not characteristic for the definitional view of atomic systems. What distinguishes the definitional view from other views of atomic systems is the fact that in the case of a definition of an atom a it is assumed in addition that nothing else defines a. This extremality condition is usually not stated explicitly in a definition; however, by referring to something as a definition one always tacitly assumes it.

The extremality condition justifies an additional reasoning principle for definitions: For an atom a defined by

$$S \begin{cases} \Gamma_1 \triangleright a \\ \vdots \\ \Gamma_k \triangleright a \end{cases}$$

one can pass from a to an arbitrary atom c whenever c can be obtained from each of the defining conditions Γ_i of a (for $1 \leq i \leq k$). That is, in addition to definitional closure one can argue by *definitional reflection*:

$$\frac{a \quad \overset{[\Gamma_1]}{c} \; \ldots \; \overset{[\Gamma_k]}{c}}{c}$$

(In linear notation: $(a, (\Gamma_1 \triangleright c), \ldots, (\Gamma_k \triangleright c)) \triangleright c$.) Note that this rule is not given by any definitional clause in S (as it is the case for definitional closure); it only becomes available by reflecting on S as a whole. Each instance of

The Definitional View of Atomic Systems in Proof-Theoretic Semantics

definitional reflection is an *elimination rule for the atom* a, where a is the *major premiss* of the rule.

In general, the formula c need not even be atomic. If atomic systems are used in the context of logical rules, or if additional rules are available that allow to manipulate higher-level rules, then definitional reflection can take the form

$$\cfrac{a \quad \overset{[\Gamma_1]}{C} \ldots \overset{[\Gamma_k]}{C}}{C}$$

for arbitrary formulas C.[5] Definitional reflection can also take this general form if a is an undefined atom, that is, if S does not contain any clauses of the form $\Gamma \triangleright a$. As the set of defining conditions of a is empty in this case, one can infer any formula C from a by definitional reflection. In other words, under the definitional view of atomic systems S a principle of *ex falso quodlibet*

$$a \vdash_S C$$

is available if at least one atom a is not defined by S.

To simplify the introduction of the rules of definitional closure and definitional reflection we considered atomic systems S with definitional clauses for only one atom a. In general, however, a *definition* can be any finite atomic system of the following form:

$$S \begin{cases} \Gamma_1^1 \triangleright a_1 & \Gamma_1^n \triangleright a_n \\ \vdots & \ldots & \vdots \\ \Gamma_{k_1}^1 \triangleright a_1 & \Gamma_{k_n}^n \triangleright a_n \end{cases}$$

Note that such a definition can in general not be divided into separate definitions for each a_i, since the definitional clauses might be entangled in the sense that a_i occurs in the defining conditions of another atom a_j. We also note that such a definition need not have base clauses $\emptyset \triangleright a_i$, and is therefore not necessarily well-founded. Moreover, the restriction of definitions to finite atomic systems guarantees that some atom, for example $\bot$, is always undefined, and that therefore *ex falso quodlibet* $\bot \vdash_S C$ is available for any definition S.

As an example for definitional reasoning using definitional closure and definitional reflection consider the following definition (the specific form of

---

[5] For rules of definitional reflection see Hallnäs (1991, 2006) and Schroeder-Heister (1993).

the distinct sets of higher-level rules Γ, Δ and Σ does not matter in what follows):

$$S^+ \begin{cases} \Gamma \triangleright a & \Gamma \triangleright b \\ \Delta \triangleright a & \Delta \triangleright b \\ & \Sigma \triangleright b \end{cases}$$

The two applications of definitional closure

$$\dfrac{\Gamma}{b}\,\langle \Gamma \triangleright b \rangle \quad \text{and} \quad \dfrac{\Delta}{b}\,\langle \Delta \triangleright b \rangle$$

show that b can be inferred from assumptions Γ as well as from assumptions Δ. Since Γ and Δ are exactly the defining conditions of a this means that b can be derived from each of the defining conditions of a. Hence definitional reflection can be applied to a, discharging the assumptions Γ and Δ:

$$\dfrac{a \quad \dfrac{[\Gamma]^1}{b}\,\langle \Gamma \triangleright b \rangle \quad \dfrac{[\Delta]^1}{b}\,\langle \Delta \triangleright b \rangle}{b}\,1 \quad (\text{def. reflection on } S^+)$$

This derivation shows that $a \vdash_{S^+} b$ holds. Note that this cannot be shown by definitional closure alone; one has to use definitional reflection in addition.

Having explained the definitional view of atomic systems, we now discuss some consequences of this view in proof-theoretic semantics.[6]

4 The definitional view in proof-theoretic semantics

Proof-theoretic semantics of logical constants can be given in different ways. One approach, which is due to Prawitz (1971, 1973, 1974, 2014), is to define a notion of validity for derivations that are constructed from arbitrary inference rules.[7] Alternatively one can define notions of proof-theoretic validity for formulas.[8] This approach, which we follow here, allows for a perspicuous formulation of the definition of validity that still captures the main ideas of the derivations-based approach.

---

[6] It should be mentioned that definitional reflection develops its full power, in particular from the computational point of view (see Hallnäs & Schroeder-Heister, 1990, 1991), when clauses for atoms with (free) variables are considered. We have here confined ourselves to atoms as sentence letters without any internal structure, as this suffices to make our point.

[7] See Schroeder-Heister (2006, 2012a).

[8] See Kreisel (1961), Gabbay (1976, 1981), Piecha et al. (2015) and Piecha (2016).

The Definitional View of Atomic Systems in Proof-Theoretic Semantics

We consider a notion of validity for formulas $A, B, C, \ldots$ that are constructed from atoms with the logical constants $\to$, $\vee$ and $\wedge$. It is intended to be a notion of validity for minimal propositional logic. In the following we first inductively define the relation of *S-validity* ($\vDash_S$), which is a relation relative to atomic systems S, by clauses (S1)-(S5). Logical validity, in short: *validity* ($\vDash$), is then defined in clause (S6) as S-validity for all atomic systems S; for Γ being a set of formulas we write $\vDash_S \Gamma$ for $\{\vDash_S A_i \mid A_i \in \Gamma\}$:

(S1) $\vDash_S a \;:\Longleftrightarrow\; \vdash_S a$,

(S2) $\vDash_S A \to B \;:\Longleftrightarrow\; A \vDash_S B$,

(S3) $\Gamma \vDash_S A \;:\Longleftrightarrow\; (\vDash_S \Gamma \;\Longrightarrow\; \vDash_S A)$,

(S4) $\vDash_S A \vee B \;:\Longleftrightarrow\; \vDash_S A \text{ or } \vDash_S B$,

(S5) $\vDash_S A \wedge B \;:\Longleftrightarrow\; \vDash_S A \text{ and } \vDash_S B$,

(S6) $\Gamma \vDash A \;:\Longleftrightarrow\; \forall S : \Gamma \vDash_S A$.

The S-validity of atoms a is defined by clause (S1) as derivability of a in an atomic system S. For the definitional view considered here this means that an atom a is S-valid if, and only if, a is derivable in S by definitional closure and definitional reflection.

Clause (S2) defines S-validity of implications $A \to B$, that is, $\vDash_S A \to B$, by *S-consequence* $A \vDash_S B$. The latter is defined by clause (S3). By combining clauses (S2) and (S3) we see that implication is explained as follows:

$$\vDash_S A \to B \;:\Longleftrightarrow\; (\vDash_S A \;\Longrightarrow\; \vDash_S B)$$

Clauses (S4) and (S5) are straightforward, and clause (S6) then gives us a proof-theoretic notion of logical validity based on atomic systems understood as definitions.

In the given notion of validity we have not considered extensions $S' \supseteq S$ of atomic systems, where an atomic system S' is an extension of an atomic system S if $S' = S$ or if S' results from S by adding atomic rules. In doing so we follow Prawitz's suggestion[9] that considering extensions is in conflict with the definitional view of atomic systems. Alternatively, one could define S-consequence using extensions by the following clause:

$$\Gamma \vDash_S A \;:\Longleftrightarrow\; \forall S' \supseteq S : (\vDash_{S'} \Gamma \;\Longrightarrow\; \vDash_{S'} A) \qquad (\text{S3}_{\text{ext}})$$

---

[9]Cf. Prawitz (2016), fn. 12, p. 18.

Thomas Piecha and Peter Schroeder-Heister

This would prevent that an S-consequence $\Gamma \vDash_S A$ holds just because some atom on which Γ depends is not valid in S. Without extensions, that is, for clause (S3), this is not the case: Consider the empty definition $S = \emptyset$. It is $\nvDash_S a$, since $\nvdash_S a$; hence $\vDash_S a$ implies $\vDash_S b$ trivially, and therefore $a \vDash_S b$ by clause (S3). In general, clause (S3$_{\text{ext}}$) guarantees monotonicity of validity in the sense that, if a is S-valid, it is S'-valid for any extension S' of S. This makes a conceptual difference depending on whether atomic systems are viewed as knowledge bases or as definitions. A knowledge base, unlike a definition, is supposed to be monotone. In fact, under the definitional view of atomic systems already atomic derivability fails to be monotone with respect to extensions of atomic systems. For example, for definition S^+ we had $a \vdash_{S^+} b$. Extending S^+ by $\Theta \triangleright a$ to

$$S^{++} \begin{cases} \Gamma \triangleright a & \Gamma \triangleright b \\ \Delta \triangleright a & \Delta \triangleright b \\ \Theta \triangleright a & \Sigma \triangleright b \end{cases}$$

would block definitional reflection for a, since b cannot be derived from Θ, which, however, is an additional defining condition of a in S^{++}. Hence $a \nvdash_{S^{++}} b$, although $a \vdash_{S^+} b$.

We can now pose the question how derivability of atoms in a definition S from assumed rules relates to S-consequence between a set of assumed formulas and atoms. First we observe that atomic rules R can be represented by formulas R^* over $\{\rightarrow, \wedge\}$ by using a translation $*$ defined as follows:

1. $a^* := a$, for atoms a.

2. $(R_1, \ldots, R_n \triangleright a)^* := R_1^* \wedge \ldots \wedge R_n^* \rightarrow a$, for a rule $R_1, \ldots, R_n \triangleright a$.

We write S^* for the set of formulas representing the rules in a given atomic system S. This device is only needed because S-consequence is defined as a relation between sets of formulas and formulas, whereas derivability is a relation between sets of rules and atoms.

Let now Δ^* be some set of formulas representing a set of rules Δ, and let S be an arbitrary atomic system. We can then ask whether under the definitional view S-validity is *stable* in the sense that the following biconditional holds for any atomic system S:

$$\Delta^* \vDash_S a \iff \Delta \vdash_S a$$

The Definitional View of Atomic Systems in Proof-Theoretic Semantics

That *atomic completeness*

$$\Delta^* \vDash_S a \implies \Delta \vdash_S a$$

does *not* hold was shown in Piecha and Schroeder-Heister (2016). The atomic system $S = \{a \triangleright a\}$ gives a simple counterexample: Obviously $\nvdash_S a$, and hence $\nvDash_S a$ by clause (S1). Therefore $a \vDash_S b$ by clause (S3). However, we have $a \nvdash_S b$. This is because any application of definitional reflection on this atomic system S with major premiss a has the form

$$\cfrac{a \quad \overset{[a]^1}{C}}{C} 1$$

and is therefore useless, since the minor premiss C already contains what is to be established.

For *atomic soundness*

$$\Delta \vdash_S a \implies \Delta^* \vDash_S a$$

the situation is a bit more complicated. If one considered S-validity with extensions by having S-consequence defined by clause (S3$_{\text{ext}}$) instead of clause (S3), then already the empty atomic system $S = \emptyset$ gives a counterexample:[10] Since, for example, a is not defined, we have $a \vdash_S b$ for any atom b by definitional reflection. Now we can extend S to $S' = S \cup \{a\}$ for which we have $\vdash_{S'} a$ and, by clause (S1), also $\vDash_{S'} a$. But clearly $\nvdash_{S'} b$, and thus $\nvDash_{S'} b$ by clause (S1). Therefore $\forall S' \supseteq S : (\vDash_{S'} a \implies \vDash_{S'} b)$ does not hold; hence $a \nvDash_S b$ by clause (S3$_{\text{ext}}$).

However, for the notion of S-validity that is under scrutiny here, we do not consider extensions of atomic systems. In this case, a possible counterexample for atomic soundness is given by the atomic system

$$S = \{(a \triangleright b) \triangleright a\}$$

The derivation

$$\mathcal{D} \begin{cases} \cfrac{a \quad \cfrac{\overline{a} \quad [a \triangleright b]^1}{b}}{b} 1 \text{ (def. reflection on } S) \end{cases}$$

---

[10] Cf. Piecha and Schroeder-Heister (2016).

shows $a \vdash_S b$, and the derivation

$$\mathscr{D}' \begin{cases} \cfrac{[a]^2 \quad \cfrac{\overline{a} \quad \cfrac{[a]^2}{b}[a \triangleright b]^1}{b}}{\cfrac{b}{a}} \text{ (def. reflection on } S) \\ 2 \cfrac{b}{a} \text{ (def. closure)} \langle (a \triangleright b) \triangleright a \rangle \end{cases}$$

which introduces the atom a in the last step by definitional closure shows $\vdash_S a$. Thus by the latter also $\vDash_S a$, by clause (S1). Now, if one allowed for derivations that are *not* normalizable, then the non-normal derivation

$$\mathscr{D}^\dagger \begin{cases} \mathscr{D}' \begin{cases} \cfrac{[a]^2 \quad \cfrac{\overline{a} \quad \cfrac{[a]^2}{b}[a \triangleright b]^1}{b}}{\cfrac{b}{a}} \text{ (def. refl. on } S) \\ 2 \cfrac{b}{a} \text{ (def. closure)} \end{cases} \\ 3 \cfrac{}{b} \end{cases} \qquad \mathscr{D}' \begin{cases} \cfrac{[a]^2 \quad \cfrac{\overline{a} \quad \cfrac{[a]^2}{b}[a \triangleright b]^1}{b}}{\cfrac{b}{a}} \text{ (def. refl. on } S) \\ 2 \cfrac{b}{a} \text{ (def. closure)} \\ \cfrac{a \quad [a \triangleright b]^3}{b} \text{ (def. refl. on } S) \end{cases}$$

would show $\vdash_S b$. This derivation results from substituting the closed derivation $\mathscr{D}'$ for the two open assumptions a in derivation $\mathscr{D}$. By clause (S1) we would then have $\vDash_S b$, and therefore $a \vDash_S b$ by clause (S3). Hence S does not give us a counterexample to atomic soundness.

That derivation $\mathscr{D}^\dagger$ is not in normal form is due to the fact that the major premiss a in the final application of definitional reflection is a maximal formula: it is introduced by an application of definitional closure in the last step of $\mathscr{D}'$, and is immediately eliminated in the last step of $\mathscr{D}^\dagger$. Moreover, $\mathscr{D}^\dagger$ is not normalizable, since b cannot be introduced by definitional closure (since there is no definitional clause for b in S).

If, on the other hand, we require that all derivations be normal, then $\mathscr{D}$ and $\mathscr{D}'$ cannot be combined into $\mathscr{D}^\dagger$. In this case there cannot be a closed derivation of b in S, that is, $\nvdash_S b$. Hence $\nvDash_S b$ by clause (S1), while $\vDash_S a$. Therefore $a \nvDash_S b$, which means that S is in this case a counterexample to atomic soundness.

This points to a deeper problem of S-validity under the definitional view of atomic systems. In the non-atomic realm of logically complex formulas or statements one usually imposes Dummett's *fundamental assumption* "that, if we have a valid argument for a complex statement, we can construct a valid argument for it which finishes with an application of one of the introduction rules governing its principal operator" (Dummett, 1991, p. 254; cf. also ch. 12). For the atomic realm, where applications of definitional closure are

The Definitional View of Atomic Systems in Proof-Theoretic Semantics

the introduction rules for defined atoms, this means that the base clause (S1) in the definition of S-validity has to be replaced by the following clause:

$$\vDash_S a \;:\Longleftrightarrow\; \vdash_S a, \text{ where } a \text{ is derived by definitional closure in the last step.} \tag{S1$'$}$$

This enforces that the fundamental assumption holds for S-validity. Alternatively, the same effect can be achieved by demanding for definitional reflection

$$\begin{array}{ccc} & [\Gamma_1] & [\Gamma_k] \\ a & c & \ldots \; c \\ \hline & c & \end{array}$$

that is, for the elimination rule for atoms a, that the major premiss a must always be the conclusion of an assumed rule; this includes the case that the major premiss is an assumed formula.[11] This forces all derivations to be in normal form, and consequently ensures that S-validity complies with the fundamental assumption.

What this tells us about the definitional view of atomic systems in proof-theoretic semantics is the following: If we impose the same assumption that we make for the validity of complex formulas also on the validity of atomic formulas, which is a natural thing to do, then S-validity is not stable on the atomic level, that is, neither atomic completeness nor atomic soundness holds. A restriction from higher-level to first-level atomic systems might rule out counterexamples to atomic soundness, but atomic completeness would still fail, as the given first-level counterexample shows.

An objection one might raise against the clause $(a \rhd b) \rhd a$ is that it is circular in that it defines a in terms of a (as is the case also with the clause $a \rhd a$) and moreover paradoxical, by defining a by $\neg a$. The latter is easily seen if we write the undefined atom b as absurdity $\bot$ and $a \rhd \bot$ as $\neg a$, so that the clause becomes $\neg a \rhd a$. However, we do not consider this a problem here, as we do not want to put restrictions on the form clauses are allowed to take. We prefer to claim *definitional freedom* in that respect. Apart from the fact that in logic programming clauses such as $\neg a \rhd a$ have always been considered and are thus not unusual at all, it gives us a most welcome tool with considerable expressive power. By means of clauses of

---

[11] This corresponds to the feature that major premisses of elimination rules for logical constants only occur as assumptions. It is sometimes considered in discussions of general elimination rules; for example, according to Tennant (2015, p. 746) "all major premises for eliminations stand proud, with no proof-work above them (that is to say, they occupy leaf-nodes of the proof-tree)".

this kind and the principle of definitional reflection we can develop a natural theory of semantical and set-theoretical paradoxes, which are characterized by non-normalizing atomic derivations.[12] The idea of a 'partial' assignment of meaning as present in self-referential definitions is at the core of the theory of definitional reasoning and its principle of definitional reflection. It is analogous to the consideration of partial recursive functions in recursive function theory (see Hallnäs, 1991).

It is important to note that the failure of stability on the atomic level says nothing about the soundness and completeness of a calculus for minimal or intuitionistic logic with respect to logical validity. Our results apply only to S-validity, that is, to validity with respect to a chosen atomic system. The fact that stability on the atomic level is lacking for approaches based on definitional reasoning shows that a satisfactory theory of proof-theoretic validity based on definitional reasoning with atoms is still a desideratum. It might be the case that a hybrid theory incorporating both a knowledge-base view and a definitional view of atomic systems might be an option, but at the present stage of research this is not much more than a speculation. In any case the theory of atomic reasoning deserves much more attention in proof-theoretic semantics than devoted to it so far.

References

Aczel, P. (1977). An introduction to inductive definitions. In J. Barwise (Ed.), *Handbook of Mathematical Logic* (pp. 739–782). Amsterdam: North-Holland.

Dummett, M. (1991). *The Logical Basis of Metaphysics*. London: Duckworth.

Gabbay, D. M. (1976). On Kreisel's notion of validity in Post systems. *Studia Logica, 35*(3), 285–295.

Gabbay, D. M. (1981). *Semantical Investigations in Heyting's Logic*. Dordrecht: Reidel.

Hallnäs, L. (1991). Partial inductive definitions. *Theoretical Computer Science, 87*, 115–142.

Hallnäs, L. (2006). On the proof-theoretic foundation of general definition theory. *Synthese, 148*, 589–602.

---

[12] See Schroeder-Heister (2012b). This approach goes back to Prawitz (1965, appendix B) and Tennant (1982).

Hallnäs, L., & Schroeder-Heister, P. (1990). A proof-theoretic approach to logic programming. I. Clauses as rules. *Journal of Logic and Computation, 1*, 261–283.

Hallnäs, L., & Schroeder-Heister, P. (1991). A proof-theoretic approach to logic programming. II. Programs as definitions. *Journal of Logic and Computation, 1*, 635–660.

Kreisel, G. (1961). *Unpublished appendix to the paper "Set theoretic problems suggested by the notion of potential totality" in: Infinitistic Methods. Proceedings of the Symposium on Foundations of Mathematics, Warsaw, 2-9 September 1959. Pergamon, Oxford (1961)*. (Reference taken from Gabbay (1976); unpublished appendix not found.)

Piecha, T. (2016). Completeness in proof-theoretic semantics. In T. Piecha & P. Schroeder-Heister (Eds.), *Advances in Proof-Theoretic Semantics* (pp. 231–251). Cham: Springer.

Piecha, T., de Campos Sanz, W., & Schroeder-Heister, P. (2015). Failure of completeness in proof-theoretic semantics. *Journal of Philosophical Logic, 44*, 321–335. (First published online August 1, 2014.)

Piecha, T., & Schroeder-Heister, P. (2016). Atomic systems in proof-theoretic semantics: two approaches. In J. Redmond, O. Pombo Martins, & Á. N. Fernández (Eds.), *Epistemology, Knowledge and the Impact of Interaction* (pp. 47–62). Cham: Springer.

Prawitz, D. (1965). *Natural Deduction: A Proof-Theoretical Study*. Stockholm: Almqvist & Wiksell. (Reprinted by Dover Publications, Mineola, N.Y., 2006.)

Prawitz, D. (1971). Ideas and results in proof theory. In J. E. Fenstad (Ed.), *Proceedings of the Second Scandinavian Logic Symposium* (pp. 235–307). Amsterdam: North-Holland.

Prawitz, D. (1973). Towards a foundation of a general proof theory. In P. Suppes et al. (Eds.), *Logic, Methodology and Philosophy of Science IV* (pp. 225–250). Amsterdam: North-Holland.

Prawitz, D. (1974). On the idea of a general proof theory. *Synthese, 27*, 63–77.

Prawitz, D. (2014). An approach to general proof theory and a conjecture of a kind of completeness of intuitionistic logic revisited. In L. C. Pereira, E. H. Haeusler, & V. de Paiva (Eds.), *Advances in Natural Deduction* (pp. 269–279). Berlin: Springer.

Prawitz, D. (2016). On the relation between Heyting's and Gentzen's approaches to meaning. In T. Piecha & P. Schroeder-Heister (Eds.), *Advances in Proof-Theoretic Semantics* (pp. 5–25). Cham: Springer.

Thomas Piecha and Peter Schroeder-Heister

Schroeder-Heister, P. (1984). A natural extension of natural deduction. *Journal of Symbolic Logic, 49*, 1284–1300.
Schroeder-Heister, P. (1993). Rules of definitional reflection. In *Proceedings of the Eighth Annual IEEE Symposium on Logic in Computer Science (Montreal 1993)* (pp. 222–232). Los Alamitos: IEEE Computer Society.
Schroeder-Heister, P. (2006). Validity concepts in proof-theoretic semantics. *Synthese, 148*, 525–571.
Schroeder-Heister, P. (2012a). Proof-theoretic semantics. In E. N. Zalta (Ed.), *The Stanford Encyclopedia of Philosophy*.
Schroeder-Heister, P. (2012b). Proof-Theoretic semantics, self-contradiction, and the format of deductive reasoning. *Topoi, 31*, 77–85.
Tennant, N. (1982). Proof and paradox. *Dialectica, 36*, 265–296.
Tennant, N. (2015). The relevance of premises to conclusions of core proofs. *Review of Symbolic Logic, 8*, 743–784.

Thomas Piecha
University of Tübingen
Germany
E-mail: `thomas.piecha@uni-tuebingen.de`

Peter Schroeder-Heister
University of Tübingen
Germany
E-mail: `psh@uni-tuebingen.de`

Two Accounts of Pairs

MARTIN PLEITZ[1]

Abstract: I argue for the desirability of a formal account that construes ordered pairs as primitive by criticizing Quine's methodological praise of the standard set theoretical substitutes. Then I contrast two construals of primitive pairs: the neo-logicist account of pairs by abstraction and a novel account of pairs by postulation that makes use of Kit Fine's procedural postulationism. I show that pairs by postulation are preferable, and valuable for plural logic variants of the neo-logicist project.

Keywords: Ordered Pairs, Kit Fine, Procedural Postulationism, Neo-Logicism, Abstraction Principle, Plural Logic, Equinumerosity, Ancestral

1 "The ordered pair as philosophical paradigm"

What are ordered pairs? On our way to an answer, let us first contrast the technical notion we will be interested in with the colloquial notion that is its likely origin. Colloquially, we speak of pairs of shoes, of pairing each knife with a fork on the table, and of a hydrogen atom being a pair of a proton and an electron. Note that according to the colloquial notion, it is always *two* things that form a pair – the components of every pair are distinct –, but not *all* two things can sensibly be paired (think of leftover socks in the drawer, or the Eiffel tower and the number three). On a more metaphysical note, according to the colloquial notion a pair of things is not itself a (third) thing; it's just two objects, in that order. This colloquial notion might be given a formal explication if the resources of plural logic were extended in some way so as to deal with *ordered* pluralities. But we need not go into that, because the colloquial notion differs in at least the three noted aspects from the technical notion of an ordered pair that is standardly used in philosophy, logic, and mathematics. According to the technical notion, the components of a pair *can* be identical (an object can be paired with itself); *any* objects can be paired; and just like its components a and b, the pair $\langle a, b \rangle$ is itself an *object*: "a single object doing the work of two" (Quine, 1960, p. 258).

---

[1] I would like to thank Kit Fine, Simon Hewitt, and Tobias Martin for helpful conversations.

When we ask: what are pairs, we will always mean ordered pairs according to this technical notion.

There is a clear consensus that all possible answers to the question are restricted by the following "instrumental property" (Kanamori, 2003, p. 289) of pairs:[2]

(**Peano's Principle**) $\langle a, b \rangle = \langle c, d \rangle$ *if and only if* $a = c$ *and* $b = d$.

Most people think it is equally clear that this principle is not only necessary but also sufficient to characterize pairs; that it is really all that needs to be said. We will hear – and endorse – a dissenting opinion in due course (cf. section 4). Willard Van Orman Quine, however, is as good an example as any of the orthodoxy, when he says with regard to (Peano's Principle) that the notion of ordered pair is "subject in effect to [a] single postulate" (Quine, 1960, p. 258).

The restriction by (Peano's Principle) leaves room for a lot of different construals of pairs. The path that is at least prima facie most appealing (and which we will ourselves take later) is to follow Peano himself and construe the pairs given by his principle as *primitive* (Kanamori, 2003, p. 289). Besides the non-reductive path, there is a multitude of reductive paths – at least when some background theory is taken as given. While reductions within number theory (e.g., understanding '$\langle a, b \rangle$' as an abbreviation of '$2^a 3^b$' (Quine, 1960, p. 260), which satisfies (Peano's Principle) because of the unique-prime-factorization theorem) give us only pairs *of natural numbers*, the reductions within set theory deliver pairs of objects of any kind (given that these are included among the urelements, if they are not pure sets to begin with). Here there are Norbert Wiener's definition $\langle a, b \rangle =_{\text{def}} \{\{\{a\}, \emptyset\}, \{\{b\}\}\}$, Felix Hausdorff's definition $\langle a, b \rangle =_{\text{def}} \{\{a, 1\}, \{b, 2\}\}$,[3] and the currently standard definition by Kazimierz Kuratowski $\langle a, b \rangle =_{\text{def}} \{\{a\}, \{a, b\}\}$ to choose from (Kanamori, 2003, 290f.). All of these satisfy (Peano's Principle), because of the principle of extensionality which is the criterion of identity for sets. On an historical note, we should point out that Frege also took (Peano's Principle) as the sole benchmark of pairhood when he gave his own "extravagant"[4] construal in

---

[2] From the historical overview (Kanamori, 2003, pp. 288–293) it emerges that it was Giuseppe Peano who first made explicit that this property is characteristic of pairs. It is for that reason – and to bring out the parallel to (Hume's Principle) – that we refer to it as '(Peano's Principle)'. It has also been called "the Peano axiom" and "the pair principle".

[3] In Hausdorff's definition, 1 and 2 are some "distinct objects alien to the situation" (Kanamori, 2003, p. 291).

[4] ...to use Boolos's word (Heck, 2012, p. 181).

Two Accounts of Pairs

Grundgesetze (Frege, 2009, pp. 210ff., 217, and 230), which is obsolete because it makes crucial use of the inconsistency-generating Basic Law V (Heck, 2012, p. 181). So did Wiener, who originally worked within the framework of Russell's type theory (hence the extra level of complexity in his construal, which Quine does away with – with a view to current set theory, rightfully – when he reports it as $\langle a, b \rangle =_{\text{def}} \{\{a, \emptyset\}, \{b\}\}$ in Quine, 1960, p. 259).

What are we to make of the fact that (Peano's Principle) allows for a multiplicity of acceptable construals? How are we to choose among them? Quine, for one, was delighted by the situation; and he used it to illustrate a general methodological maxim of his (Quine, 1960, pp. 257–262): If in a theory we already endorse there is a substitute that has the requisite instrumental properties, then we should use it instead of the intuitive notion we initially were interested in. As Quine is aiming only for extensional adequacy, he is unperturbed by any *artifacts* of the substitute; i.e., by any of its properties beside the instrumental ones that make it suitable to stand in for the initial notion. In fact, he reckons these among the "waste cases, what the computing-machine engineers call don't cares" (Quine, 1960, pp. 182, 259). An illustration of these ideas with pairs commends itself to Quine because with (Peano's Principle) there is "a single explicit standard whereby to judge whether a version is suitable" (Quine, 1960, p. 258). The Quinean "don't cares" for pairs are, e.g., that a pair with identical components is a singleton set on Kuratowski's construal (but not on the Hausdorff's), and more generally, that every pair is *a set*.[5] But is it right not to care? Observe that there are several preconditions for the Quinean view: the aforementioned merely extensional orientation, an ideal of ontological austerity that makes it desirable to minimize the sorts of abstract objects we countenance, and a prior commitment to an ontology of sets. But those of us who care about the nature of things and who are less enamored of ontological austerity or more reluctant to commit to sets cannot be content with the Quinean view. (At least when we are doing metaphysics, that is, because *within set theory* a set theoretical construal of pairs is of course entirely sufficient.) For us there are in fact several problems with the Quinean view.

The first problem with the Quinean view becomes evident when we drop the merely extensional orientation, but keep the aim of giving a set theoretical construal of pairs. For *which* set is a given pair? We can transpose the

---

[5] On the number theoretical construal $\langle a, b \rangle =_{\text{def}} 2^a 3^b$, the Quinean "don't cares" are, e.g., that the only pairs of numbers that are prime numbers are $\langle 0, 1 \rangle$ and $\langle 1, 0 \rangle$, and more generally, that every pair of numbers is itself *a number*.

problem that Paul Benacerraf famously described for the identification of numbers with sets (Benacerraf, 1983, pp. 272f., 278) to the present identification of pairs with sets:

> Imagine Felix and Kazimierz, sons of two militant Quineans – children who have been taught ontology not in the vulgar (old-fashioned) way by characterizing objects as belonging to different primitive kinds, but in the more direct way of using sets in a systematic way. Where we talk of the pair of a and b, Felix was told to use the set $\{\{a, 1\}, \{b, 2\}\}$, while Kazimierz was told to use the set $\{\{a\}, \{a, b\}\}$. But now disagreements arise between Felix and Kazimierz, whether a pair with identical components is a singleton set etc.[6]

What we have done here is of course to turn Quine's "don't cares" into proper disadvantages, giving us motivation for a structuralist or otherwise non-reductive account.[7]

The second problem with the Quinean view depends on our interest in the nature of things. One need not oppose *all* use of set theory in philosophy (cf. Simons, 2005) to be skeptical of its overuse, according to which all abstract objects (and maybe even some concrete entities) are really set theoretical constructions. Even in view of the possibility of objects of one sort standing in for objects of another sort – and without any qualms about the *existence* of sets –, we should keep the metaphysical distinctions between natural numbers, sets, pairs, and other objects clear. This might go against the grain of the official practice in current mathematics and perhaps philosophy. But it is really just a common sense position, which is implicitly taken, e.g., when ordered pairs are distinguished from *unordered pairs* (roughly, sets of two elements) to motivate the instrumental property that any set theoretical construal needs to exemplify. And it fits nicely with the historical fact that numbers, sets, and pairs came to the attention of mathematicians at different historical moments[8] and were each at first construed as primitive objects.[9] So regardless of whether we accept sets or not (in metaphysics), it

---

[6] Here, I am of course deliberately echoing the famous passage by Benacerraf, and also my own earlier adaptation of it in the context of the idea of identifying a symbol of a formal language of arithmetic with its Gödel number (Pleitz, 2010, p. 210).

[7] The point is not new. Set theoretical construals of pairs have already been criticized for their arbitrariness (Armstrong, 1986, p. 87) and for their conventionality (Forrest, 1986, p. 91). Cf. Sider (1996).

[8] Kanamori observes: "to focus the historical background it should be noted that ordered pairs are not explicit in Descartes and in the early work on analytic geometry. Hamilton (1837) may have been the first to objectify ordered pairs in his reconstrual of the complex numbers as ordered 'couples' of real numbers." (Kanamori, 2003, p. 288)

[9] Kanamori is perhaps unduly surprised that Nicolas Bourbaki still used primitive pairs in

Two Accounts of Pairs

is good to have a primitive account of pairs because our interest in the nature of things outweighs the value we put in ontological parsimony.

So we do care about the differences among kinds of abstract objects. But why care about *pairs* in particular? First of all, given the dialectics of the debate about the analysis of abstract objects, it will be instructive to *accept* the paradigmatic status that Quine bequeathed to pairs,[10] if only to reject the methodological maxim he saw them as illustrating.[11] Another reason is that pairs do important work in certain contexts. Within current set theory, a relation (and more particularly a function) is standardly understood as a set (or class) of ordered pairs.[12] Now, in set theory this role can of course be played by pairs that are construed in some set theoretical way. But in formal accounts that (for whatever reason) accept pluralities but abstain from sets, the addition of primitive pairs can be crucial to enable a corresponding construal of a relation (or function) as a *plurality* of ordered pairs; see e.g., Linnebo (2010, p. 151).[13] Examples that we will be concerned with specifically in the present paper are certain variants of neo-logicism that are formulated in a logic that has plural quantifiers but no second-order quantifiers.

In sum, we care for pairs not only for the general reason that we want to understand how their nature differs from that of sets, number, etc., but also for the more pressing particular reason that within certain theoretical contexts there is indeed no easy substitute for primitive pairs. And it is again because of our interest in their nature, and to evade the Benacerraf Problem,

---

the 1954 edition of their book on set theory (Kanamori, 2003, p. 292).

[10] Pairs are also well-suited to play a paradigmatic role within my own broader project of developing a theory of reification. I construe *reification* as the close association of a non-object with an object that encodes some relevant information about the non-object, and take it to be the root of such diverse kinds of abstract objects as sets, cardinal numbers, ordinal numbers, facts, propositions, and arguably, even meaningful expressions. In the present paper, we are concerned with a particularly simple example of reification, where the ordered plurality of objects a and b (a non-object because it's *two* objects) is associated closely with the pair $\langle a, b \rangle$ (an object) via the criterion of identity that is implicit in (Peano's Principle).

[11] Thus we take a cue from Kanamori, who closes his overview of the history of ordered pairs by reflecting on Quine's view: "The ordered pair [...] is now carrying the weight of what 'analysis' is to mean in philosophy." (Kanamori, 2003, p. 293)

[12] Although Frege construed relations and functions in a quite different way, in *Gundgesetze* he made a similar use of ordered pairs on a technical level (Frege, 2009, pp. 210ff.); cf. Heck (1995) and Heck (2012, pp. 180–186).

[13] The technical fact in the background is that although plural logic can mimic the monadic part of second order logic (roughly, quantification over concepts), it cannot represent its dyadic part (roughly, quantification over relations). Cf., e.g., Boolos (1998, p. 11) and Linnebo (2014, section 2.2). The problem is also relevant for David Lewis's attempt to reduce mathematics to mereology in Lewis (1991).

that we would prefer pairs to be *primitive*; we have ample motivation to look for a non-reductive account of pairs. There are several ways of spelling out a non-reductive account; primitive pairs can be construed in proof theoretical, structuralist, neo-logicist, or postulationist terms. In the following we will sketch and critique the neo-logicist construal and develop a novel postulationist alternative.[14]

2 Pairs by abstraction

Before we can introduce and assess the neo-logicist construal of pairs, we need to set the stage – very quickly, for reasons of space.[15] The project of *logicism* was to reduce arithmetic to logic and definitions. Despite Frege's immense and in many regards successful work, his logicism famously failed because he based it on Basic Law V, which entails that there is an extension (roughly, a set) for every condition that can be expressed in the language, and thus gives rise to Russell's paradox.

(**Basic Law V**) $\forall F \forall G \, (\{x|F(x)\} = \{x|G(x)\} \leftrightarrow \forall z \, (F(z) \leftrightarrow G(z)))$

For a time, everybody gave up. But since the 1980s, there has been a logicist renaissance. It is typical of *neo-logicism* that *abstraction principles* play a central role. In general, an abstraction principle is of the following form:

(**Abstraction Principle in General**) $\forall \alpha \forall \beta \, (\S \alpha = \S \beta \leftrightarrow \alpha \sim \beta)$

Here, α and β are of the same syntactic category (typically, they are predicates), '$\S$' is a term-forming operator, and '$\sim$' expresses an equivalence suited to that syntactic category. Thus the principle says, roughly, that the

---

[14] Neil Tennant has developed a proof theoretical account of a pairing function (Tennant, 2009, sections 2 through 10). As he uses a free logic and a proof theoretical presentation, a detailed discussion of his proposal would take us too far afield. But I would like to acknowledge that reading his thorough study of pairs construed as primitive was a valuable inspiration in writing the present paper. The structuralist account, on the other hand, is too close to the neo-logicist one for an interesting contrast. The Benacerrafian argument for a non-reductive account would of course lead naturally to a contrual of primitive pairs within ante rem structuralism (Shapiro, 1997, pp. 83ff.), understanding the plurality of all pairs as the structure characterized by the theory with (Peano's Principle) as its sole axiom, given some base ontology of objects that are not pairs, corresponding to the urelements of (impure) set theory. But this (to anticipate) is quite similar to the neo-logicist account which is based on (Peano's Principle) as an abstraction principle. As we have no room to bring out whatever subtle difference there might be between the role of an axiom in structuralism and the role of an abstraction principle in neo-logicism, we will say no more about this here.

[15] With regard to neo-logicism, cf., Tennant (2014) for an overview; Potter (2000) for historical motivation; as well as Wright (1983) and Hale and Wright (2001) as exemplary texts.

abstract of α is identical to the abstract of β if and only if α and β are equivalent. The most important neo-logicist abstraction principle is:

(**Hume's Principle**) $\forall F \forall G\ (\sharp F = \sharp G \leftrightarrow F \approx G)$

Here, the term-forming operator '$\sharp$' reads 'the number of ...' and '$\approx$' expresses equinumerosity. Thus (Hume's Principle) says that the number of Fs is identical to the number of Gs if and only if the Fs and Gs are equinumerous, i.e., stand in one-to-one correspondence. Now it turns out that on the background of second-order logic and given suitable definitions, (Hume's Principle) entails the axioms of Peano arithmetic. As most of this was already shown by Frege in *Grundgesetze* (cf., e.g., Wright, 1983, p. 130ff.; Boolos, 1998, part II; Heck, 2011; Zalta, 2016) this fact has been dubbed "Frege's theorem" (Boolos, 1998, pp. 209, 277). On the basis of this technical result, Crispin Wright and others have started to carry out the philosophical program of a Neo-Fregean account of mathematics.

Neo-logicism has many commendable aspects, but we will move right on to certain problems that beset it. The *Caesar Problem* was noticed already by Frege, who was the first to consider abstraction as a way of introducing objects, but rejected this way because an abstraction principle settles the truth of only those identity statements that have a certain form (using terms formed with the operator '§' in both places of the identity predicate). He was worried because (Hume's Principle) cannot on its own settle whether Julius Caesar is the number of some concept (Frege, 1988, pp. 66ff., 74ff.). The *Bad Company Problem* is specific to the dialectical situation after the discovery of Russell's paradox. It is based on the observation that the inconsistent (Basic Law V) has just as much claim to being an abstraction principle as (Hume's Principle), putting a burden on the friends of the latter to explain what distinguishes it from the bad company it keeps. Then there is a problem of existential commitment, which we will call the *Anselm Problem*.[16] It is the problem of reconciling the claimed analyticity of (Hume's Principle), which is crucial because of the epistemological goals of neo-logicism, with the fact that it entails the existence of objects – the cardinal numbers (Field, 1984; Boolos, 1998, pp. 301ff.). From the Anselm Problem, we may want to extricate the more particular *Cardinality Problem*, which is that it is an *infinity* of numbers that are entailed to exist; (Hume's Principle) can only be satisfied on an infinite domain, which is a quite drastic requirement on

---

[16] While the labels 'Caesar problem' and 'Bad Company problem' are widely used in the literature about neo-logicism, the label 'Anselm problem' is my own addition (with a nod to Field, 1984, pp. 509ff.).

the world to be made by what is alleged to be a conceptual truth. None of these problems speaks decisively against neo-logicism, but they all need to be addressed.

We will now turn to those variants of neo-logicism that are especially relevant for our investigation of pairs. Besides the above variant that is formulated in second-order (singular) logic, it is also worthwhile to study variants that are formulated in (first order) plural logic. This is well-motivated because a good case can be made (and *is* made in the introduction of Hewitt [manuscript]) for the claim that a number statement (e.g., that Jupiter has 67 moons) pace Frege is not about a concept (*being a moon of Jupiter*) but about a plurality of objects (the moons of Jupiter). But it has far-reaching consequences for the abstraction principle at the heart of the neo-logicist project. Let us contrast the second-order logic to the plural logic variant of (Hume's Principle):

(Hume's Principle in 2nd-Order Logic) $\forall F \forall G \, (\sharp F = \sharp G \leftrightarrow F \approx_{\text{sec.}} G)$

(Hume's Principle in Plural Logic) $\forall xx \forall yy \, (\sharp xx = \sharp yy \leftrightarrow xx \approx_{\text{pl.?}} yy)$

Besides the difference in the kind of item that numbers are abstracted from, which is made explicit here by a difference in syntactic category (predicates vs. plural terms), there is an equally important difference that concerns the analysis of the notion of equinumerosity. In (dyadic) second-order logic, there are straightforward ways of saying that there is a one-to-one correspondence; abbreviated here as '$\approx_{\text{sec.}}$'. Not so in plural logic (Linnebo, 2014, section 4.2); hence the question mark in '$\approx_{\text{pl.?}}$'. One way to supply the lack is to introduce pairs by abstraction. For note that (Peano's Principle) is logically less demanding than (Hume's Principle) – its right hand side can already be formulated in first order (singular) logic with identity. The idea is to enable the introduction of numbers by abstraction by the simultaneous introduction of pairs by abstraction. This path is considered by Shapiro and Weir (2007, pp. 132ff.) and taken by Hewitt (Manuscript, section 3).[17] Others have also considered introducing pairs by abstraction, but fordifferent reasons and not with a view to plural logic variants of neo-logicism in particular.[18]

---

[17] Francesca Boccuni also employs plural logic in a foundational account of arithmetic (Boccuni, 2010, 2013), but her work is farther removed from the present topic – what she develops is a plural logic variant of *logicism*, not a plural logic variant of *neo-logicism*.

[18] Tennant gives a critical discussion of pairs by abstraction as a contrast to his own proof theoretical account (Tennant, 2009, subsection 6.2). His work on the pairing function is due not to an interest in a plural logic variant of neo-logicism, but in a neo-logicist account of addition and multiplication (Tennant, 2009, sections 11 through 14). Shapiro (2000, pp. 337f.)

Two Accounts of Pairs

To arrive at a neo-logicist account of pairs, similar to the neo-logicist account of numbers, we need only understand (Peano's Principle) as playing the role of (Hume's Principle), i.e., of an abstraction principle. Recall:

(**Peano's Principle**) $\forall a \forall b \forall c \forall d (\langle a, b \rangle = \langle c, d \rangle \leftrightarrow a = c \land b = d)$

This has indeed the right form to qualify as an abstraction principle – the usual pair brackets are a term forming operator, and the right hand side expresses an equivalence, each taking ordered pluralities with two members as arguments.[19] Note that this particular abstraction principle has as good a claim to analyticity as any,[20] so *pairs by abstraction* appear to be a very natural addition to the neo-logicist universe.

It is not surprising that the problems that beset the neo-logicist account of numbers recur for the neo-logicist account of pairs. The Caesar Problem for pairs by abstraction is that (Peano's Principle) does not settle whether Julius Caesar (or any other object that we intuitively do not understand as a pair) is identical with some pair of objects. The Bad Company Problem for pairs by abstraction is just the same as for numbers by abstraction, because the abstraction principles that allow to introduce pairs and numbers, respectively, share the bad company of (Basic Law V). The Anselm Problem for pairs by abstraction is that despite its alleged analyticity, (Peano's Principle) commits us to the existence of further objects if there are at least two objects. And there is also again a more particular Cardinality Problem for pairs by abstraction, because (Peano's Principle) cannot be satisfied on a finite domain of more than one object.[21] The recurrence of these problems might be reason enough to look for an alternative account (which we will do in section 5); but let us first look more closely at two (potentially) problematic aspects.

---

uses pairs by abstraction to illustrate abstracting over ordered pluralities with two members in Shapiro (2000) itself and in Hale (2001), but in both texts the aim is not to introduce pairs but other objects: *ratios* and *reals*.

[19] Stewart Shapiro construes the right hand side as expressing a *four-place* relation, so that strictly speaking it cannot be an equivalence relation. He points out himself though that it has the requisite structural properties of an equivalence "when the variables are taken two at a time" (Shapiro, 2000, p. 337). And I think that in the context of neo-logicism we should be more liberal with regard to what counts as an equivalence, because most abstraction principles abstract over non-objectual items anyway: The usual second-order variant of (Hume's Principle) abstracts over *concepts*, and the plural logic variant abstracts over *pluralities*. So why not construe (Peano's Principle) as abstracting over *ordered pluralities with two members*?

[20] Its consistency (on domains of a certain cardinality) follows already from the existence of the set theoretical and number theoretical substitutes that satisfy the principle.

[21] (Peano's Principle) can only be satisfied either on a domain of exactly one object (which would then be a pair with itself as both components) or on a domain of infinitely many objects (Boolos, 1998, p. 208).

3 Existential commitment

Let us look more closely at the question of existential commitment. Recall that we mentioned that according to our *colloquial* notion, not any two objects can sensibly be paired. On the technical notion, in contrast, any two objects can be paired, and on the neo-logicist account this is spelled out as the existence claim that for any objects a and b, the pair of a and b exists. Now many will see that as fitting for the technical notion – but from a metaphysical point of view it is surely too much to build this far-reaching existential commitment already into the *language* we use to talk about pairs. For note that by using the descriptive function 'the pair of a and b' as our only device to talk about pairing, we cannot even *say* of any objects that they fail to form a pair.[22] Let us change this! Specifically, we will no longer treat the usual term-forming operator '$\langle a,b \rangle$' as primitive, and rather base our account on a ternary predicate, 'PAIRS$(x{:}a,b)$', which we read as 'x pairs a and b'. Its meaning will be governed by the following principle of (**Pair Identity**):[23]

$\forall x \forall y \forall a \forall b \forall c \forall d$ (PAIRS$(x{:}a,b) \land$ PAIRS$(y{:}c,d)) \to (x=y \leftrightarrow a=c \land b=d)$

Note that this principle is (trivially) satisfied on a domain of several objects none of which form a pair; and more generally, it makes no existential claim. Such claims can be supplied separately. Let us distinguish:[24]

(**Pair Universalism**) $\forall a \forall b \exists x$ PAIRS$(x{:}a,b)$

(**Pair Moderatism**) $\exists a \exists b \exists c \exists d$ ($\exists x$ PAIRS$(x{:}a,b) \land \neg \exists y$ PAIRS$(y{:}c,d))$

(**Pair Nihilism**) $\neg \exists a \exists b \exists x$ PAIRS$(x{:}a,b)$

This partition of the spectrum is finer than the dichotomy of *realism* and *anti-realism* about pairs (i.e., respectively, '$\exists a \exists b \exists x$ PAIRS$(x{:}a,b)$', which is entailed both by (Pair Universalism)[25] and by (Pair Moderatism), and its negation, (Pair Nihilism)). By formulating an intermediate position between

---

[22] This is so at least as long as we are using classical predicate logic (where '$\neg \exists x \ x = \langle a, b \rangle$' by existential generalization entails '$\exists y \ \neg \exists x \ x = y$', which contradicts the reflexivity of identity). In a free logic where not all terms need to have existential import the use of the pair-forming operator need not stand in the way of denying that some objects form a pair. That is why Neil Tennant, who is also skeptical of too far-reaching ontological commitment, uses free logic in his account of pairing in Tennant (2009, section 3). We will shortly see a way to circumvent this departure from classical logic.

[23] Contrary to current conventions, I note all the universal quantifiers prefixing the formula, because I will later want to draw attention to them.

[24] The three labels are inspired by (van Inwagen, 1990, pp. 72ff.)

[25] We make the classical presupposition that the domain is not empty.

universalism and nihilism about pairs we show how to express the existential claim that is connected to the colloquial notion of pairing – some objects form pairs, others don't.

Given (Pair Identity), the usual pair-forming operator can be defined by $\langle a,b \rangle =_{\text{def}} \imath x\, \text{PAIRS}(x{:}a,b)$, and given (Pair Universalism), it will deliver a value for all arguments. Thus it becomes evident that, in effect, (Pair Identity) is a conditionalized variant of (Peano's Principle), and (Peano's Principle) the conjunction of (Pair Identity) and (Pair Universalism). By factoring (Peano's Principle) into these two parts, we can assess them separately. Clearly, (Pair Identity) is a conceptual truth – someone who does not endorse it fails to grasp the concept of pairing –, and it is plausible that this is the root of the widely shared conviction that (Peano's Principle) is analytic. But separated out as (Pair Universalism), the existentially committing remainder of (Peano's Principle) loses its claim to analyticity. Pace Anselm, the analysis of a concept alone should never entail that there is an object that falls under it. (By thus reconciling the intuition of analyticity with the conviction that existential claims cannot be analytic, we sharpen the Anselm Problem for pairs by abstraction.)

Of course, endorsing that (Pair Universalism) is not conceptually true does not commit us to rejecting it as false. In general, metaphysics may well go beyond conceptual analysis. So: Can we say something about the three options concerning the ontology of pairs? We find some prima facie support for each. (Pair Universalism) is implicit in the widely used technical notion of pair[26] and might thus be supported by an indispensability argument; (Pair Moderatism) is implicit in the colloquial notion of pair and thus likely to have common sense on its side; and (Pair Nihilism) is in harmony with an ideal of ontological austerity, making it attractive to nominalists. Besides what the other options have going for them, there is a historical fact that casts some doubt on (Pair Universalism): In marked contrast to geometrical shapes and natural numbers (although in parallel to sets), pairs despite their being so basic are abstract objects that have made a relatively recent entry on the scientific stage – only in the nineteenth century (Kanamori, 2003, p. 288).

One way to proceed in this dialectical situation would be to try and find the one true answer to the question 'When do objects a and b form a pair?'[27] This would be in clear analogy to Øystein Linnebo's question

---

[26] (Pair Universalism) corresponds to the "Rule of Totality" in Tennant (2009, section 4ff.).

[27] In the context of a comparison with the questions of Linnebo and van Inwagen, it would

'When do some things form a set?' and Peter van Inwagen's "special composition question" 'When do some objects compose a whole?' (Linnebo, 2010, pp. 144ff.; van Inwagen, 1990, pp. 21ff.).[28] But in the case of pairs, this route is not very promising. While in the case of sets, objective aspects can be pointed out that are necessary and sufficient for the forming of the objects in question (considerations of "size" for sets and being an organism for composition), this is intuitively not so in the case of pairs. When we consider following common sense in taking only some objects to be paired, it becomes evident that there is nothing intrinsic to the objects in question that explains their being paired – rather, the fact which objects are paired is always relative to some purpose.[29] The contrast to the formation of sets and of composite objects thus brings out that intuitively there is no fact of the matter with regard to which objects are paired; there is an element of conventionality that is particular to pairing. We will take this to motivate a desideratum of *ontological neutrality* for an account of pairs.[30]

4 Well-foundedness

Let us now draw attention to an aspect of the metaphysics of pairs that we have so far slid over: The question of well-foundedness. A relation R is *well-founded* (on a certain domain) if and only if there are no infinitely-descending R-chains (on that domain). In graph theoretical terms, there are two ways in which a relation can fail to be well-founded: either it contains a branch that goes 'down' without end, or a circular branch. For example, the relation of being smaller than is well-founded on the natural numbers, but it is not well-founded on the integers or on any interval of the reals, and neither is the relation of being later than on the dial of a clock.

---

be misleading to ask simply 'When do *some objects* form a pair?'. Even if we specified that the number of objects denoted by 'some objects' must be one or two, there would be an ambiguity with regard to their order. But in the presence of (Pair Moderatism), it may come about that some specific objects a_0 and b_0 form a pair, but the objects b_0 and a_0 do not.

[28] Asking when objects a and b form a pair does not mean to be partial to (Pair Moderatism), because the relevant aspect might be something shared by *all* objects or by *none*, justifying (Pair Universalism) or (Pair Nihilism), respectively.

[29] There are some limit cases where the relevant purpose arguably *is* intrinsic to the objects that are paired: when two lovers decide to form a couple. But given that we are interested in abstract objects here, such examples would take us too far afield.

[30] We should note that the ontological neutrality required here is part of an account of the *metaphysics* of pairs – in contrast to the ontological neutrality that we require (against Anselm) of all conceptual analysis. Cf. the distinction between "analytic" and "ontological intuitions" in Tennant (2009, sections 7 and 8).

Two Accounts of Pairs

Neil Tennant deserves credit for pointing out what is usually overlooked (Tennant, 2009, sections 8 and 9): Intuitively, we think of pairs as well-founded objects, in the following sense: The relation that is the transitive closure of the relation of being a component is well-founded, so that no pair can have itself as a component, and the relation of having as component always bottoms out in some objects that are not pairs. However, (Peano's Principle) does not ensure well-foundedness, and a fortiori the weaker principle of (Pair Identity) is also compatible with non-well-foundedness.[31] Observe that the standard substitutes for pairs are indeed well-founded – in the set theoretical construals due to the well-foundedness of the relation of elementhood and in the above-mentioned number theoretical construal due to the well-foundedness of the relation of being a divisor of. (This might explain the fact that well-foundedness despite its intuitiveness is usually not required explicitly.) So the question of well-foundedness constitutes another metaphysical problem for pairs by abstraction, beside the desirability of ontological neutrality.

5 Pairs by postulation

We will now develop an alternative to the neo-logicist construal of pairs. The basic idea is to use Kit Fine's procedural postulationism, which he has used so far to give accounts of numbers and of sets, apply it to the ontology of pairs, and compare the ensuing *pairs by postulation* to the pairs by abstraction of neo-logicism. We will also sketch technical applications within a postulationist alternative to the neo-logicist introduction of numbers that make essential use of features specific to pairs by postulation (in the next section).

The basic idea of procedural postulationism is to construe the introduction of objects by postulation *in an explicitly processual way*. Although the process does not occur in physical time, it shares important features. In particular, as the postulational introduction of objects expands the domain *from within* (Fine, 2005, pp. 102f.; Fine, 2012, pp. 15f.), the framework shares some structural features with *tensed time*. We should therefore embed the postulational account in a tense logical framework with a variable domain.[32]

---

[31] This is shown by the (almost) trivial model that has exactly one object in the domain, which satisfies the monadic predicate 'PAIRS$(x{:}x{,}x)$', so that when it is denoted 'a', the sentence '$a = \langle a,a \rangle$' comes out as true. Cf. Tennant (2009, section 9).

[32] In being explicit about tense, we deviate from Fine's own presentation, which is couched in terms of modal logic and more cautious with regard to the metaphysical picture. But this is

The domain being variable allows to model that objects exist only relative to some stages of the introduction process. With regard to pairs this means that we can indeed have ontological neutrality, in the sense that a scenario is coherent where (Pair Nihilism) is true relative to some initial stages, (Pair Moderatism) is true relative to many intermediary stages, and (Pair Universalism) is true relative to some late, limit stage. For this, the quantifiers in these existential claims must be understood as relativized to the domain of the respective stage. In this regard, they differ crucially from the quantifiers in the conditionalized criterion of (Pair Identity). As we construe this principle as a conceptual truth, we should not understand its quantifiers as relative to *any* particular domain of objects. Our logical notation will need some way to make this explicit; here we will simply prefix the formulas that are understood as conceptually necessary universal claims with a modal operator '$\Box_C$'.[33]

Fine's procedural postulationism is a novel approach in the philosophy of mathematics which he so far has applied to numbers and sets. Crucially, a postulate for Fine is not static but dynamic – it specifies a procedure for extending the domain (Fine, 2005, pp. 91f.). The basic form of a postulate is '$!x.C(x)$', for 'Make an object that is C!'. Postulates can be composed (read '$\alpha;\beta$' as 'Do α, then do β!'), conditionalized (read '$p \to \alpha$' as 'If p, then do α!'), executed simultaneously for everything in the domain (read, e.g., '$\forall x \, !y.R(x,y)$' as 'for every x, make an object that bears R to x!') and, most importantly, iterated (read 'α^*' as 'Repeat doing α until finished!'). E.g., to introduce numbers Fine uses the postulates (Fine, 2005, pp. 92f.):[34]

ZERO: $!x.\text{NUM}(x)$

SUCCESSOR: $\forall x(\text{NUM}(x) \to !y.(\text{NUM}(y) \land \text{SUCC}(y,x)))$

NUMBER: ZERO; SUCCESSOR*

Fine requires that the postulation of numbers must be guided by certain "constraints" that fix the understanding of the notions involved (expressed by 'NUM' and 'SUCC'). He lays down, e.g., that only numbers stand in the

---

likely no more than a difference of emphasis.

[33] Ultimately, I would prefer a different approach, where an additional pair of *conceptual quantifiers* is used to formalize conceptual statements, distinguishing them more clearly from statements that are meant to be about objects and formalized with the standard *objectual quantifiers*. But as we are not giving a full formal treatment here, an additional operator will likely be less confusing.

[34] This differs from our envisioned postulational variant of the *neo-logicist* introduction of numbers. In Fine's words: "This is arithmetic à la Dedekind. It is also possible to provide a postulational treatment of arithmetic à la Frege." (Fine, 2005, p. 92)

Two Accounts of Pairs

successor relation and each one has at most one successor (Fine, 2005, p. 107). The details do not matter here, but we should point out that it is these postulational constraints that we will construe as conceptual truths.

Although we are giving only the barest sketch of a logical framework here, we should point out some of its basic features. The main aims are to distinguish conceptual and objectual truth and to represent the process of the postulation of objects from within. As mentioned, we achieve the former here with the operator '$\Box_C$'. For the latter, we use tense operators that share some feature with those of a metric tense logic. For example, for every postulate α there is a future tense operator $\ulcorner[\vec{\alpha}]\urcorner$, read: $\ulcorner$directly after executing α it will be the case that$\urcorner$, and a past tense operator $\ulcorner[\overleftarrow{\alpha}]\urcorner$, read: $\ulcorner$directly before executing α it was the case that$\urcorner$. Under certain conditions,[35] we thus gain the operators of a generalized metric tense logic, with each postulate inducing a discrete metric.[36] This allows us to express, e.g., that postulating the first number will extend the domain: $\neg \exists x\, \text{NUM}(x) \to [!z.\overrightarrow{\text{NUM}(z)}]\, \exists x\, \text{NUM}(x)$

Because of the interaction between postulates and operators, the tense logic of postulation is a species of dynamic logic (cf., e.g., Harel, Kozen, & Tiuryn, 2000). As we aim to represent the expansion of the domain from within, the internal stance of the object language is ultimately privileged. But of course something should also be said about the external stance of model theory: The stages correspond to the moments in a branching-time structure, and there is a relation of accessibility for each postulate (alternatively, we can think of the postulates as labels on transitions between stages). If we think of objects by postulation as persistent, the domain will *grow* from stage to stage along the relation of accessibility.

Now we can proceed to the postulational introduction of pairs. It will be guided by a single postulational constraint, the conditionalized criterion of (**Pair Identity$_C$**) as a conceptual truth:

$\Box_C \forall x \forall y \forall a \forall b \forall c \forall d\, (\text{PAIRS}(x{:}a,b) \land \text{PAIRS}(y{:}c,d) \to (x{=}y \leftrightarrow a{=}c \land b{=}d))$

Note that adding the modal operator of conceptual necessity changes the meaning of the quantifiers – the principle is not made true by what is the case with objects in the current domain, rather it is true for purely conceptual reasons and thus *restricts* what can be the case with the current domain.

---

[35] More specifically, the condition is that there is exactly one way of executing the postulate in question.

[36] Some care is needed with regard to past tense operators formed from postulates that contain the iteration asterisk '*'. I would like to thank Kit Fine for pointing this out to me.

Now, for objects a and b that are in the current domain, we can introduce a pair by postulating: $!z.\,\text{PAIRS}(z{:}a,b)$. Then '$\exists x\,\text{PAIRS}(z{:}a,b)$' will be true relative to the subsequent stage (and as pairs are persistent, it will exist relative to *all* subsequent stages). With the help of the postulational tense operators we can express, e.g., that postulating the pair of certain objects a and b will extend the domain:

$$\neg\exists x\,\text{PAIRS}(x{:}a,b) \to \overrightarrow{[!z.\text{PAIRS}(z{:}a,b)]}\,\exists x\,\text{PAIRS}(x{:}a,b)$$

Observe that we are thus able to speak about pairs that currently do not exist yet, which will be relevant for the technical applications sketched in the next section.

The postulational account of pairs solves the three problems we saw to beset the neo-logicist account and satisfies the two desiderata we developed from metaphysical intuitions about pairs. Firstly, the postulational account solves the Caesar Problem because Caesar like any other concrete object exists relative to *all* postulational stages or relative to *none*, but a pair like any other object by postulation exists relative to *only some* stages. Secondly, it solves the Bad Company Problem because it introduces pairs in a well-founded process, which is well-known to ensure the consistency of all abstraction principles (cf. Leitgeb, 2017; Linnebo, 2009; Studd, 2015). Note that in contrast to similar accounts, the postulational account gives a *rationale* for the well-foundedness of the process of the introduction of objects. Thirdly, it solves the Anselm Problem *and* delivers the desired ontological neutrality. The only part of the account with the status of a conceptual truth is the conditionalized criterion of pair identity, which does not entail anything about the existence of pairs. Otherwise the account distributes all conceivable existential claims about pairs – from nihilism via all conceivable variants of moderatism to universalism – over separate stages. (Thus it solves the Cardinality Problem, too.) We can understand this distribution as a specific way of spelling out the aspect of the technical notion of pair that *any* objects a and b can be paired: Rather than claim the existence of all conceivable pairs, we are taking the modal character ('*can* be') of this aspect seriously,[37] and are even able to reconcile it with the corresponding aspect of the colloquial notion, that not all objects actually *do* form a pair. Finally, the postulational account ensures the well-foundedness of pairs in a natural way by its procedural tensed structure: As any pair starts to exist later than its components (which must be denoted in the postulate), no pair can be one

---

[37] Our claim that any objects a and b *can* form a pair is in exact parallel to Øystein Linnebo's claim that any plurality *can* form a set (e.g., Linnebo, 2009, p. 157).

of its own components, or a component of one of its components, and so on. And as the postulational process has to start at some 'nihilist' stage relative to which only non-pairs exist, there can be no structure where it is pairs 'all the way down'.[38]

6 Two procedural analyses with pairs

In addition to the philosophical advantages, the postulational account of pairs is also interesting for technical reasons. We will sketch two technical applications that are relevant to the plural logic variant of neo-logicism, or rather, to a postulational alternative of it: procedural analyses of the notions of equinumerosity and of the ancestral. Both applications employ a feature specific to the postulational account – that the existence of pairs is relative to stages –, and could not be given in the same way on the basis of the other accounts of pairs.

First, we give a procedural analysis of *equinumerosity*, which we have seen to be lacking in the plural logic variant of neo-logicism. We use '$P_l(x,yy)$' to abbreviate '$\exists z\, \exists y \prec yy\, \text{PAIRS}(z{:}x,y)$', for '$x$ is paired with one of the yy', and we use '$P_r(xx,y)$' to abbreviate '$\exists z\, \exists x \prec xx\, \text{PAIRS}(z{:}x,y)$', for 'one of the xx is paired with y'. Now let us define a postulate to pair off one of the xx with one of the yy in the following way:

PAIR-OFF: $!z.(\exists x \prec xx\, \exists y \prec yy\, (\neg P_l(x,yy) \wedge \neg P_r(xx,y) \wedge \text{PAIRS}(z{:}x,y)))$

Then equinumerosity can be defined for every initial stage:[39]

(EQ) $xx \approx_{\text{pl.}} yy \leftrightarrow_{\text{def}} [\overrightarrow{\text{PAIR-OFF}^*}](\forall x \prec xx\, P_l(x,yy) \wedge \forall y \prec yy\, P_r(xx,y))$

Read: 'The xx and the yy are equinumerous if and only if after iterated pairing off of one of the xx with one of the yy none of the xx are left unpaired and none of the yy are left unpaired.' In effect, this is a procedural analysis of a bijection: The iterated postulate guarantees that the relation is left-unique and right-unique, and the statement in the scope of the tense operator defined from the iterated postulate guarantees that it is left-total and right-total. In contrast to the standard analysis of equinumerosity the present

---

[38] To my knowledge, there is one other account of pairs beside the present one that takes a measure to ensure well-foundedness: Tennant's proof theoretical definition of a pairing function. But Tennant guarantees well-foundedness in a rather complicated and arguably unnatural way (Tennant, 2009, section 10).

[39] Here, an initial stage is one where no pairs have yet been postulated. For the condition to give the desired results it would be enough to require that with respect to the stage in question none of the xx is paired off with any of the yy and vice versa.

account does not need second-order quantification. Also, it is ontologically more parsimonious, because it is formulated from the viewpoint of a stage where the pairs used in the procedure do not yet exist. We might even consider construing it in an entirely hypothetical way: '*Were* we to postulate pairs, then the following would hold: ...'.

The *ancestral* of a relation – its transitive closure – is another notion important for the logicist project. It can also be defined with recourse to pairing. We use 'P(x,y)' to abbreviate '$\exists z$ PAIRS$(z{:}x,y)$', read: 'x is paired with y'. Then let us use the following postulates:

GRAPH$_R$: $\forall x \forall y (R(x,y) \rightarrow\ !z.\text{PAIRS}(z{:}x,y))$

CLOSE: $\forall a \forall b \forall c\ (a{\neq}b \wedge b{\neq}c \wedge P(a,b) \wedge P(b,c) \rightarrow\ !z.\text{PAIRS}(z{:}a,c))$

CLOSURE$_R$: GRAPH$_R$; CLOSE*

As pairs are among the postulated objects, there are no pairs in the domain of an initial stage. Therefore we can define the ancestral of a relation R at any initial stage like this:

(**Ancestral**) $R^*(x,y) \leftrightarrow_{\text{def}} [\overrightarrow{\text{CLOSURE}_R}]\ \exists z\ \text{PAIRS}(z{:}x,y)$

Read: 'y follows x in the R-chain if and only if after making the graph of R and iterating closing it off, there will be a pair of x and y.' In contrast to Frege's analysis of the ancestral, the present account again does not need second-order quantification and is ontologically parsimonious. I take it that it is also closer to what we intuitively mean when we talk about the transitive closure of a relation.

7 Conclusion

We have compared two accounts of pairs construed as primitive, pairs by abstraction and pairs by postulation, and found ample reason to prefer the second account. It evades the problems that beset the neo-logicist construal, gets the metaphysics of pairs right, and promises to be valuable for a postulationist variant of (or alternative to) the neo-logicist construal of numbers. The comparison is also meant as a miniature case study of both neo-logicism and its postulationist variant (or alternative).

References

Armstrong, D. M. (1986). In defence of structural universals. *Australasian*

Journal of Philosophy, 64, 85–88.

Benacerraf, P. (1983). What numbers could not be. In P. Benacerraf & H. Putnam (Eds.), *Philosophy of Mathematics* (pp. 272–294). Cambridge: Cambridge University Press.

Boccuni, F. (2010). Plural Grundgesetze. *Studia Logica, 96(2)*, 315–330.

Boccuni, F. (2013). Plural logicism. *Erkenntnis, 78(5)*, 1051–1067.

Boolos, G. (1998). *Logic, Logic, and Logic*. Cambridge (Massachusetts): Harvard University Press.

Field, H. (1984). Is mathematical knowledge just logical knowledge? *Philosophical Review, 93*, 509–552.

Fine, K. (2005). Our knowledge of mathematical objects. In T. Z. Gendler & J. Hawthorne (Eds.), *Oxford Studies in Epistemology* (pp. 89–109). Oxford: Clarendon Press.

Fine, K. (2012). Mathematics: Discovery or invention? *Think, 11(32)*, 11–27.

Forrest, P. (1986). Neither magic nor mereology: A reply to Lewis. *Australasian Journal of Philosophy, 64*, 89–91.

Frege, G. (1988). *Grundlagen der Arithmetik*. Hamburg: Meiner. (Edited by Christian Thiel)

Frege, G. (2009). *Grundgesetze der Arithmetik. Begriffsschriftlich abgeleitet. Band I und II*. Paderborn: Mentis. Transcribed into modern notation and edited by Thomas Müller, Bernhard Schröder and Rainer Stuhlmann-Laeisz.

Hale, B. (2001). Reals by abstraction. In B. Hale & C. Wright (Eds.), *The Reason's Proper Study. Essays Towards a Neo-Fregean Philosophy of Mathematics* (pp. 399–420). Oxford: Oxford University Press.

Hale, B., & Wright, C. (2001). *The Reason's Proper Study. Essays Towards a Neo-Fregean Philosophy of Mathematics*. Oxford: Oxford University Press.

Hamilton, W. R. (1837). Theory of conjugate functions, or algebraic couples. With a preliminary and elementary essay on algebra as the science of pure time. *Transactions of the Royal Irish Academy, 17*, 293–422.

Harel, D., Kozen, D., & Tiuryn, J. (2000). *Dynamic Logic*. Cambridge (Massachusetts) and London: MIT Press.

Heck, R. G. J. (1995). Definition by induction in Frege's *Grundgesetze der Arithmetik*. In W. Demopolous (Ed.), *Frege's Philosophy of Mathematics* (pp. 1931–1967). Cambridge: Harvard University Press.

Heck, R. G. J. (2011). *Frege's Theorem*. Oxford: Oxford University Press.

Heck, R. G. J. (2012). *Reading Frege's Grundgesetze*. Oxford: Clarendon.
Hewitt, S. (Manuscript). *Flat Logicism*.
van Inwagen, P. (1990). *Material Beings*. Ithaca / London: Cornell University Press.
Kanamori, A. (2003). The empty set, the singleton, and the ordered pair. *The Bulletin of Symbolic Logic, 9.3*, 273–298.
Leitgeb, H. (2017). Abstraction grounded: A note on abstraction and truth. In P. Ebert & M. Rossberg (Eds.), *Abstractionism. Essays in Philosophy of Mathematics* (pp. 269–282). Oxford: Oxford University Press.
Lewis, D. (1991). *Parts of Classes*. Oxford: Blackwell.
Linnebo, Ø. (2009). Bad company tamed. *Synthese, 170.3*, 371–391.
Linnebo, Ø. (2010). Pluralities and sets. *Journal of Philosophy, 107.3*, 144–164.
Linnebo, Ø. (2014). Plural quantification. In E. N. Zalta (Ed.), *The Stanford Encyclopedia of Philosophy (Fall 2014 Edition)*. (Available online at https://plato.stanford.edu/archives/fall2014/entries/plural-quant/)
Pleitz, M. (2010). Curves in Gödel-space. *Studia Logica, 96*, 193–218.
Potter, M. (2000). *Reason's Nearest Kin*. Oxford: Oxford University Press.
Quine, W. V. O. (1960). *Word and Object*. Cambridge: MIT Press.
Shapiro, S. (1997). *Philosophy of Mathematics. Structure and Ontology*. Oxford: Oxford University Press.
Shapiro, S. (2000). Frege meets Dedekind: A neo-logicist treatment of real numbers. *Notre Dame Journal of Formal Logic, 41.4*, 335–364.
Shapiro, S., & Weir, A. (2007). "Neo-logicist" logic is not epistemically innocent. In R. T. Cook (Ed.), *The Arché Papers on the Mathematics of Abstraction* (pp. 119–146). Dordrecht: Springer.
Sider, T. (1996). Naturalness and arbitrariness. *Philosophical Studies, 81*, 283–301.
Simons, P. (2005). Against set theory. In J. C. Marek & M. E. Reicher (Eds.), *Experience and Analysis. Proceedings of the 2004 Wittgenstein Symposium* (pp. 143–152). Vienna: öbv&hpt.
Studd, J. (2015). Abstraction reconceived. *British Journal for the Philosophy of Science, 67.2*, 579–615.
Tennant, N. (2009). Natural logicism via the logic of orderly pairing. In S. Lindström, E. Palmgren, K. Segerberg, & V. Stoltenberg-Hansen (Eds.), *Logicism, Intuitionism, Formalism: What has become of them?* (pp. 91–125). Dordrecht: Springer.
Tennant, N. (2014). Logicism and neologicism. In E. N. Zalta

(Ed.), *The Stanford Encyclopedia of Philosophy (Fall 2014 Edition)*. (URL: http://plato.stanford.edu/archives/fall2014/entries/logicism/)

Wright, C. (1983). *Frege's Conception of Numbers as Objects*. Aberdeen: Aberdeen University Press.

Zalta, E. N. (2016). Frege's theorem and foundations for arithmetic. In E. N. Zalta (Ed.), *The Stanford Encyclopedia of Philosophy (Winter 2016 Edition)*. (URL: https://plato.stanford.edu/archives/win2016/entries/frege-theorem/)

Martin Pleitz
Department of Philosophy, Münster University
Germany
E-mail: martinpleitz@web.de

A General Framework for Logics of Questions

VÍT PUNČOCHÁŘ[1]

Abstract: This paper provides an overview of basic inquisitive semantics and its generalization proposed in (Punčochář, submitted). It is shown that the generalization allows to model questions over any of a large class of non-classical logics and so avoids paradoxes of material implication and irrelevance in the logic of questions. Moreover, it is advocated that the general framework does not lose any of the characteristic features of basic inquisitive semantics that are needed for modeling of questions.

Keywords: Inquisitive semantics, Logic of questions, Substructural logics

1 Introduction

The goal of this paper is (a) to present inquisitive semantics and logic (see, e.g., Ciardelli, 2016; Ciardelli, Groenendijk, & Roelofsen, 2013; Ciardelli & Roelofsen, 2011) as an adequate logical model of questions, (b) to explain how it is possible to generalize the basic framework of inquisitive semantics without a loss of its essential features that make it suitable for modeling of questions. The paper relies heavily on (Punčochář, submitted) where all the technical details are worked out. Here we intend to provide a comprehensible explanation of the motivation that is behind the basic and generalized inquisitive semantics, and of the way in which the latter generalizes the former. For this reason we will present all the results without proofs. Most of the results presented in Section 2 can be found with proofs in (Ciardelli, 2016). The proofs of the results presented in Section 3 can be found in (or can be easily reconstructed from) (Punčochář, submitted).

---

[1] The work on this paper was supported by grant no. 16-07954J of the Czech Science Foundation.

2 Basic inquisitive semantics

Inquisitive semantics is a framework that has been recently developed with the aim of providing a uniform formal representation of statements and questions that would allow us to model adequately logical relations among these kinds of sentences. The main idea behind this framework can be seen as an extension or generalization of a standard view according to which the meaning of a sentence is identified with the truth conditions of the sentence. A classical philosophical expression of this view was formulated by Ludwig Wittgenstein in his *Tractatus Logico-Philosophicus*:

> "To understand a proposition means to know what is the case if it is true." (Wittgenstein, 1922, paragraph 4.024)

So, according to this criterion, one understands a sentence A if one is able to classify possible states of affairs (possible worlds) into those in which A is true and those in which A is false. The standard formal representation of a sentential meaning (i.e., a proposition) as a set of possible worlds (those worlds in which the sentence is true) stems from this picture. However, this picture is obviously applicable only to declarative sentences and not to questions, for questions do not have truth conditions and truth values. Inquisitive semantics introduces a more general notion of sentential meaning that is applicable to declarative sentences as well as to questions.

Definition 1 *A possible world is any function that assigns to every atomic formula a truth value (either TRUE, or FALSE). An information state (or just state) is any set of possible worlds. A proposition is any nonempty downward closed set of information states.*[2]

So, in inquisitive semantics a proposition P is defined as any downward closed set of sets of possible worlds. The set $\bigcup P$, i.e. the union of P, represents the *informative content* of the proposition which is just a proposition in the traditional sense (a set of possible worlds). The informative content locates to some degree the actual world in the space of all possible worlds. But the proposition P has also an *inquisitive content* that can be informally understood as a request to locate the actual world more precisely within the informative content. The members of P are intuitively viewed as information states that contain enough information to settle the request. If

---

[2] A set of information states P is downward closed, if it is closed under substates, that is if $a \in P$ and $b \subseteq a$ implies $b \in P$.

Logics of Questions

we have two states a and b such that a is a proper subset of b, then the state a excludes more possibilities than b, and so a can be regarded as stronger or containing more information than b. This explains why propositions are defined as *downward closed* sets of information states, since if an information state counts as containing enough information to settle the request to locate the actual world more precisely, then every stronger state must also count as containing enough information to settle the request.

Definition 2 *Let P be a proposition. We say that P is declarative if it contains its own informative content, i.e. $\bigcup P \in P$. If P is not declarative, we say that it is inquisitive.*

Informally, the inquisitive content of a sentence can be understood as the issue that is raised by uttering the sentence. A sentence expresses a declarative proposition if the information it provides resolves the issue it raises. The difference between declarative and inquisitive propositions is illustrated with Figure 1, where (a) represents a declarative proposition and (b) an inquisitive proposition. Let us restrict ourself only to two atomic formulas, say p and q. Then for example FT represents the world that assigns *FALSE* to p and *TRUE* to q. For the sake of simplicity, only the maximal states of the two propositions are depicted, so (a) represents the declarative proposition consisting of all subsets of $\{TT, TF, FT\}$ and (b) the inquisitive proposition $\{\{TT, FT\}, \{TT, TF\}, \{TT\}, \{TF\}, \{FT\}, \emptyset\}$.

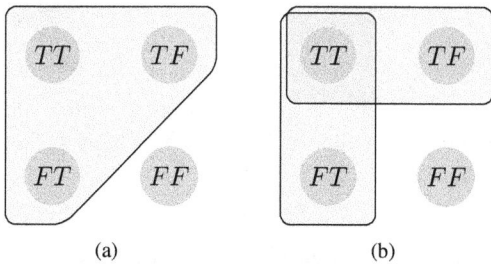

Figure 1: Declarative vs. inquisitive proposition

Now let us introduce the standard formal propositional language L which consists of formulas that are built from atomic formulas using negation ($\neg$), implication ($\rightarrow$), conjunction ($\wedge$), and disjunction ($\vee$). If we extend this language with one additional binary connective ? that is called inquisitive disjunction,[3] we obtain the language $L^?$. The formulas from L and $L^?$ will be called L-formulas and $L^?$-formulas, respectively.

The formula $p?q$ represents the question *whether p or q*. A polar question $[?]p$, read as *whether p*, is equivalent to *whether p or $\neg p$*. In general, one can define $[?]\varphi =_{def} \varphi?\neg\varphi$. Formal inquisitive semantics is based on the so called support relation between information states and $L^?$-formulas that is defined by the following Kripke-style semantic clauses:

C1 $a \vDash p$ iff p is true in every world of a.

C2 $a \vDash \neg\varphi$ iff for any nonempty $b \subseteq a$, $b \nvDash \varphi$.

C3 $a \vDash \varphi \to \psi$ iff for any $b \subseteq a$, if $b \vDash \varphi$, then $b \vDash \psi$.

C4 $a \vDash \varphi \wedge \psi$ iff $a \vDash \varphi$ and $a \vDash \psi$.

C5 $a \vDash \varphi \vee \psi$ iff there are b, c such that $b \vDash \varphi$, $c \vDash \psi$ and $a = b \cup c$.

C6 $a \vDash \varphi?\psi$ iff $a \vDash \varphi$ or $a \vDash \psi$.

"$a \vDash \varphi$" is read as "the state a supports φ". The proposition expressed by φ, denoted as $||\varphi||$, can be defined as the set of those states that support φ. This terminology is justified by the following fact.

Fact 1 *Every $L^?$-formula expresses a proposition in the sense of Def. 1. This means that it holds for any $L^?$-formula φ that*

(a) $\emptyset \vDash \varphi$,

(b) *if $a \vDash \varphi$ and $b \subseteq a$, then $b \vDash \varphi$.*

The empty set $\emptyset$ can be regarded as an inconsistent state that (leaves no open possibility and) supports everything. The next fact states that all L-formulas "behave classically".

Fact 2 *For any L-formula α and any information state a, $a \vDash \alpha$ iff α is true (according to classical logic) in every world of a.*

Let $|\alpha|$ be the set of possible worlds in which α is true according to classical logic. Then the previous fact says that for any L-formula α, $a \in ||\alpha||$ iff $a \subseteq |\alpha|$. The following fact is a straightforward consequence of the previous one.

---

[3]We decided to use a nonstandard notation here. The symbol ? is usually not used for inquisitive disjunction.

Logics of Questions

Fact 3 *Every L-formula expresses a declarative proposition.*

So, one can view the language L as a language of declarative sentences. It is easy to check that, for instance, if p, q are different atomic formulas, then $p?q$ expresses a proposition that is inquisitive. Indeed,

$$\bigcup ||p?q|| = |p| \cup |q| \notin ||p|| \cup ||q|| = ||p?q||,$$

which shows that the proposition $||p?q||$ is not declarative. If there are only two atomic formulas, p and q, the proposition $||p?q||$ corresponds to Figure 1-(b), while Figure 1-(a) represents $||p \vee q||$.

The support relation can be viewed as a unifying notion suitable for both declarative sentences and questions. If α represents a declarative sentence, the meaning of $a \models \alpha$ is that a contains enough information to establish that α is true, or briefly that a implies α. If φ represents a question, the meaning of $a \models \varphi$ is that a contains information that is sufficient to resolve φ. Note that if α, β represent declarative sentences (express declarative propositions), the semantic clause for inquisitive disjunction amounts to the following: an information state resolves the question *whether α or β* iff it implies α or β (which differs from the claim that it implies $\alpha \vee \beta$). In this interpretation α and β represent direct answers to the question $\alpha?\beta$.

In inquisitive semantics the semantic notions of logical validity, logical consequence, and logical equivalence are defined in terms of the support relation. Logical validity is defined as universal support, logical consequence as preservation of support, and logical equivalence as support by the same states. These notions determine the basic inquisitive logic ($InqB$).

Definition 3 *Let $\Delta \cup \{\varphi, \psi\}$ be a set of $L^?$-formulas. φ is logically valid in InqB ($\models_{InqB} \varphi$) if it is supported by every information state. φ is a consequence of Δ in InqB ($\Delta \models_{InqB} \varphi$) if every information state that supports every formula from Δ supports also φ. φ and ψ are logically equivalent in InqB ($\varphi \equiv_{InqB} \psi$) if they are supported by the same states, i.e., if $\varphi \models_{InqB} \psi$ and $\psi \models_{InqB} \varphi$.*

The logical notions introduced in Definition 3 generalize corresponding notions of classical logic (CL). Most importantly, the consequence relation of $InqB$ generalizes the consequence relation of CL in the following sense.

Fact 4 *If $\Delta \cup \{\alpha\}$ is a set of L-formulas, $\Delta \models_{InqB} \alpha$ iff $\Delta \models_{CL} \alpha$.*

Besides the standard cases, in which the premises, as well as the conclusion, represent declarative sentences (i.e., express declarative propositions), one can also consider arguments consisting purely of questions, or hybrid cases in which premises contain a question or the conclusion is a question. It was demonstrated by Ciardelli (2016) that all these cases have a reasonable meaning. Let Δ be a set of L-formulas that represent a description of a context in which an argument is formulated. If α is a statement (expresses a declarative proposition) and φ a question (expresses an inquisitive proposition), then $\Delta, \alpha \vDash_{InqB} \varphi$ is interpreted as follows: in the context Δ, the statement α resolves the question φ. Moreover, if ψ is a question, then $\Delta, \psi \vDash_{InqB} \varphi$ might be interpreted as: in the context Δ, every information that resolves the question ψ, resolves also the question φ. And $\Delta, \psi \vDash_{InqB} \alpha$ is interpreted as: in the context Δ, the question ψ presupposes the statement α.

Let us consider three simple examples. (a) It holds that $p?q \vDash_{InqB} p \vee q$. This says that the question *whether p or q* presupposes the statement *p or q*. (b) It holds that $p \to q, p \vDash_{InqB} [?]q$. This might be interpreted as follows: in a context, in which *p implies q*, the statement *p* resolves the question *whether q*. (c) $(p \wedge q) \to r, (r \wedge p) \to q, [?]r \vDash_{InqB} p \to [?]q$. This might be interpreted in this way: in a context, in which *p and q implies r*, and *r and p implies q*, every information that resolves the question *whether r*, resolves also the conditional question *whether q if p*. (The reason is that in such a context, if r is true then $p \to q$ is true, and if r is false then $p \to \neg q$ is true. Note that *if p then q* and *if p then not q* are the two possible direct answers to the conditional question *whether q if p*.)

A crucial feature of the logic $InqB$ is that every question is logically equivalent to the inquisitive disjunction of the direct answers to the question. More precisely, for any $L^?$-formula φ, we define a finite set of L-formulas $\mathcal{R}(\varphi)$ by the following recursive equations:

- $\mathcal{R}(p) = \{p\}$,
- $\mathcal{R}(\neg \varphi) = \{\bigwedge_{\alpha \in \mathcal{R}(\varphi)} \neg \alpha\}$,
- $\mathcal{R}(\varphi \to \psi) = \{\bigwedge_{\alpha \in \mathcal{R}(\varphi)} \alpha \to f(\alpha); f : \mathcal{R}(\varphi) \to \mathcal{R}(\psi)\}$,
- $\mathcal{R}(\varphi \wedge \psi) = \{\alpha \wedge \beta; \alpha \in \mathcal{R}(\varphi), \beta \in \mathcal{R}(\psi)\}$,
- $\mathcal{R}(\varphi \vee \psi) = \{\alpha \vee \beta; \alpha \in \mathcal{R}(\varphi), \beta \in \mathcal{R}(\psi)\}$,
- $\mathcal{R}(\varphi ? \psi) = \mathcal{R}(\varphi) \cup \mathcal{R}(\psi)$.

Notice that for any L-formula α, $\mathcal{R}(\alpha) = \{\alpha\}$. The equation for implication says that for any function f that assigns elements of $\mathcal{R}(\psi)$ to the elements of $\mathcal{R}(\varphi)$, the set $\mathcal{R}(\varphi \to \psi)$ contains the conjunction $\bigwedge_{\alpha \in \mathcal{R}(\varphi)} \alpha \to f(\alpha)$. The resolutions of φ represent direct answers to φ. For example, $\mathcal{R}(p?q) = \{p, q\}$, and $\mathcal{R}(p \to [?]q) = \{p \to q, p \to \neg q\}$.

Fact 5 *If $\mathcal{R}(\varphi) = \{\alpha_1, \ldots, \alpha_n\}$, then $\varphi \equiv_{InqB} \alpha_1? \ldots ?\alpha_n$.*

As a consequence of this fact, we obtain the following feature of inquisitive semantics.

Fact 6 *Every proposition expressible by an $L^?$-formula is identical with the union of a finite set of declarative propositions.*

The notion of a resolution allows for a specific reduction of basic inquisitive logic of questions to classical propositional logic that plays the role of a background logic of declarative sentences. The reduction of the logical validity of $InqB$ to the logical validity of CL has the following form.

Fact 7 $\vDash_{InqB} \varphi$ *iff there is $\alpha \in \mathcal{R}(\varphi)$ such that $\vDash_{CL} \alpha$.*

The reduction can be stated also for the consequence relation and logical equivalence.

Fact 8 $\varphi_1, \ldots, \varphi_n \vDash_{InqB} \psi$ *iff for every $\alpha_1 \in \mathcal{R}(\varphi_1), \ldots, \alpha_n \in \mathcal{R}(\varphi_n)$ there is $\beta \in \mathcal{R}(\psi)$ such that $\alpha_1, \ldots, \alpha_n \vDash_{CL} \beta$.*

Notice that this result corresponds to the informal interpretation of the generalized consequence relation. If a given argument involves only questions, Fact 8 corresponds to saying that the argument is logically valid iff any set of statements that resolve the questions in premises resolves also the question in conclusion. As a consequence, two questions are regarded as logically equivalent iff any statement that resolves one of them resolves also the other.

Fact 9 $\varphi \equiv_{InqB} \psi$ *iff for every $\alpha \in \mathcal{R}(\varphi)$ there is $\beta \in \mathcal{R}(\psi)$ such that $\alpha \vDash_{CL} \beta$, and for every $\beta \in \mathcal{R}(\psi)$ there is $\alpha \in \mathcal{R}(\varphi)$ such that $\beta \vDash_{CL} \alpha$.*

In the next section we will present a general semantics that allows for modeling of questions in the style of inquisitive semantics while classical logic, as the background logic of declarative sentences, can be replaced by any of a large class of non-classical logics. The idea to replace classical logic, as a logic of declarative sentences, by an alternative logical system is well-motivated. If the background logic of declarative sentences is classical, the system is not immune to the well-known paradoxes of material implication and irrelevance. This has unwanted consequences even for arguments involving questions as is illustrated with the following simple argument forms that are valid in $InqB$:

AF1 $\neg(q \to p) \vDash_{InqB} [?]q$, *Example:* The statement *it is not the case that if John is older than Peter, then Peter is older than John* resolves the question *whether John is older than Peter*.

AF2 $p \land \neg p \vDash_{InqB} [?]q$, *Example:* The statement *Peter is smart and he is not smart* resolves the question *whether it will be raining tomorrow*.

AF3 $[?]p \vDash_{InqB} q \lor \neg q$. *Example:* The question *whether we will win the match* presupposes the statement *god exists or god does not exist*.

It is clear that the unintuitive evaluation of these arguments is not the result of the way in which questions are modelled in inquisitive semantics but rather of the classical logic as the background logic of declarative sentences.

Before we introduce the general framework, it will be useful to make the following observation. The basic framework for inquisitive semantics is usually introduced via the semantic clauses C1-C6.[4] For the generalization of inquisitive semantics that we are going to introduce in the next section it will be useful to reformulate the semantic clauses in the following way:

C1* $a \Vdash p$ iff p is true in every world of a,

C2* $a \Vdash \neg\varphi$ iff for any b, if $a \cap b \neq \emptyset$, then $b \nVdash \varphi$,

C3* $a \Vdash \varphi \to \psi$ iff for any b, if $b \Vdash \varphi$, then $a \cap b \Vdash \psi$,

C4* $a \Vdash \varphi \land \psi$ iff $a \Vdash \varphi$ and $a \Vdash \psi$,

C5* $a \Vdash \varphi \lor \psi$ iff there are b, c such that $b \Vdash \varphi$, $c \Vdash \psi$ and $a \subseteq b \cup c$,

C6* $a \Vdash \varphi?\psi$ iff $a \Vdash \varphi$ or $a \Vdash \psi$.

---

[4] This is not exactly true. The semantic clause $C5$ for disjunction $\lor$ is often omitted and disjunction is sometimes introduced in a non-equivalent way by $\varphi \lor \psi =_{def} \neg(\neg\varphi \land \neg\psi)$, as for example in (Ciardelli, 2016, p. 47). The clause $C5$ is standardly used in the related framework of propositional dependence logic (see Yang, 2014). This semantic clause for disjunction was systematically studied also in (Punčochář, 2016a). Negation is often (equivalently) defined in basic inquisitive semantics as follows: $\neg\varphi =_{def} \varphi \to \bot$, where $\bot$ is regarded as a primitive symbol of the language with the semantic clause: $a \vDash \bot$ iff $a = \emptyset$.

Logics of Questions

The following fact can be easily proved by induction on the complexity of φ. The main feature of the semantics that is used in the proof is that support is downward persistent.

Fact 10 *The semantic clauses C1-C6 are equivalent to the semantic clauses C1*-C6* in the following sense: for every $L^?$-formula φ and every state a, it holds that $a \vDash \varphi$ iff $a \Vdash \varphi$.*

3 Generalized inquisitive semantics

In this section we will present a generalized inquisitive semantics that has been developed in (Punčochář, submitted). It generalizes the framework introduced in (Punčochář, 2016b). The non-inquisitive part can be viewed as a modified and extended version of the framework developed by Došen (1989). It can be compared also to (Wansing, 1993).

The semantic structures are called *informational models*. An informational model is any structure of this type:

$$\mathcal{M} = \langle S, +, \cdot, 0, 1, C, V \rangle$$

that satisfies several conditions specified below. S is an arbitrary nonempty set, the elements of S are called *information states* or just *states* (let us stress that now information states are not defined as sets of possible worlds but are regarded as primitive entities); $+$ and $\cdot$ are binary operations on S, called respectively *addition* and *fusion* of states; 0 and 1 are two distinguished elements of S, called respectively the *trivially inconsistent state* and the *logical state*; C is a binary relation on S, called a *compatibility relation*; and V is a *valuation*.

Moreover, the following conditions are required to be satisfied: $\langle S, + \rangle$ is a join-semilattice, i.e. addition is idempotent, commutative, and associative. As a consequence, addition determines a partial order on S defined in the standard way: $a \leq b$ iff $a + b = b$. We assume that 0 is the least element with respect to this ordering,[5] i.e. $0 + a = a$, for every state a. Fusion is distributive over addition in both directions, i.e. $a \cdot (b+c) = (a \cdot b) + (a \cdot c)$ and $(b + c) \cdot a = (b \cdot a) + (c \cdot a)$. Moreover, we assume that $1 \cdot a = a$ and $0 \cdot a = 0$. The relation C satisfies the following conditions: (a) the trivially inconsistent state is compatible with no state, i.e. there is no a such that $0Ca$, (b) compatibility is symmetrical, i.e. if aCb, then bCa, (c)

---

[5]Let us stress that 1 does not have to be the greatest element.

compatibility is monotone, i.e. if aCb and $a \leq c$, then cCb, and finally (d) if $(a+b)Cc$, then aCc or bCc. The valuation V assigns to every atomic formula an ideal in $\mathcal{M}$. An ideal I in $\mathcal{M}$ is a subset of S satisfying: (a) $0 \in I$, (b) if $a \in I$ and $b \leq a$, then $b \in I$, (c) if $a \in I$ and $b \in I$, then $a + b \in I$.

Let us introduce a language of substructural logics L_s the formulas of which are called L_s-formulas and are built from atomic formulas, and the constants $\bot$ (explosive contradiction) and t (logical truth) using the connectives $\neg, \rightarrow, \wedge, \otimes$, and $\vee$. The connective $\otimes$ is usually called *intensional conjunction*. This will be our language of declarative sentences. When we add to L_s inquisitive disjunction ? we will get the language $L_s^?$, a language of declarative sentences and questions.

Given an informational model $\mathcal{M} = \langle S, +, \cdot, 0, 1, C, V \rangle$, we define the support relation between the states of $\mathcal{M}$ and $L_s^?$-formulas by the following recursive clauses:

D1 $a \vDash p$ iff $a \in V(p)$,

D2 $a \vDash \bot$ iff $a = 0$,

D3 $a \vDash t$ iff $a \leq 1$,

D4 $a \vDash \neg \varphi$ iff for any b, if aCb, then $b \nvDash \varphi$,

D5 $a \vDash \varphi \rightarrow \psi$ iff for any b, if $b \vDash \varphi$, then $a \cdot b \vDash \psi$,

D6 $a \vDash \varphi \wedge \psi$ iff $a \vDash \varphi$ and $a \vDash \psi$,

D7 $a \vDash \varphi \otimes \psi$ iff there are b, c such that $b \vDash \varphi$, $c \vDash \psi$ and $a \leq b \cdot c$,

D8 $a \vDash \varphi \vee \psi$ iff there are b, c such that $b \vDash \varphi$, $c \vDash \psi$ and $a \leq b + c$,

D9 $a \vDash \varphi ? \psi$ iff $a \vDash \varphi$ or $a \vDash \psi$.

We say that an $L_s^?$-formula φ is $\mathcal{M}$-valid if $1 \vDash \varphi$ in $\mathcal{M}$. An $L_s^?$-formula φ is an $\mathcal{M}$-consequence of a set of $L_s^?$-formulas Δ if every state of $\mathcal{M}$ that supports every formula from Δ supports also φ. $L_s^?$-formulas φ and ψ are $\mathcal{M}$-equivalent if they are supported by the same states in $\mathcal{M}$.

The nature of this semantic framework will be illustrated with the following example that can also serve as a counterexample to the argument forms AF1-AF3 mentioned in the previous section. Consider informational model $\mathcal{M}_e = \langle S, +, \cdot, 0, 1, C, V \rangle$ defined as follows. $S = \{0, 1, 2, 3\}$, the operations $+$ and $\cdot$ are defined by the following tables:

Logics of Questions

+	0	1	2	3
0	0	1	2	3
1	1	1	3	3
2	2	3	2	3
3	3	3	3	3

·	0	1	2	3
0	0	0	0	0
1	0	1	2	3
2	0	2	3	3
3	0	3	3	3

Addition $+$ determines the partial order $\leq$, in which 0 is the least element and 3 is the greatest element. The states 1 and 2 are two incomparable elements between 0 and 3. The compatibility relation is defined in this way:

$$C = \{\langle 1, 2\rangle, \langle 2, 1\rangle, \langle 1, 3\rangle, \langle 3, 1\rangle, \langle 2, 3\rangle, \langle 3, 2\rangle, \langle 3, 3\rangle\}.$$

We will restrict ourselves to two atomic formulas p and q and define $V(p) = \{0, 2\}$ and $V(q) = \{0, 1\}$. It takes some time but it can be verified that all the conditions from the definition of informational models are satisfied and, consequently, that $\mathcal{M}_e$ is indeed an informational model. This model (without fusion) can be visualized by Figure 2 in which the ordering is determined by the vertical position of the elements ($a < b$ iff a is below b) and the dashed lines represent the compatibility relation.

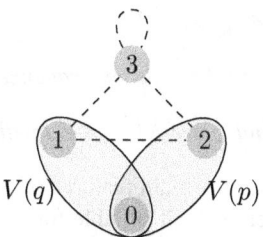

Figure 2: A visualization of $\mathcal{M}_e$

It holds, for example, that $2 \not\models [?]q$, since (i) 2 is not in $V(q)$, i.e. $2 \not\models q$, and (ii) there is a state (namely the state 1) that is compatible with 2 and is in $V(q)$, so $2 \not\models \neg q$. Moreover, we will show that $2 \models \neg(q \to r)$ in $\mathcal{M}_e$. This means that there is no b, such that $2Cb$ and b supports $q \to p$. In particular, $1 \not\models q \to p$ and $3 \not\models q \to p$. This is true, since 1 supports q but neither $1 \cdot 1 = 1$, nor $3 \cdot 1 = 3$ supports p. As a consequence, the state 2 in $\mathcal{M}_e$ provides a counterexample to the argument form AF1: $[?]q$ is not an $\mathcal{M}_e$-consequence of $\neg(q \to p)$.

It holds that $2 \vDash p \land \neg p$, since 2 is in $V(p)$ but no state that is compatible with 2 is in $V(p)$. We have already shown that $2 \nvDash [?]q$, so 2 in $\mathcal{M}_e$ provides a counterexample to the argument form AF2: $[?]q$ is not an $\mathcal{M}_e$-consequence of $p \land \neg p$.

Since $2 \vDash p$, it also holds that $2 \vDash [?]p$. But $2 \nvDash q \lor \neg q$. So, the state 2 in $\mathcal{M}_e$ provides also a counterexample to the argument form AF3: $q \lor \neg q$ is not an $\mathcal{M}_e$-consequence of $[?]p$.

The whole basic inquisitive semantics can be viewed as semantics of one particular informational model $\mathcal{M}_{InqB} = \langle S, +, \cdot, 0, 1, C, V \rangle$, where the set of states S is the set of all sets of possible worlds, the addition $+$ is union, the fusion $\cdot$ is intersection, the trivially inconsistent state 0 is the empty set $\emptyset$, the logical state 1 is the set of all possible worlds, two states a and b are compatible (aCb) iff $a \cap b \neq \emptyset$, and $a \in V(p)$ iff p is true in every world of a. $\varphi \otimes \psi$ is $\mathcal{M}_{InqB}$-equivalent to $\varphi \land \psi$, $\bot$ is $\mathcal{M}_{InqB}$-equivalent to $p \land \neg p$, and t is $\mathcal{M}_{InqB}$-equivalent to $\neg\bot$. So the additional symbols of the language $L_s^?$ are not needed in this models. However, these equivalences do not hold generally in every informational model. Due to Fact 10, the following result should be obvious.

Fact 11 *Let $\Delta \cup \{\varphi, \psi\}$ be a set of $L^?$-formulas. Then it holds:*

(a) $\vDash_{InqB} \varphi$ *iff φ is $\mathcal{M}_{InqB}$-valid,*

(b) $\Delta \vDash_{InqB} \varphi$ *iff φ is an $\mathcal{M}_{InqB}$-consequence of Δ,*

(c) $\varphi \equiv_{InqB} \psi$ *iff φ and ψ are $\mathcal{M}_{InqB}$-equivalent.*

Now we will show that the main features of the basic framework of inquisitive semantics, which make it suitable for modeling of questions, hold throughout the class of all informational models. Let us fix an arbitrary informational model $\mathcal{M}$. With respect to $\mathcal{M}$ one can define semantic notions that correspond to the notions introduced in the previous section for the basic framework.

Definition 4 *A proposition in $\mathcal{M}$ is any nonempty downward closed set of information states of $\mathcal{M}$. We say that a proposition P in $\mathcal{M}$ is declarative if P is an ideal in $\mathcal{M}$. If P is not declarative, it is called inquisitive.*

If we define $||\varphi||$ in $\mathcal{M}$, i.e. the proposition expressed by an $L_s^?$-formula φ in the model $\mathcal{M}$, as the set of states that support φ in $\mathcal{M}$, we obtain the following generalization of Fact 1.

Logics of Questions

Fact 12 *Every $L_s^?$-formula expresses a proposition in $\mathcal{M}$.*

The notion of a declarative proposition introduced in Definition 4, when applied to the model $\mathcal{M}_{InqB}$, does not coincide exactly with the corresponding notion from Definition 2, which defines propositions, algebraically speaking, as principal ideals in $\mathcal{M}_{InqB}$. If we assume that the set of atomic formulas is infinite, there are ideals in $\mathcal{M}_{InqB}$ that are not principal. However, the two notions coincide on the propositions expressed by $L_s^?$-formulas.

Fact 13 *An $L_s^?$-formula φ expresses a declarative proposition in $\mathcal{M}_{InqB}$ iff $\|\varphi\|$ in $\mathcal{M}_{InqB}$ is a principal ideal (i.e. contains its own union and thus is a declarative proposition in the sense of Def. 2).*

The following claim generalizes Fact 3.

Fact 14 *Every L_s-formula expresses a declarative proposition in $\mathcal{M}$.*

As in the basic framework, the connective ? generates formulas that typically express inquisitive propositions. Assume that a and b are two states of $\mathcal{M}$ such that neither $a \leq b$, nor $b \leq a$. If $V(p) = \{c; c \leq a\}$ and $V(q) = \{c; c \leq b\}$, then both a and b support $p?q$, but $a + b$ does not support $p?q$. In such case, $\|p?q\|$ is not an ideal in $\mathcal{M}$.

The logic of all informational models, for the language of declarative sentences L_s, is very weak. It is a non-distributive, non-commutative, non-associative version of Full Lambek with a paraconsistent negation, i.e. a basic substructural logic. A Hilbert system for this logic is presented and completeness proved in (Punčochář, submitted).

Since the logic of all informational models is very weak, the set of its extensions is large. Let us denote the set of L_s-formulas that are valid in a class of informational models $\mathcal{C}$ as $Log(\mathcal{C})$.

Definition 5 *A set of L_s-formulas λ is called a logic of declarative sentences if there is a class of informational models $\mathcal{C}$ such that $\lambda = Log(\mathcal{C})$.*

The set of logics of declarative sentences encompasses most of the "logics" studied in the literature. In particular, besides classical logic, it contains intuitionistic logic and its extensions, relevant logics, and many-valued logics, including fuzzy logics. In our framework, every logic of this class can be enriched with questions in the style of inquisitive semantics. Let us denote the set of $L_s^?$-formulas that are valid in a class of informational models $\mathcal{C}$ as $Log^?(\mathcal{C})$ and the class of models of some given set of L_s-formulas Δ as $Mod(\Delta)$.

Definition 6 *Let λ be a logic of declarative sentences. The inquisitive extension of λ, denoted as $\lambda^?$, is the set of all $L_s^?$-formulas that are valid in every model of λ. In symbols, $\lambda^? = Log^?(Mod(\lambda))$.*

In particular, the basic inquisitive logic is the inquisitive extension of classical logic.

Fact 15 $CL^? = InqB$.

For any logic of declarative sentences and its inquisitive extension, we will define the semantic concepts of logical validity, logical consequence, and logical equivalence. Then we will show that the inquisitive extension of any logic of declarative sentences λ is related to λ in the same way as basic inquisitive logic to classical logic.

Definition 7 *Let λ be a logic of declarative sentences and $\Delta \cup \{\alpha, \beta\}$ a set of L_s-formulas. α is λ-valid ($\vDash_\lambda \alpha$) if for every informational model $\mathcal{M}$ in $Mod(\lambda)$, α is $\mathcal{M}$-valid. α is a λ-consequence of Δ ($\Delta \vDash_\lambda \alpha$) if for every informational model $\mathcal{M}$ in $Mod(\lambda)$, α is an $\mathcal{M}$-consequence of Δ. α and β are λ-equivalent ($\alpha \equiv_\lambda \beta$) iff for every informational model $\mathcal{M}$ in $Mod(\lambda)$, α and β are $\mathcal{M}$-equivalent.*

For the inquisitive extension of λ, these notions are defined in an analogous way.

Definition 8 *Let λ be a logic of declarative sentences and $\Delta \cup \{\varphi, \psi\}$ a set of $L_s^?$-formulas. φ is $\lambda^?$-valid ($\vDash_{\lambda^?} \varphi$) if for every informational model $\mathcal{M}$ in $Mod(\lambda)$, φ is $\mathcal{M}$-valid. φ is a $\lambda^?$-consequence of Δ ($\Delta \vDash_{\lambda^?} \varphi$) if for every informational model $\mathcal{M}$ in $Mod(\lambda)$, φ is an $\mathcal{M}$-consequence of Δ. φ and ψ are $\lambda^?$-equivalent ($\varphi \equiv_{\lambda^?} \psi$) iff for every informational model $\mathcal{M}$ in $Mod(\lambda)$, φ and ψ are $\mathcal{M}$-equivalent.*

The consequence relation of $\lambda^?$ coincides with the consequence relation of λ on L_s-formulas, that is on the language of declarative sentences. This generalizes Fact 4.

Fact 16 *If $\Delta \cup \{\alpha\}$ is a set of L_s-formulas, then $\Delta \vDash_{\lambda^?} \alpha$ iff $\Delta \vDash_\lambda \alpha$.*

The concept of resolution for the language $L_s^?$ extends the concept of resolution for $L^?$ in a straightforward way.

- $\mathcal{R}(p) = \{p\}$,

- $\mathcal{R}(\bot) = \{\bot\}$,
- $\mathcal{R}(t) = \{t\}$,
- $\mathcal{R}(\neg\varphi) = \{\bigwedge_{\alpha \in \mathcal{R}(\varphi)} \neg\alpha\}$,
- $\mathcal{R}(\varphi \to \psi) = \{\bigwedge_{\alpha \in \mathcal{R}(\varphi)} \alpha \to f(\alpha); f : \mathcal{R}(\varphi) \to \mathcal{R}(\psi)\}$,
- $\mathcal{R}(\varphi \wedge \psi) = \{\alpha \wedge \beta; \alpha \in \mathcal{R}(\varphi), \beta \in \mathcal{R}(\psi)\}$,
- $\mathcal{R}(\varphi \otimes \psi) = \{\alpha \otimes \beta; \alpha \in \mathcal{R}(\varphi), \beta \in \mathcal{R}(\psi)\}$,
- $\mathcal{R}(\varphi \vee \psi) = \{\alpha \vee \beta; \alpha \in \mathcal{R}(\varphi), \beta \in \mathcal{R}(\psi)\}$,
- $\mathcal{R}(\varphi ? \psi) = \mathcal{R}(\varphi) \cup \mathcal{R}(\psi)$.

Let us fix any logic of declarative sentences λ. The following two results generalize Facts 5 and 6.

Fact 17 *If $\mathcal{R}(\varphi) = \{\alpha_1, \ldots, \alpha_n\}$, then $\varphi \equiv_{\lambda^?} \alpha_1 ? \ldots ? \alpha_n$.*

Fact 18 *Let $\mathcal{M}$ be any informational model (of λ). Every proposition in $\mathcal{M}$ that is expressible by an $L_s^?$-formula is identical with the union of a finite set of declarative propositions in $\mathcal{M}$.*

Finally, the following three results generalize Facts 7-9 and show in which sense the semantic notions of $\lambda^?$ are determined by the corresponding notions of λ.

Fact 19 $\vDash_{\lambda^?} \varphi$ *iff there is $\alpha \in \mathcal{R}(\varphi)$ such that $\vDash_\lambda \alpha$.*

Fact 20 $\varphi_1, \ldots, \varphi_n \vDash_{\lambda^?} \psi$ *iff for every $\alpha_1 \in \mathcal{R}(\varphi_1), \ldots, \alpha_n \in \mathcal{R}(\varphi_n)$ there is $\beta \in \mathcal{R}(\psi)$ such that $\alpha_1, \ldots, \alpha_n \vDash_\lambda \beta$.*

Fact 21 $\varphi \equiv_{\lambda^?} \psi$ *iff for every $\alpha \in \mathcal{R}(\varphi)$ there is $\beta \in \mathcal{R}(\psi)$ such that $\alpha \vDash_\lambda \beta$, and for every $\beta \in \mathcal{R}(\psi)$ there is $\alpha \in \mathcal{R}(\varphi)$ such that $\beta \vDash_\lambda \alpha$.*

4 Conclusion

To sum up, we have described basic and generalized inquisitive semantics. We have shown that the generalized version enables us to model questions over a large class of non-classical logics. Due to this it provides means to avoid paradoxes of material implication and irrelevance in the logic of questions without the loss of any of the characteristic features of basic inquisitive semantics that are needed for an adequate modeling of questions.

References

Ciardelli, I. (2016). *Questions in Logic*. Doctoral dissertation. University of Amsterdam.

Ciardelli, I., Groenendijk, J., & Roelofsen, F. (2013). Inquisitive semantics: A new notion of meaning. *Language and Linguistics Compass*, 7, 459–476.

Ciardelli, I., & Roelofsen, F. (2011). Inquisitive logic. *Journal of Philosophical Logic*, 40, 55–94.

Došen, K. (1989). Sequent systems and groupoid models, part 2. *Studia Logica*, 48, 41–65.

Punčochář, V. (2016a). A generalization of inquisitive semantics. *Journal of Philosophical Logic*, 45, 399–428.

Punčochář, V. (2016b). Algebras of information states. *Journal of Logic and Computation*, DOI: 10.1093/logcom/exw021.

Punčochář, V. (submitted). Substructural inquisitive logics.

Wansing, H. (1993). Informational interpretation of substructural propositional logics. *Journal of Logic, Language and Information*, 2, 285–308.

Wittgenstein, L. (1922). *Tractatus Logico-philosophicus*. London: Routledge & Kegan Paul.

Yang, F. (2014). *On Extensions and Variants of Dependence Logic*. Doctoral dissertation. University of Helsinki.

Vít Punčochář
Institute of Philosophy, Czech Academy of Sciences
The Czech Republic
E-mail: vit.puncochar@centrum.cz

Non-Classical PDL on the Cheap

IGOR SEDLÁR[1]

Abstract: A four-valued version of PDL is obtained by defining a DeMorgan negation as a negative modality using a special atomic program. Simple examples indicating applications of the formalism in formal verification of epistemic programs and epistemic planning are provided. Decidability and weak completeness are established.

Keywords: DeMorgan negation, Epistemic planning, Formal verification, Propositional dynamic logic

1 Introduction

Propositional dynamic logic PDL is a well-known formalism for specifying and verifying properties of regular programs (Fischer & Ladner, 1979; Harel, Kozen, & Tiuryn, 2000). In the Kripke-style semantics for PDL, programs are represented by state transitions ('state y is a possible outcome of executing program α in state x') where states of the computer are taken to be complete and consistent possible worlds. Technically speaking, in PDL states correspond to functions from the set of formulas to the set of classical truth values $\{0, 1\}$: every formula is either false (0) or true (1), but not both.

Several generalisations of this approach have been suggested. In this article we shall focus on the one put forward by Belnap and Dunn (Belnap, 1977a, 1977b; Dunn, 1976). Belnap–Dunn states correspond to functions from formulas to *subsets* of $\{0, 1\}$; and are seen as *bodies of information about the world* rather than possible states of the world (possible worlds). On this view, the four possible truth values $\emptyset, \{0\}, \{1\}$ and $\{0, 1\}$ correspond to four possible answers to queries about a formula with respect to a fixed body of information: the body of information does not provide any information about the formula ($\emptyset$); it provides information that the formula is false and no information that it is true ($\{0\}$); it provides information that the formula is true and no information that it is false ($\{1\}$); it provides conflicting information about the formula ($\{0, 1\}$). The notion of a computer

---

[1]This work was supported by the long-term strategic development financing of the Institute of Computer Science (RVO:67985807).

program operating on Belnap–Dunn states is natural since such programs can be seen as algorithmic transformations of database-like bodies of information, central to areas such as epistemic planning.[2]

Sedlár (2016) outlines a version of PDL over an extension of the Belnap–Dunn logic studied by Odintsov and Wansing (2010). This article introduces a simplification of the approach. It is shown that a version of PDL corresponding to the Belnap–Dunn notion of state can be defined by a very simple modification of standard PDL. Building on the approach of Fagin, Halpern, and Vardi (1995), we extend PDL with a *modal DeMorgan negation* connective '$\sim$', interpreted semantically by the Routley star operator (Routley & Routley, 1972). We construe the Routley star as a serial, symmetric and functional atomic program. Decidability and weak completeness of the resulting system are established.

We note that (Sedlár, 2016) and the present article can be seen as an addition to the small but growing literature on non-classical PDL. We should mention Teheux (2014) who formulates PDL over finitely-valued Łukasiewicz logics to model the Rényi–Ulam searching game with errors. Baltag and Smets (2006) present a dynamic logic for reasoning about information flow in quantum programs and Bergfeld and Sack (2015) discuss a probabilistic logic for quantum programs.

The article is organised as follows. Section 2 outlines the basics of standard PDL. Section 3 discusses a modal logic with De Morgan negation and Section 4 extends this logic to PDL with De Morgan negation, NPDL. Section 5 outlines potential applications of NPDL. Section 6 provides a brief summary of the article.

2 Standard PDL

This section outlines standard PDL using complete and consistent possible worlds as representations of the states programs operate on. For more details, see (Harel et al., 2000).

Language $\mathcal{L}$ consists of two classes of expressions, namely, programs ($\alpha \in P$) and formulas ($\phi \in F$), defined as follows:

$$\alpha ::= a \mid \alpha;\alpha \mid \alpha \cup \alpha \mid \alpha^* \mid \phi? \qquad \phi ::= p \mid \neg\phi \mid \phi \to \phi \mid [\alpha]\phi$$

---

[2] To give another example of a generalisation of the classical notion of state, if some formulas correspond to statements involving imprecise or graded predicates, then functions from the set of formulas to the real interval [0, 1] are the appropriate formalisation.

Non-Classical PDL on the Cheap

($a \in AP$, a countable set of atomic programs and $p \in AF$, a countable set of atomic formulas). Formulas of the form $[\alpha]\phi$ are read 'It is necessary that after executing α, ϕ will hold'. The operator ';' is seen as program composition ('Execute α, then execute β'), '$\cup$' as non-deterministic choice ('Choose either α or β nondeterministically and execute it'), '*' as iteration ('Execute α a nondeterministically chosen finite number of times') and '?' as test ('Test whether ϕ is the case; proceed if true, fail if false'). As usual, $\phi \wedge \psi \stackrel{def}{=} \neg(\phi \to \neg\psi)$, $\phi \vee \psi \stackrel{def}{=} \neg(\neg\phi \wedge \neg\psi)$ and $\langle\alpha\rangle\phi \stackrel{def}{=} \neg[\alpha]\neg\phi$. Models for this language (*'dynamic models'*) are multi-dimensional Kripke models $M = \langle S, R, V \rangle$, where every α is assigned a binary relation $R(\alpha)$ on the set of states S and every ϕ is assigned a subset $V(\phi)$ of S as follows: $V(p)$ is arbitrary; $V(\neg\phi)$ is the complement of $V(\phi)$; $V(\phi \to \psi) = (S - V(\phi)) \cup V(\psi)$; $V([\alpha]\phi)$ is the set of $x \in S$ such that, for all y, if $\langle x, y \rangle \in R(\alpha)$, then $y \in V(\phi)$ ('the set of such states x that every successful (terminating, halting) execution of α in x results in a state that satisfies ϕ'); $R(\alpha; \beta)$ ($R(\alpha \cup \beta)$) is the composition (union) of $R(\alpha)$ and $R(\beta)$; $R(\alpha^*)$ is the reflexive transitive closure of $R(\alpha)$; and $R(\phi?)$ is the identity relation on $V(\phi)$. (Note that $x \in V(\langle\alpha\rangle\phi)$ iff there is y such that $R(\alpha)xy$ and $y \in V(\phi)$, i.e. $\langle\alpha\rangle\phi$ means that it is possible that after executing α, ϕ will hold.) V could have been defined equivalently as a function from S to functions from F to $\{0, 1\}$ (as discussed in the introduction), but we have chosen this simpler formulation. Sometimes the notation 'S_M', 'R_M' and 'V_M' is used to make the relevant M explicit.

The following metalogical notions are used, as defined here, throughout the article. A formula ϕ is valid in model M iff $V_M(\phi) = S_M$ (notation: $M \models \phi$). A formula ϕ follows from (is entailed by) Γ as the set of local assumptions and Δ as the set of global assumptions (notation: $[\Delta], \Gamma \models \phi$) iff

$$\bigcap_{\psi \in \Gamma} V_M(\psi) \subseteq V_M(\phi)$$

for all M such that $M \models \Delta$ (i.e. $M \models \chi$ for all $\chi \in \Delta$). A formula ϕ follows from Γ globally (locally) iff $[\Gamma], \emptyset \models \phi$ ($[\emptyset], \Gamma \models \phi$). We also use the notation $[\Gamma] \models \phi$ ($\Gamma \models \phi$) for global (local) entailment. A formula ϕ is valid in a class of models $\mathcal{C}$ iff $M \models \phi$ for all $M \in \mathcal{C}$; ϕ is valid in PDL iff it is valid in the class of all dynamic models (notation: $\models \phi$). A formula ϕ is satisfiable iff there is $\mathcal{M}$ such that $M \not\models \neg\phi$.

The language $\mathcal{L}$ is able to express a number of standard programming constructs. For example, *if ϕ then α else β* translates to $(\phi?; \alpha) \cup (\neg\phi?; \beta)$

241

Igor Sedlár

and *while* ϕ *do* α to $(\phi?; \alpha)^*; \neg\phi?$. *Correctness* of programs (given some desired functionality) can be seen as a relation between specific inputs and outputs of terminating executions of the program: given an input specified by a formula ϕ, every terminating (successful) execution of the program terminates in a state satisfying ψ. In this case, we say that a program α is *partially correct* with respect to precondition ϕ and postcondition ψ. Partial correctness assertions are expressible in $\mathcal{L}$ as formulas of the form

$$\phi \to [\alpha]\psi \qquad (1)$$

PDL can be used to check whether specific partial correctness assertions follow from some given global assumptions Δ (thought of as 'invariant' assumptions holding in the initial state and every possible outcome of every possible computation) and some local assumptions Γ (required to hold only in the initial state of the computation): $[\Delta], \Gamma \models^? \phi \to [\alpha]\psi$. The key observation is that, in most practically interesting cases, such questions are decidable.

Proposition 1 *Let* $\Gamma, \Delta \subseteq F_\mathcal{L}$ *be finite and* $\phi \in F_\mathcal{L}$. *If* $\{a_1, \ldots, a_n\}$ *is the set of all atomic programs appearing in some formula in* $\Delta \cup \Gamma$ *or in* ϕ, *then*

$$[\Delta], \Gamma \models \phi \iff \models [(a_1 \cup \ldots \cup a_n)^*] \bigwedge \Delta \to \left(\bigwedge \Gamma \to \phi\right)$$

Proof. Let us write the claim on the right-hand side as $\models \Delta^* \to (\bigwedge \Gamma \to \phi)$. The right-to-left implication is trivial. If $M \models \Delta$ and $x \in V_M(\bigwedge \Gamma)$, then obviously $x \in V_M(\Delta^* \land \bigwedge \Gamma)$. So $x \in V_M(\phi)$ follows from the assumption.

To establish the converse implication, assume that $\Delta^* \to (\bigwedge \Gamma \to \phi)$ is not valid. This means that there are some M and $x \in S_M$ such that $x \in V_M(\Delta^* \land \bigwedge \Gamma)$ and $x \notin V_M(\phi)$. Now define $M_x = \langle S_x, R_x, V_x \rangle$ as follows: $S_x = \{y \mid \langle x, y \rangle \in R((a_1 \cup \ldots \cup a_n)^*)\}$; $R_x(\alpha) = R(\alpha) \cap S_x^2$ for all α; $V_x(p) = V(p) \cap S_x$ and $V_x(\phi)$ for complex ϕ are built up recursively as in the definition of dynamic model. It is plain that $M_x \models \Delta$, but $M_x \not\models \bigwedge \Gamma \to \phi$. (The key fact, easily established by induction on the complexity of α, is that if every atomic program appearing in α is in $\{a_1, \ldots, a_n\}$, then $R(\alpha)zz'$ only if $R_x(\alpha)zz'$, for all $z, z' \in S_x$.) □

Theorem 1 *The validity problem for* PDL *($\models^? \phi$) is decidable (EXPTIME-complete).*

Proof. See (Harel et al., 2000), chapters 6 and 8. □

Non-Classical PDL on the Cheap

Hence, for any finite sets of global assumptions Δ, local assumptions Γ and any partial correctness assertion $\phi \to [\alpha]\psi$, there is an algorithm running in time exponential to the size of the input and deciding whether the assertion follows from the assumptions. For infinite Δ, the situation gets worse; it is even sufficient to consider the set $SI(\phi)$ of substitution instances of some fixed ϕ.

Theorem 2 *The question whether $[SI(\phi)], \emptyset \models \psi$ is undecidable (Π_1^1-complete).*

Proof. See (Harel et al., 2000), chapter 8.3. □

Because of the 'infinitary' iteration operator $*$, PDL is not compact.[3] Hence, there is no hope of providing a strongly complete axiomatization. However, weak completeness is another story.

Theorem 3 *Let $H(\mathsf{PDL})$ be the Hilbert-style system extending any axiomatisation of classical propositional logic in $\{\neg, \to\}$ by schemas*

1. $[\alpha](\phi \to \psi) \to ([\alpha]\phi \to [\alpha]\psi)$

2. $[\alpha \cup \beta]\phi \leftrightarrow ([\alpha]\phi \land [\beta]\phi)$

3. $[\alpha;\beta]\phi \leftrightarrow [\alpha][\beta]\phi$

4. $[\psi?]\phi \leftrightarrow (\psi \to \phi)$

5. $[\alpha^*]\phi \leftrightarrow (\phi \land [\alpha][\alpha^*]\phi)$

6. $(\phi \land [\alpha^*](\phi \to [\alpha]\phi)) \to [\alpha^*]\phi$

and the Necessitation rule $\phi/[\alpha]\phi$. Let theoremhood in $H(\mathsf{PDL})$ be defined as usual. Then ϕ is a theorem of $H(\mathsf{PDL})$ iff $\models \phi$.

Proof. See (Harel et al., 2000), chapter 7. □

---

[3]Every finite subset of $\{\langle \alpha^* \rangle \phi\} \cup \{\neg \phi, \neg \langle \alpha \rangle \phi, \neg \langle \alpha; \alpha \rangle \phi, \ldots\}$ is satisfiable but the set itself is not satisfiable.

Igor Sedlár

3 Modal logic with De Morgan negation

This section outlines the framework of Fagin et al. (1995). The language $\mathcal{L}_n^\sim$ is a fragment of $\mathcal{L}$ without program operators ';', '∪', '∗' and '?', with only a finite $AP_n \subseteq AP$ of cardinality n, extended by a new unary connective '∼'. (Informally, members of AP_n are seen as agents, but this is not important for our purposes.) *Star models* are $\mathcal{M} = \langle S, R, \star, V \rangle$, where S and R are as before (the range of R is AP_n), $\star$ is a unary operation of period two (i.e. $(x^\star)^\star = x$) on S and V is defined as before with the addition of

$$V(\sim\phi) = \{x \mid x^\star \notin V(\phi)\} \qquad (2)$$

Let us denote the set of $\mathcal{L}_n^\sim$-formulas valid in every star model as NK. It is plain that '∼' does not adhere to many of the laws satisfied by classical negation '¬'. For instance, not every formula of the form $\sim\phi \lor \phi$ or $(\phi \land \sim\phi) \to \psi$ is in NK, but $\sim\sim\phi \leftrightarrow \phi$ and all of the De Morgan laws are in NK.[4] Hence, '∼' turns out to be what is usually called a De Morgan negation (Dunn, 1993).

Theorem 4 *Let $H(\text{NK})$ be a Hilbert-style system that extends any axiomatization of classical propositional logic in $\{\neg, \to\}$ by schemas*

1. $[\alpha](\phi \to \psi) \to ([\alpha]\phi \to [\alpha]\psi)$

2. $\sim\sim\phi \leftrightarrow \phi$

3. $\sim(\phi \land \psi) \leftrightarrow (\sim\phi \lor \sim\psi)$

4. $\sim(\phi \lor \psi) \leftrightarrow (\sim\phi \land \sim\psi)$

and the Necessitation rule $\phi/[\alpha]\phi$. Then ϕ is a theorem of $H(\text{NK})$ iff $\mathcal{M} \models \phi$ for every star model $\mathcal{M}$.

Proof. See (Fagin et al., 1995), where a slightly different but equivalent axiomatization is used. □

Theorem 5 *Membership in NK (the problem of validity in every star model) is decidable (PSPACE-complete).*

Proof. See (Fagin et al., 1995). □

---

[4] Standard Kripke models can be seen as a special case where $x^\star = x$.

Note that this framework is, in effect, a fragment of PDL. Syntactically, the set of programs is limited to $AP_{n+1} = \{a_1, \ldots, a_n, b\}$ and $\sim\phi$ defined as $[b]\neg\phi$. Semantically, $R(b)$ is required to be a total function of period two, i.e., a relation that is *serial* $((\forall x)(\exists y)(Rxy))$, *symmetric* $((\forall xy)(Rxy \Rightarrow Ryx))$ and *functional* $((\forall xyz)((Rxy \,\&\, Rxz) \Rightarrow y = z))$.

4 PDL with De Morgan negation

Now consider the full language $\mathcal{L}$ again. Let us fix an atomic program $a \in AP$ and denote it as '$\star$'. Define $\sim\phi$ as $[\star]\neg\phi$ ('It is necessary that after executing $\star$, ϕ will not hold'). A *Routley model* is a dynamic model where $R(\star)$ is a total function of period two (serial, symmetric and functional). We reiterate that, when considering Routley models, we can meaningfully say $y = x^\star$ instead of $R(\star)xy$. Formula ϕ is said to be valid in NPDL iff it is valid in the class of Routley models.

Proposition 2 *Any formula of one of the following forms is valid in* NPDL*:*

1. $\langle\star\rangle\phi \leftrightarrow [\star]\phi$
2. $\phi \to [\star]\langle\star\rangle\phi$

Proof. If $R(\star)$ is functional, then, for every x, there is at most one y such that $R(\star)xy$. So if $x \in V(\langle\star\rangle\phi)$, then $y \in V(\phi)$ for every y such that $R(\star)xy$. Hence, $\langle\star\rangle\phi \to [\star]\phi$ is valid in NPDL. The converse $[\star]\phi \to \langle\star\rangle\phi$ is valid since $R(\star)$ is serial and $\phi \to [\star]\langle\star\rangle\phi$ is valid because $R(\star)$ is symmetric. □

Corollary 1 *Any formula of one of the following forms is valid in* NPDL*:*

1. $\phi \to [\star][\star]\phi$
2. $[\star]\langle\star\rangle\phi \to \phi$

Proof. The first fact is obvious. To see that the second one holds as well, observe that

$$\begin{aligned}
(\forall\phi)(M \models \phi \to [\star]\langle\star\rangle\phi) &\Longrightarrow (\forall\phi)(M \models \neg\phi \to [\star]\langle\star\rangle\neg\phi) \\
&\Longrightarrow (\forall\phi)(M \models \neg[\star]\langle\star\rangle\neg\phi \to \neg\neg\phi) \\
&\Longrightarrow (\forall\phi)(M \models \langle\star\rangle[\star]\phi \to \phi) \\
&\Longrightarrow (\forall\phi)(M \models [\star][\star]\phi \to \phi) \\
&\Longrightarrow (\forall\phi)(M \models [\star]\langle\star\rangle\phi \to \phi)
\end{aligned}$$

Igor Sedlár

□

Corollary 2 *Any formula of one of the following forms is valid in* NPDL:

(a) $\sim\sim\phi \leftrightarrow \phi$

(b) $\sim(\phi \land \psi) \leftrightarrow (\sim\phi \lor \sim\psi)$

(c) $\sim(\phi \lor \psi) \leftrightarrow (\sim\phi \land \sim\psi)$

Proof. (a) boils down to $[\star]\langle\star\rangle\phi \leftrightarrow \phi$; (b) is equivalent to $\langle\star\rangle(\neg\phi \lor \neg\psi) \leftrightarrow (\langle\star\rangle\neg\phi \lor \langle\star\rangle\neg\psi)$; and (c) is equivalent to $[\star](\neg\phi \land \neg\psi) \leftrightarrow ([\star]\neg\phi \land [\star]\neg\psi)$.
□

If $\sim\phi$ is read as 'ϕ is false', then states in Routley models can be seen as Belnap–Dunn states. To be more specific, there are four possibilities for every x:

- $x \in V(\phi)$ and $x^\star \in V(\phi)$, so $x \in V(\phi \land \neg\sim\phi)$;
- $x \notin V(\phi)$ and $x^\star \notin V(\phi)$, so $x \in V(\neg\phi \land \sim\phi)$;
- $x \in V(\phi)$ and $x^\star \notin V(\phi)$, so $x \in V(\phi \land \sim\phi)$;
- $x \notin V(\phi)$ and $x^\star \in V(\phi)$, so $x \in V(\neg\phi \land \neg\sim\phi)$.

For every Routley model M and $x \in S_M$, let f_x^M be a function from formulas to subsets of $\{0, 1\}$ defined by setting $1 \in f_x^M(\phi)$ iff $x \in V(\phi)$ and $0 \in f_x^M(\phi)$ iff $x \in V(\sim\phi)$. The four possibilities specified above correspond to:

- $f_x^M(\phi) = \{1\}$,
- $f_x^M(\phi) = \{0\}$,
- $f_x^M(\phi) = \{0, 1\}$ and
- $f_x^M(\phi) = \emptyset$,

respectively. (Note that defining $\sim_1, \ldots, \sim_n$ in terms of $\star_1, \ldots, \star_n$ would enable to simulate 2^n-valued states for every finite n.[5])

Informally, the present framework treats the Routley star operator (Routley & Routley, 1972) as a program operating on database-like bodies of

---

[5]Thanks to Kit Fine for a suggestion along these lines.

information.[6] For the sake of simplicity, let us represent these bodies of information by functions from formulas to subsets of $\{0, 1\}$.[7] For every such f, define $f^\star$ by

$$f^\star(\phi) = \{\xi \in \{0, 1\} \mid f(\phi) \cup \{\xi\} \neq \{0, 1\}\}$$

In other words, $f^\star$ assigns to ϕ those 'classical' truth values that can be consistently added to the 'classical' truth values assigned to ϕ by f. Observe that $0 \in f(\phi)$ iff $1 \notin f^\star(\phi)$. In general,

$$\{0\} \xrightarrow{f^\star} \{0\} \qquad \{1\} \xrightarrow{f^\star} \{1\} \qquad \{0,1\} \xrightarrow{f^\star} \emptyset \qquad \emptyset \xrightarrow{f^\star} \{0,1\}$$

Hence, in a sense, the Routley star program corresponds to what has been called *conflation* in the literature on bilattices (Arieli & Avron, 1996).

Routley models comply with this interpretation, as witnessed by the following fact.

Proposition 3 *Any formula of one of the following forms is valid in* NPDL*:*

1. $(\phi \wedge [\star]\phi) \to [\star](\phi \wedge [\star]\phi)$
2. $(\phi \wedge [\star]\neg\phi) \to [\star](\neg\phi \wedge [\star]\phi)$
3. $(\neg\phi \wedge [\star]\phi) \to [\star](\phi \wedge [\star]\neg\phi)$
4. $(\neg\phi \wedge [\star]\neg\phi) \to [\star](\neg\phi \wedge [\star]\neg\phi)$

Proof. For all ϕ, ψ, $\phi \wedge [\star]\psi$ (both locally and globally) entails $[\star][\star]\phi \wedge [\star]\psi$ by Corollary 1, which entails $[\star](\psi \wedge [\star]\phi)$ by elementary modal logic. □

Theorem 6 *The* NPDL *validity problem is decidable.*

Theorem 7 *Let $H(\mathrm{NPDL})$ be $H(\mathrm{PDL})$ extended by the schemas $\langle\star\rangle\phi \leftrightarrow [\star]\phi$ and $\phi \to [\star]\langle\star\rangle\phi$. Then ϕ is valid in* NPDL *iff it is a theorem of $H(\mathrm{NPDL})$.*

---

[6] Some find the Routley star in need of a convincing informal interpretation. For example, Dunn (1976) expresses worries whether the Routley star can be seen as something more that just a purely technical device without a reasonable informal interpretation.

[7] This is a simplifying assumption as, of course, there are Routley models M that contain a pair of states $x \neq y$ such that $f_x^M = f_y^M$. We can say that states are characterized by these functions up to $\mathcal{L}$-equivalence, where x and y are $\mathcal{L}$-equivalent iff $x \in V(\phi) \iff y \in V(\phi)$ for all $\phi \in F_\mathcal{L}$.

Igor Sedlár

The proofs of both of these theorems add little technical novelty to the standard proofs for PDL. The only thing that requires modification is the definition of the set used in filtration of the canonical model. In the proof for PDL, the so-called Fisher–Ladner closure $FL(\phi)$ of a given non-provable ϕ is used. However, it is easy to show that while $R(\star)$ in the canonical model is functional, $R(\star)^{FL(\phi)}$ in the filtrated model may fail to be a function. The solution to this problem lies in using a special finite superset $FLR(\phi)$ of $FL(\phi)$, also called the Fisher–Ladner–Routley closure of ϕ. The interested reader is referred to Appendix 6.

5 Some examples

This section outlines some simple examples that demonstrate the expressivity (and possible applications) of NPDL. In these examples we use 'ϕ is supported (by x)' and 'There is information (in x) that ϕ is true' interchangeably. Instead of calling x a database-like body of information, we refer to it as a database. We say that ϕ is *decided* (by x) if ϕ or $\sim\phi$ is supported (by x) and that ϕ is *undecided* (by x) if ϕ is not decided (by x). We write ϕ^+ instead of $\phi \vee \sim\phi$ and ϕ^- instead of $\neg(\phi^+)$.

Being a 'special case' of PDL, NPDL can be used to check correctness of programs. In the context of NPDL, however, programs are seen as operating on bodies of information about the world, not the world itself. We may call these programs *epistemic programs*.[8]

Example 1 If states are seen as databases, then perhaps the simplest kinds of epistemic programs correspond to *adding* and *removing* ϕ from the database. Other simple examples are *testing* whether the database supports or decides a specific ϕ. The test programs are representable as $\phi?$ and $(\phi^+)?$, respectively. Representing addition and removal of information is trickier, but it can be done as follows. We may assume that, for a specific fixed ϕ, two (distinct) atomic programs a, b correspond to adding and removing ϕ, respectively. This assumption can be formalised by taking the formulas $\neg\phi \to [a]\phi$ and $\phi \to [b]\neg\phi$ as global assumptions. The role of global assumptions is perhaps best understood in contrast to the test programs. For instance, if we want to check whether $\phi?$ is partially correct with respect to precondition ψ and postcondition $\phi \wedge \psi$, then we ask whether $\psi \to [\phi?](\phi \wedge \psi)$ is valid in NPDL (the answer is, obviously, 'yes'). Now assume that we

---

[8]This terminology is related to but somewhat different in meaning from the one used by Baltag and Moss (2004).

Non-Classical PDL on the Cheap

want to check whether $a;b$ is partially correct with respect to $\neg\phi$ as both the precondition and postcondition. Obviously, $\neg\phi \to [a;b]\neg\phi$ is not valid in NPDL, but this is not the intuitively correct answer. However,

$$[\neg\phi \to [a]\phi, \phi \to [b]\neg\phi] \models \neg\phi \to [a;b]\neg\phi$$

holds.[9]

Example 2 An interesting special case of a postcondition is ϕ^+. Formulas of the form $\psi \to [\alpha]\phi^+$ say that every terminating execution of α where the input is a ψ-supporting database is a database that decides ϕ. Note that, in a sense, partial correctness claims play an important role in 'strategic' assessments of programs. For example, if the *goal* is to have a database that decides ϕ and

$$[\Gamma] \models \psi \to [\alpha]\phi^+ \quad \text{but} \quad [\Gamma] \not\models \psi \to [\beta]\phi^+,$$

(where Γ is some set of relevant global assumptions) then running α on a ψ-database is a *better choice* than running β. This is related to *epistemic planning*, i.e. the activity of choosing the appropriate course of actions given specific epistemic goals. Thus, NPDL can be seen as a formalism for checking the correctness of epistemic plans.

Epistemic planning, a 'special case' of automated planning[10] (Ghallab et al., 2004), has recently been formalised using the framework of dynamic epistemic logic (Bolander & Andersen, 2011). However, this approach has the disadvantage that checking the correctness of epistemic plans is undecidable even in the single-agent case if 'knowledge' is modelled by an epistemic logic weaker that S5 (Aucher & Bolander, 2013). Another advantage of NPDL as a formalism for epistemic planning is that it is able to distinguish between *different 'kinds' of inconsistency*. For instance, assume that p represents some irrelevant piece of information and q represents some rather important one. It is natural to assume that the presence of conflicting information about p in a database requires a different kind of action (probably

---

[9]If $x \in V_M(\neg\phi)$ and $M \models \neg\phi \to [a]\phi$, then $R(a)xy$ only if $y \in V_M(\phi)$. But then, if also $M \models \phi \to [b]\neg\phi$, then $R(b)yz$ only if $z \in V_M(\neg\phi)$. Therefore, if $x \in V_M(\neg\phi)$, then $R(a;b)xz$ only if $z \in V_M(\neg\phi)$. The argument also illustrates why it is not sufficient to take $\neg\phi \to [a]\phi$ and $\phi \to [b]\neg\phi$ as local assumptions. If they were local, we could not apply $\phi \to [b]\neg\phi$ to y.

[10]We note that the 'plans as programs' approach, related to the present discussion, is the prevalent approach in deductive planning and there are applications of PDL in the area, see (Ghallab, Nau, & Traverso, 2004, Ch. 12) and (Rosenschein, 1981; Stephan & Biundo, 1993).

Igor Sedlár

do nothing) than inconsistency concerning q. Planning formalisms based on normal modal logic (such as dynamic epistemic logic) recognize only one inconsistent database – the empty set. Hence, epistemic planning based on these approaches cannot diversify plans according to the 'seriousness' of the inconsistency involved.

6 Conclusion

This article explored the possibility to adapt PDL to modelling programs that operate on possibly incomplete and inconsistent bodies of information. It turns out that there is quite a simple way to do this, namely, defining a De Morgan negation in terms of a special atomic program $\star$, seen as the Routley star operation. The only extra assumptions needed pertain to $R(\star)$, which is assumed to be serial, symmetric and functional (function of period two). The resulting extension of PDL, NPDL, is decidable and has a sound and weakly complete axiomatisation (any axiomatisation of PDL plus the obvious axioms defining symmetry, seriality and functionality of $R(\star)$), as shown by simple modifications of the standard proofs for PDL. Last, but not least, simple examples of expressivity indicate that NPDL (and similar formalisms, e.g., the Belnapian PDL of Sedlár (2016)) might find applications in epistemic planning and related areas.

Appendix

A. Decidability and completeness of NPDL

This appendix contains (the interesting parts of) proofs of Theorems 6 and 7. The proofs are very close to similar proofs for standard PDL, the only difference being the assumptions concerning $R(\star)$. Of course, these assumptions need to be complied with when constructing the canonical model and defining filtrations.

We start by proving Theorem 6.

Definition 1 *Let $FL(\phi)$ be the Fisher-Ladner closure of ϕ (Harel et al., 2000, ch. 6.1). The* Fisher-Ladner-Routley closure *of ϕ, $FLR(\phi)$, is defined as*

$$FLR(\phi) = FL(\phi) \cup \{[\star]\psi \mid \psi \in FL(\phi) \text{ and } \psi \neq [\star]\chi, \text{ for all } \chi\}$$

Non-Classical PDL on the Cheap

Hence, $FLR(\phi)$ is $FL(\phi)$ extended by $[\star]\psi$ for every $\psi \in FL(\phi)$ such that ψ does not begin with '$[\star]$'. It is plain that $FLR(\phi)$ is finite for all ϕ (as $FL(\phi)$ is). Our reason for working with FLR instead of FL is that we want the filtrated $R(\star)$ to be a function of period two, but this is not the case if the canonical model is filtered trough $FL(\phi)$. We show later (Lemma 4) that the problem does not arise when FL is replaced by FLR.

Our main task now is to check that FLR has all the important properties of FL. In fact, the original proofs need to be modified only slightly. To be closer to the original proofs, we assume in this appendix that, as in (Harel et al., 2000), '$\bot$' and '$\to$' are primitive and the rest of the Boolean connectives are defined using these two in the usual manner.

Lemma 1 *If $\psi \to \chi \in FLR(\phi)$, then $\psi \in FLR(\phi)$ and $\chi \in FLR(\phi)$.*

Proof. If $\psi \to \chi \in FLR(\phi)$, $\psi \to \chi \in FL(\phi)$ by the definition of FLR. But then, by Lemma 6.1. of (Harel et al., 2000, ch. 6.1), $\{\psi, \chi\} \subseteq FL(\phi) \subseteq FLR(\phi)$. □

Lemma 2

1. *If $[\alpha]\psi \in FLR(\phi)$, then $\psi \in FLR(\phi)$*

2. *If $[\chi?]\psi \in FLR(\phi)$, then $\chi \in FLR(\phi)$*

3. *If $[\alpha \cup \beta]\psi \in FLR(\phi)$, then $[\alpha]\psi \in FLR(\phi)$ and $[\beta]\psi \in FLR(\phi)$*

4. *If $[\alpha; \beta]\psi \in FLR(\phi)$, then $[\alpha][\beta]\psi \in FLR(\phi)$ and $[\beta]\psi \in FLR(\phi)$*

5. *If $[\alpha^*]\psi \in FLR(\phi)$, then $[\alpha][\alpha^*]\psi \in FLR(\phi)$*

Proof. The only case to check is $[\star]\psi \in (FLR(\phi) \setminus FL(\phi))$. However, this case arises only if $\psi \in FL(\phi)$, ergo, if $\psi \in FLR(\phi)$. □

In the following lemma, $\#X$ denotes the cardinality of set X and $|\phi|$ denotes the length (number of symbols) of ϕ, excluding parentheses.

Lemma 3 *For any ϕ, $\#FLR(\phi) \leq (2 \times |\phi|)$*

Proof. By Lemma 6.3. of (Harel et al., 2000, ch. 6.1), $\#FL(\phi) \leq |\phi|$. The worst-case scenario is that no $\psi \in FL(\phi)$ begins with a '$[\star]$'. In that case, $\#FLR(\phi) = 2 \times \#FL(\phi)$. □

Igor Sedlár

In what follows, a *non-standard* Routley model is a non-standard dynamic model in the sense of (Harel et al., 2000, ch. 6.3), where $R(\star)$ is serial, symmetric and functional. A *standard* Routley model is just a Routley model. M^ϕ for non-standard M is defined just as in the case of standard M. It is plain that M^ϕ is standard even if M is non-standard.

Definition 2 *Let M be a (standard or non-standard) Routley model. Define M^ϕ, the filtration of M trough ϕ, as follows:*

- $x \equiv_\phi y$ iff for all $\psi \in FLR(\phi)$, $(x \models \psi$ iff $y \models \psi)$
- $[x] = \{y \mid x \equiv_\phi y\}$
- $S^\phi = \{[x] \mid x \in S\}$
- $R_a^\phi = \{\langle [x], [y] \rangle \mid R_a xy\}$ *for all* $a \in AP$
- $V^\phi(p) = \{[x] \mid x \in V(p)\}$ *for all* $p \in AF$

R^ϕ *for compound programs and* $\models^\phi$ *are defined on* S^ϕ *as in the definition of Routley models.*

Lemma 4 $R_\star^\phi$ *is serial, symmetric and functional.*

Proof. Seriality and symmetry are straightforward. Functionality needs a bit more work. Assume that (a) $R_\star^\phi[x][y]$, $R_\star^\phi[x][z]$, but (b) $y \not\equiv_\phi z$. (b) means that there is $\psi \in FLR(\phi)$ such that $y \models \psi$ and $z \not\models \psi$. (a) means that there are x', x'', y', z' such that

- $x', x'' \in [x]$, $y' \in [y]$ and $z' \in [z]$
- $R_\star x'y'$ and $R_\star x''z'$

Together with (b), this means that

- $y' \models \psi$ and, therefore, $x' \models \langle \star \rangle \psi$, i.e. (Corollary 1), $x' \models [\star]\psi$
- $z' \not\models \psi$ and, therefore, $x'' \not\models [\star]\psi$

These two claims imply that $[\star]\psi \notin FLR(\phi)$ even if $\psi \in FLR(\phi)$. Now, either $\psi = [\star]\chi$ for some χ or not. If not, then $\psi \in FL(\phi)$ and, by the definition of FLR, $[\star]\psi \in FLR(\phi)$, a contradiction. If $\psi = [\star]\chi$ for some χ, then we can reason as follows. By symmetry of $R_\star$, $y' \models \psi$ implies $x' \models \chi$. By Lemma 2, $\chi \in FLR(\phi)$ and, hence, $x'' \models \chi$. However, $x'' \not\models [\star][\star]\chi$ ($x'' \not\models [\star]\psi$) implies, by Corollary 1, that $x'' \not\models \chi$. A contradiction. □

Lemma 4 implies that M^ϕ is a standard Routley model, even if M happens to be non-standard.

Lemma 5 (Filtration Lemma) *Let M be a (standard or non-standard) Routley model. Then*

1. *For all $\psi \in FLR(\phi)$, $x \models \psi$ iff $[x] \models^\phi \psi$*

2. *For all $[\alpha]\psi \in FLR(\phi)$,*

   (a) If $R_\alpha xy$, then $R_\alpha^\phi [x][y]$

   (b) If $R_\alpha^\phi [x][y]$ and $x \models [\alpha]\psi$, then $y \models \psi$

Proof. The proof is an exact copy of the proofs of Lemmas 6.4 and 6.6. (Harel et al., 2000, ch. 6.2 and 6.3). As an inspection of the original proofs shows, this copy-pasting is possible because of Lemmas 1 and 2. □

Theorem 8 *If ϕ is satisfied in a (standard or non-standard) Routley model, then it is satisfied in some standard Routley model with at most $2^{(2\times|\phi|)}$ states.*

Proof. Lemma 5 implies that $(M,x) \models \phi$ for any (standard or non-standard) M only if $(M^\phi, [x]) \models^\phi \phi$. M^ϕ is a standard Routley model by Lemma 4. Finally, S^ϕ has no more states than the number of truth assignments to formulas in $FLR(\phi)$, which is by Lemma 3 at most $2^{(2\times|\phi|)}$. □

Theorem 6 follows immediately.

To prove Theorem 7, it is sufficient to establish the following claim; the theorem then follows by the standard argument (Harel et al., 2000, ch. 7). The $H(\text{NPDL})$-canonical model M^c is defined in the usual way; $X \models^c \phi$ is defined as $\phi \in X$.

Lemma 6 *M^c is a non-standard Routley model.*

Proof. The fact that M^c is non-standard ($R^c(\alpha^*)$ is a superset of, not necessarily identical to, the reflexive transitive closure of $R(\alpha)$) is established as in PDL, as are the facts that $R^c(\alpha)$ for compound α and $\models^c$ behave as they should. The only new thing to prove is that $R^c(\star)$ is serial, symmetric and functional.

This follows from including the $\star$-axioms by standard modal reasoning. We formulate only the argument concerning functionality explicitly. Assume that $R_\star^c XY$ and $R_\star^c XZ$ but $Y \neq Z$. Hence, there is $\phi \in Y$ and

$\phi \notin Z$. Consequently, $\neg\phi \in Z$, $\langle\star\rangle\phi \wedge \langle\star\rangle\neg\phi \in X$. It follows that $\langle\star\rangle\bot \in X$ ($\langle\star\rangle\chi \leftrightarrow [\star]\chi$ is an axiom). But $R_\star^c$ is serial and, therefore, $\bot \in Y$ for some maximal consistent Y. A contradiction. □

References

Arieli, O., & Avron, A. (1996). Reasoning with logical bilattices. *Journal of Logic, Language and Information*, 5(1), 25–63.

Aucher, G., & Bolander, T. (2013). Undecidability in epistemic planning. In F. Rossi (Ed.), *Proceedings of the Twenty-Third International Joint Conference on Artificial Intelligence* (pp. 27–33). AAAI Press.

Baltag, A., & Moss, L. S. (2004). Logics for epistemic programs. *Synthese*, 139(2), 165–224.

Baltag, A., & Smets, S. (2006). LQP: the dynamic logic of quantum information. *Mathematical Structures in Computer Science*, 16(03), 491.

Belnap, N. (1977a). How a computer should think. In G. Ryle (Ed.), *Contemporary Aspects of Philosophy*. Oriel Press Ltd.

Belnap, N. (1977b). A useful four-valued logic. In J. M. Dunn & G. Epstein (Eds.), *Modern Uses of Multiple-Valued Logic* (pp. 5–37). Dordrecht: Springer Netherlands.

Bergfeld, J. M., & Sack, J. (2015). Deriving the correctness of quantum protocols in the probabilistic logic for quantum programs. *Soft Computing*.

Bolander, T., & Andersen, M. B. (2011). Epistemic planning for single and multi-agent systems. *Journal of Applied Non-Classical Logics*, 21(1), 9–34.

Dunn, J. M. (1976). Intuitive semantics for first-degree entailments and "coupled trees". *Philosophical Studies*, 29, 149–168.

Dunn, J. M. (1993). Star and perp: Two treatments of negation. *Philosophical Perspectives*, 7, 331–357.

Fagin, R., Halpern, J. Y., & Vardi, M. (1995). A nonstandard approach to the logical omniscience problem. *Artificial Intelligence*, 79(2), 203–240.

Fischer, M. J., & Ladner, R. E. (1979). Propositional dynamic logic of regular programs. *Journal of Computer and System Sciences*, 18, 194–211.

Ghallab, M., Nau, D., & Traverso, P. (2004). *Automated Planning: Theory and Practice*. San Francisco: Morgan Kaufmann.

Harel, D., Kozen, D., & Tiuryn, J. (2000). *Dynamic Logic*. MIT Press.

Odintsov, S., & Wansing, H. (2010). Modal logics with Belnapian truth values. *Journal of Applied Non-Classical Logics*, *20*(3), 279–301.

Rosenschein, S. J. (1981). Plan synthesis: A logical perspective. In *IJCAI'81 Proceedings of the 7th International Joint Conference on Artificial Intelligence* (pp. 331–337). San Franscisco: Morgan Kaufmann Publishers Inc.

Routley, R., & Routley, V. (1972). The semantics of first degree entailment. *Noûs*, *6*(4), 335–359.

Sedlár, I. (2016). Propositional dynamic logic with Belnapian truth values. In L. Beklemishev, S. Demri, & A. Máté (Eds.), *Advances in Modal Logic. Vol. 11* (pp. 503–519). London: College Publications.

Stephan, W., & Biundo, S. (1993). A new logical framework for deductive planning. In *IJCAI'93 Proceedings of the 13th International Joint Conference on Artificial Intelligence* (pp. 32–38). San Francisco: Morgan Kaufmann Publishers Inc.

Teheux, B. (2014). Propositional dynamic logic for searching games with errors. *Journal of Applied Logic*, *12*(4), 377–394.

Igor Sedlár
Institute of Computer Science, Czech Academy of Sciences
The Czech Republic
E-mail: `sedlar@cs.cas.cz`

LP, K3, and FDE as Substructural Logics

LIONEL SHAPIRO[1]

Abstract: Building on recent work, I present sequent systems for the non-classical logics LP, K3, and FDE with two main virtues. First, derivations closely resemble those in standard Gentzen-style systems. Second, the systems can be obtained by reformulating a classical system using nonstandard sequent structure and simply removing certain structural rules (relatives of exchange and contraction). I clarify two senses in which these logics count as "substructural."

Keywords: Non-classical logics, Substructural logics, Many-valued logics, Many-sided sequent systems, Structural rules

1 Introduction

The non-classical propositional logics LP, K3, and FDE, standardly presented using many-valued semantics, have long enjoyed cut-free sequent proof systems.[2] The paracomplete and paraconsistent logic FDE is in fact introduced in Anderson and Belnap (1963) using a two-sided sequent system; the extensions of this system to the paraconsistent LP and the paracomplete K3 are straightforward (Avron, 1991; Beall, 2011). Three-sided systems for LP and K3 are given by Ripley (2012) and Hjortland (2013).[3] Most recently, several authors have independently proposed alternative sequent systems for LP and K3 (Fjellstad, 2016; Shapiro, 2016) or all three logics (Wintein, 2016). These systems, which are closely related, are motivated by a variety of formal and philosophical considerations.

---

[1] For invaluable help with these topics, I am indebted to Dave Ripley. I also benefited greatly from discussion with Jc Beall and the participants of Logica 2016, where a version of the earlier paper on which I here elaborate was given.

[2] FDE is the logic of "first-degree entailments" of Anderson and Belnap (1963, 1975). LP is the "logic of paradox" of Priest (1979); its propositional fragment is due to Asenjo (1966). The "Strong Kleene" logic K3 derives from Kleene (1952). For an overview, see Priest (2008).

[3] These are based on the methods of Baaz, Fermüller, and Zach (1993), which yield a four-sided system for FDE (as Hjortland notes).

Lionel Shapiro

My purpose here is to present a more explicit version of the *substructural* approach proposed in Shapiro (2016), and to show how it can be improved by incorporating the key insight of Fjellstad and Wintein. The result is a class of sequent systems for these sub-classical logics whose use is especially intuitive: derivations look almost like familiar classical derivations of the same sequents. To obtain these systems, I start with a Gentzen-style system for classical logic and introduce new sequent structure, together with a set of structural rules. By including different subsets of those rules, I arrive at sequent systems for FDE, LP, K3 or (when all are included) classical logic CL. The systems coincide in their initial sequents, their operational rules, and in how derivable sequents correspond to true consequence claims. They differ only in their structural rules.

The paper is organized as follows. §2 introduces the consequence relations FDE, LP, and K3 and the respective two-sided sequent systems. §3 motivates and elaborates the three-sided substructural approach from Shapiro (2016). In the process, two senses are distinguished in which a logic can count as substructural. Finally, §4 modifies this approach to yield four-sided substructural systems that amount to variants of the systems of Wintein and Fjellstad. The resulting systems possess the advantages of the three-sided substructural systems without the disadvantage these share with two-sided systems—namely, the need for rules involving more than one connective.

2 Two-sided semantics and sequent systems

2.1 Many-valued semantics

To start, I present the four propositional logics semantically (cf. Priest, 2008, §8.4). Truth-values will be the members of $\{t, n, b, f\}$, partially ordered as reflected in this lattice:

$$\begin{array}{c} t \\ \nearrow \nwarrow \\ n \quad\quad b \\ \nwarrow \nearrow \\ f \end{array}$$

Definition 1 *For each of our logics X, an X-valuation is a function from the set of sentences to $\{t, n, b, f\}$. An FDE-valuation has as its range the full set $\{t, n, b, f\}$, while a LP-valuation has range $\{t, b, f\}$, a K3-valuation has range $\{t, n, f\}$, and a CL-valuation a function has range $\{t, f\}$. Every X-valuation is subject to the following conditions:*

LP, K3, and FDE as Substructural Logics

- *if $v(\alpha) = t$ then $v(\neg\alpha) = f$,*
  if $v(\neg\alpha) = f$ then $v(\alpha) = t$,
  if $v(\alpha) = n$, or $v(\alpha) = b$, then $v(\neg\alpha) = v(\alpha)$,

- *$v(\alpha \wedge \beta)$ is the greatest lower bound of $v(\alpha)$ and $v(\beta)$,*

- *$v(\alpha \vee \beta)$ is the least upper bound of $v(\alpha)$ and $v(\beta)$.*

For each logic X, we can now define its multiple-conclusion consequence relation by taking the values t and b as designated:

Definition 2 *For any sets Γ and Δ of sentences, $\Gamma \vDash_X \Delta$ just in case there is no X-valuation such that each sentence in Γ receives either t or b, and each sentence in Δ receives either f or n.*

It's easy to check that

- $\Gamma \vDash_{LP} \Delta, \alpha \vee \neg\alpha$, whereas this consequence (excluded middle) fails for FDE and K3, which are thus *paracomplete*,

- $\Gamma, \alpha \wedge \neg\alpha \vDash_{K3} \Delta$, whereas this consequence (explosion) fails for FDE and LP, which are thus *paraconsistent*.

2.2 Sequent systems

The following proof system $\mathbf{S2}_{FDE}$ (for 'two-sided') is a variant of one of Anderson and Belnap's systems (1963; 1975, pp. 179–80). Sequents are of form $\Gamma \triangleright \Delta$, where (as throughout this paper) upper-case Greek letters stand for sets of sentences.[4] However, as I'll later be generalizing the sequent structure, I write the rules using $\mathfrak{A} \triangleright \mathfrak{B}$ instead. In the present case, $\mathfrak{A}(\alpha)$ abbreviates some Γ, α, which in turn stands for the set $\Gamma \cup \{\alpha\}$.

1. *Initial sequents*

   For atomic α: (Id) $\alpha \triangleright \alpha$ (NegId) $\neg\alpha \triangleright \neg\alpha$

2. *Operational rules*

   Conjunction and disjunction rules

$$(\wedge L) \frac{\mathfrak{A}(\alpha, \beta) \triangleright \mathfrak{B}}{\mathfrak{A}(\alpha \wedge \beta) \triangleright \mathfrak{B}} \qquad (\wedge R) \frac{\mathfrak{A} \triangleright \mathfrak{B}(\alpha) \qquad \mathfrak{A} \triangleright \mathfrak{B}(\beta)}{\mathfrak{A} \triangleright \mathfrak{B}(\alpha \wedge \beta)}$$

[4] I follow Ripley (2012) in using $\triangleright$ as sequent separator.

$$(\vee L) \frac{\mathfrak{A}(\alpha) \triangleright \mathfrak{B} \quad \mathfrak{A}(\beta) \triangleright \mathfrak{B}}{\mathfrak{A}(\alpha \vee \beta) \triangleright \mathfrak{B}} \qquad (\vee R) \frac{\mathfrak{A} \triangleright \mathfrak{B}(\alpha, \beta)}{\mathfrak{A} \triangleright \mathfrak{B}(\alpha \vee \beta)}$$

Negated connective rules

$$(\neg \wedge L) \frac{\mathfrak{A}(\neg\alpha) \triangleright \mathfrak{B} \quad \mathfrak{A}(\neg\beta) \triangleright \mathfrak{B}}{\mathfrak{A}(\neg(\alpha \wedge \beta)) \triangleright \mathfrak{B}} \qquad (\neg \wedge R) \frac{\mathfrak{A} \triangleright \mathfrak{B}(\neg\alpha, \neg\beta)}{\mathfrak{A} \triangleright \mathfrak{B}(\neg(\alpha \wedge \beta))}$$

$$(\neg \vee L) \frac{\mathfrak{A}(\neg\alpha, \neg\beta) \triangleright \mathfrak{B}}{\mathfrak{A}(\neg(\alpha \vee \beta)) \triangleright \mathfrak{B}} \qquad (\neg \vee R) \frac{\mathfrak{A} \triangleright \mathfrak{B}(\neg\alpha) \quad \mathfrak{A} \triangleright \mathfrak{B}(\neg\beta)}{\mathfrak{A} \triangleright \mathfrak{B}(\neg(\alpha \vee \beta))}$$

$$(\neg\neg L) \frac{\mathfrak{A}(\alpha) \triangleright \mathfrak{B}}{\mathfrak{A}(\neg\neg\alpha) \triangleright \Delta} \qquad (\neg\neg R) \frac{\mathfrak{A} \triangleright \mathfrak{B}(\alpha)}{\mathfrak{A} \triangleright \mathfrak{B}(\neg\neg\alpha)}$$

3. *Structural rule*

$$(\text{Weak}) \frac{\mathfrak{A}(\Gamma) \triangleright \mathfrak{B}(\Delta)}{\mathfrak{A}(\Gamma, \Gamma') \triangleright \mathfrak{B}(\Delta, \Delta')}$$

To obtain $\mathbf{S2}_{LP}$ and $\mathbf{S2}_{K3}$ simply add, respectively, the initial sequents LEM or Explosion, where α is atomic:

$$(\text{LEM}) \; \emptyset \triangleright \alpha, \neg\alpha \qquad (\text{Explosion}) \; \alpha, \neg\alpha \triangleright \emptyset$$

Including LEM as well as Explosion yields the system $\mathbf{S2}_{CL}$.

Proposition 1 (Beall, 2011) *For each logic X, the sequent $\Gamma \triangleright \Delta$ is derivable in $\mathbf{S2}_X$ iff $\Gamma \vDash_X \Delta$.*

3 Three-sided substructural systems

3.1 Motivation

Shapiro (2016) argues that the systems $\mathbf{S2}_X$ are unsatisfactory when evaluated from the perspective according to which it is desirable to preserve as much as possible of the familiar ways of deriving consequence claims using a classical multiple-conclusion sequent system.[5] Such a system includes rules that give negation its usual "flip-flop behavior" (Beall, 2016).

---

[5]Fjellstad too argues that it is "easier to adopt the sequent calculus as a tool for reasoning and thus perhaps also the logic as such" when "the rules are relatively familiar" (2016).

LP, K3, and FDE as Substructural Logics

$$(\neg L) \frac{\Gamma \triangleright \Delta, \alpha}{\Gamma, \neg \alpha \triangleright \Delta} \qquad (\neg R) \frac{\Gamma, \alpha \triangleright \Delta}{\Gamma \triangleright \Delta, \neg \alpha}$$

When we consider the material conditional $\supset$, where $\alpha \supset \beta$ is $\neg \alpha \vee \beta$, this behavior corresponds to central aspects of a conditional's behavior. Conditional proof, the rule of right-$\supset$ introduction, is derivable using $\neg R$. Likewise, the $\supset$-left introduction rule that yields the *modus ponens* sequent $\alpha, \alpha \supset \beta \triangleright \beta$ is derivable using $\neg L$.

$$\cfrac{\cfrac{\Gamma, \alpha \triangleright \Delta, \beta}{\Gamma \triangleright \Delta, \beta, \neg \alpha} \neg R}{\Gamma \triangleright \Delta, \alpha \supset \beta} \vee R \qquad \cfrac{\cfrac{\Gamma \triangleright \alpha, \Delta}{\Gamma, \neg \alpha \triangleright \Delta} \neg L \qquad \Gamma, \beta \triangleright \Delta}{\Gamma, \alpha \supset \beta \triangleright \Delta} \vee L$$

How much of this familiar inferential behavior must we give up when employing one of our non-classical logics? Beall argues that we renounce it entirely if we adopt FDE as our logic. Since that's what he proposes doing, he concludes that "there is no logical negation." What he means, he clarifies, is that while there is a "logical connective called *negation*, ... logic imposes no interesting constraint on it ... aside from what logic demands of its interaction with other logical connectives." By the same token, he should hold, FDE contains no "logical conditional."

One aim of Shapiro (2016) was to give sequent systems for LP and K3 that preserve negation's flip-flop behavior together with familiar rules for conjunction and disjunction.[6] The lesson, when applied to FDE, is that Beall's conclusion is problematic.[7] As I'll now show, the way to preserve familiar classical derivations of consequence claims of our non-classical logics is to modify the standard structure of a consequence relation—in a way that goes beyond Beall's own embrace of multiple-conclusion sequents.

3.2 Systems distinguished by negation rules

The systems in Shapiro (2016) weren't designed for their formal virtues, but to make that paper's philosophical point. Here I give a slightly modified formulation, which recovers a version of the subformula property: for each

---

[6] The presentation there focuses on LP, but the system for K3 is given in footnote 20.

[7] This is so even if we understand him as demanding of a "logical negation" that it coexist with other connectives obeying familiar operational rules. In other words, his conclusion is problematic even when interpreted so as not to already be refuted by a many-sided system based on Baaz et al. (1993), whose derivations look very unlike classical ones. Such systems do display "stand-alone negation behavior" involving flip-flops.

formula in a sequent's derivation, either it, or a formula of which it is the negation, is a subformula of some sentence appearing in that sequent.[8]

Sequents will take the form $\Gamma; \Sigma \triangleright \Delta$. The two-sided $\Gamma \triangleright \Delta$ is to be understood as a convenient notation for $\Gamma; \emptyset \triangleright \Delta$.[9] The system $\mathbf{S3}_{FDE}$ consists of the following initial sequents and rules. Where one of the rules from §2.2 is included, $\mathfrak{A}(\alpha)$ on the left may be either $\Gamma, \alpha; \Sigma$ or $\Gamma; \Sigma, \alpha$ and $\mathfrak{B}(\alpha)$ on the right is Δ, α.

1. *Initial sequents*

    For atomic α: (Id) $\alpha \triangleright \alpha$ (NegId*) $\neg \alpha; \alpha \triangleright \emptyset$

2. *Operational rules*

    Negation rules

    $$(\neg L) \frac{\Gamma; \Sigma \triangleright \Delta, \alpha}{\Gamma; \Sigma, \neg \alpha \triangleright \Delta} \qquad (\neg R) \frac{\Gamma; \Sigma, \alpha \triangleright \Delta}{\Gamma; \Sigma \triangleright \Delta, \neg \alpha}$$

    All above conjunction, disjunction and negated connective rules

3. *Structural rule*

    Weak

We then obtain $\mathbf{S3}_{K3}$ by including an additional rule $\neg L^*$, and $\mathbf{S3}_{LP}$ by including instead its dual rule $\neg R^*$. The system $\mathbf{S3}_{CL}$ includes both rules.

$$(\neg L^*) \frac{\Gamma; \Sigma \triangleright \Delta, \alpha}{\Gamma, \neg \alpha; \Sigma \triangleright \Delta} \qquad (\neg R^*) \frac{\Gamma, \alpha; \Sigma \triangleright \Delta}{\Gamma; \Sigma \triangleright \Delta, \neg \alpha}$$

Shapiro (2016, note 16) describes how to extend the three-valued semantics for LP to the three-sided sequents of these systems. I present this semantics in §4.2, generalized to four-sided sequents following Fjellstad and Wintein. Using it and Proposition 1, I show that $\mathbf{S3}_X$ is sound and complete with respect to X-consequence as defined above.

---

[8] In place of a left introduction rule for negation, Shapiro (2016) uses the elimination rule that is the inverse of a right introduction rule. Initial sequents are of form Id, where α isn't restricted to atomic sentences.

[9] Shapiro (2016) uses a slightly different format. Sequents there include three-sided ones $\Gamma; \Sigma \vdash \Delta$ as well as two-sided "ordinary sequents" $\Gamma \vdash \Delta$. The notations $\emptyset; \Gamma \vdash \Delta$ and $\Gamma; \emptyset \vdash \Delta$ are *both* stipulated to stand for $\Gamma \vdash \Delta$, which behaves like $\Gamma; \emptyset \triangleright \Delta$ in $\mathbf{S3}_X$ below. This necessitates restricting $\neg L$ and $\neg R$ below to instances with nonempty Γ. Furthermore, in the system for LP, it makes possible a weakening rule that is more general than Weak in that it covers the move from $\Gamma \vdash \Delta$ to $\Gamma'; \Gamma \vdash \Delta$. The sequent $\emptyset; \Gamma \triangleright \Delta$ of $\mathbf{S3}_X$ corresponds to no sequent in the systems of Shapiro (2016).

LP, K3, and FDE as Substructural Logics

Proposition 2 $\Gamma \triangleright \Delta$ *is derivable in* $S3_X$ *iff* $\Gamma \vDash_X \Delta$.

Proof. This follows from Corollary 1 in §4.2 below.

The systems $S3_{FDE}$, $S3_{LP}$ and $S3_{K3}$ count as "substructural" in one common sense of that term: each renders inadmissible one or more versions of the standard structural rules of weakening, exchange, contraction or cut.

Fact 1 *The following rule is inadmissible in* $S3_{FDE}$, $S3_{LP}$ *and* $S3_{K3}$.

$$\text{(Exch)} \frac{\Gamma, \alpha; \Sigma, \beta \triangleright \Delta}{\Gamma, \beta; \Sigma, \alpha \triangleright \Delta}$$

Proof. From Proposition 2, since adding Exch derives $\beta \triangleright \alpha, \neg\alpha$, which is invalid in K3, and $\neg\alpha, \alpha \triangleright \neg\beta$, which is invalid in LP.

$$\frac{\dfrac{\dfrac{\dfrac{\alpha \triangleright \alpha}{\alpha; \beta \triangleright \alpha} \text{Weak}}{\beta; \alpha \triangleright \alpha} \text{Exch}}{\beta \triangleright \alpha, \neg\alpha} \neg R} \qquad \frac{\dfrac{\dfrac{\dfrac{\neg\alpha; \alpha \triangleright \emptyset}{\neg\alpha, \beta; \alpha \triangleright \emptyset} \text{Weak}}{\neg\alpha, \alpha; \beta \triangleright \emptyset} \text{Exch}}{\neg\alpha, \alpha \triangleright \neg\beta} \neg R}$$

Fact 2 *The following rules are inadmissible in* $S3_{FDE}$ *and* $S3_{LP}$.

$$\text{(Cont1)} \frac{\Gamma, \alpha; \Sigma, \alpha \triangleright \Delta}{\Gamma, \alpha; \Sigma \triangleright \Delta} \qquad \text{(Cut)} \frac{\Gamma; \Sigma \triangleright \Delta, \alpha \quad \Gamma; \Sigma, \alpha \triangleright \Delta}{\Gamma; \Sigma \triangleright \Delta}$$

Proof. Again, adding Cont1 or Cut derives Explosion.

$$\frac{\dfrac{\dfrac{\dfrac{\alpha \triangleright \alpha}{\alpha; \neg\alpha \triangleright \emptyset} \neg L}{\alpha, \neg\alpha; \neg\alpha \triangleright \emptyset} \text{Weak}}{\alpha, \neg\alpha \triangleright \emptyset} \text{Cont1}} \qquad \frac{\dfrac{\alpha \triangleright \alpha}{\alpha, \neg\alpha \triangleright \alpha} \text{Weak} \quad \dfrac{\dfrac{\neg\alpha; \alpha \triangleright \emptyset}{\alpha, \neg\alpha; \alpha \triangleright \emptyset} \text{Weak}}{\alpha, \neg\alpha \triangleright \emptyset}}{\alpha, \neg\alpha \triangleright \emptyset} \text{Cut}$$

Fact 3 *The following rule is inadmissible in* $S3_{FDE}$ *and* $S3_{K3}$.

$$\text{(Cont2)} \frac{\Gamma, \alpha; \Sigma, \alpha \triangleright \Delta}{\Gamma; \Sigma, \alpha \triangleright \Delta}$$

Proof. Adding Cont2 derives LEM.[10]

$$\frac{\dfrac{\dfrac{\dfrac{\alpha \triangleright \alpha}{\alpha; \alpha \triangleright \alpha} \text{Weak}}{\emptyset; \alpha \triangleright \alpha} \text{Cont2}}{\emptyset \triangleright \alpha, \neg\alpha} \neg R}$$

---

[10] In fact, Cont2 is inadmissible in $S3_{LP}$ too, since $\alpha; \alpha \triangleright \alpha$ is derivable but $\emptyset; \alpha \triangleright \alpha$ is not. However, Cont2 will be admissible in the revised system $S3'_{LP}$ I introduce next.

3.3 Systems distinguished by shift rules

The term 'substructural' is sometimes used in a different sense. In this sense, the failure of a standard structural rule doesn't suffice to make the logic specified by a sequent system count as substructural. Rather, the logic must be obtainable from classical logic *solely* by restricting structural rules. Došen, who coined the term, introduces it thus:

> Our proposal is to call logics that can be obtained ... by restricting structural rules, *substructural logics*.... Canonically, we should assume the same rules for logical constants in the logic whose structural rules we restrict, i.e. classical logic, and in the resulting substructural logic.... We don't insist that the Gentzen formulation of classical logic whose structural rules we restrict in order to obtain a substructural logic should be the standard one. It could as well be a nonstandard formulation with sequents whose left-hand and right-hand sides are not sequences of formulae of L but some other sort of structure involving formulae of L.... (Došen, 1993, p. 6)

Notice that $\mathbf{S3}_{FDE}$, $\mathbf{S3}_{LP}$ and $\mathbf{S3}_{K3}$ differ from classical logic, and from each other, in their negation rules. Hence they don't qualify as substructural in Došen's sense. But they can be modified so as to meet his condition.

$\mathbf{S3}'_{FDE}$ is just $\mathbf{S3}_{FDE}$ without the rules $\neg\wedge R$, $\neg\vee R$ and $\neg\neg R$, which are easily seen to be admissible in light of Lemma 1 of the Appendix. Each of the systems $\mathbf{S3}'_{LP}$ and $\mathbf{S3}'_{K3}$ is then obtained by simply adding a *structural shift* rule, which is a relative of the usual structural rules of contraction and exchange. For $\mathbf{S3}'_{LP}$, we add the rule RightShift, while for $\mathbf{S3}'_{K3}$ we add the dual rule LeftShift. Adding both shift rules yields $\mathbf{S3}'_{CL}$.

$$(\text{RightShift}) \ \frac{\Gamma, \alpha; \Sigma \rhd \Delta}{\Gamma; \Sigma, \alpha \rhd \Delta} \qquad (\text{LeftShift}) \ \frac{\Gamma; \Sigma, \alpha \rhd \Delta}{\Gamma, \alpha; \Sigma \rhd \Delta}$$

Using RightShift and $\neg R$ we can immediately derive $\neg R^*$, while using $\neg L$ and Left Shift we can derive $\neg L^*$.

Proposition 3 $\Gamma \rhd \Delta$ *is derivable in* $\mathbf{S3}'_X$ *iff* $\Gamma \vDash_X \Delta$.

Proof. This follows from Corollary 1 in §4.2 below.

This means that FDE, LP, and K3 can be obtained from classical logic (as formulated using a sequent framework in which the left-hand sides are pairs of sets of sentences) by removing one or more of this framework's structural rules. Classical logic thus relates to LP, K3, and FDE in a manner analogous to the way it relates to *affine logic*, *distribution-free relevant logic*, and *linear logic*. In a sequent system that uses multisets of sentences, these three logics can be obtained from classical logic by dropping (respectively) contraction, weakening, and both contraction and weakening.[11]

We have seen how $\mathbf{S3}_{FDE}$ and $\mathbf{S3}'_{FDE}$ partly recover the flip-flop behavior of negation in the form of rules ¬L and ¬R. They also partly recover the classical behavior of $\supset$, as the following are both derivable.[12]

$$(\supset\!\mathrm{L}*)\ \frac{\Gamma;\Sigma \triangleright \alpha, \Delta \quad \Gamma;\Sigma, \beta \triangleright \Delta}{\Gamma;\Sigma, \alpha \supset \beta \triangleright \Delta} \qquad (\supset\!\mathrm{R}*)\ \frac{\Gamma;\Sigma, \alpha \triangleright \beta, \Delta}{\Gamma;\Sigma \triangleright \alpha \supset \beta, \Delta}$$

So is the *modus ponens* sequent $\alpha \supset \beta; \alpha \triangleright \beta$.

$$\frac{\neg\alpha; \alpha \triangleright \beta \qquad \beta; \alpha \triangleright \beta}{\alpha \supset \beta; \alpha \triangleright \beta}\ \mathsf{VL1}$$

Consequently, many classical derivations go through virtually unchanged. For example, here is a derivation using the conditional rules admissible in all the above three-sided systems:

$$\frac{\dfrac{\dfrac{\dfrac{\alpha \supset \beta; \alpha \triangleright \beta}{\alpha \supset \beta; \gamma, \alpha \triangleright \gamma, \beta}\ \text{Weak} \quad \dfrac{\alpha \supset \beta; \alpha \triangleright \beta}{\alpha \supset \beta; \gamma, \alpha \triangleright \beta}\ \text{Weak}}{\alpha \supset \beta; \gamma, \gamma \supset \alpha \triangleright \beta}\ \supset\!\mathrm{L}*}{\alpha \supset \beta; \gamma \supset \alpha \triangleright \gamma \supset \beta}\ \supset\!\mathrm{R}*}{\alpha \supset \beta \triangleright (\gamma \supset \alpha) \supset (\gamma \supset \beta)}\ \supset\!\mathrm{R}*$$

4 Four-sided substructural systems

Nonetheless, with a view to preserving as much classical behavior as possible, the present systems still aren't optimal. As did $\mathbf{S2}_X$, they require negated connective rules.[13] Also, while $\alpha \supset \beta; \alpha \triangleright \beta$ is derivable in $\mathbf{S3}'_X$, it isn't directly derivable using any $\supset$-L-like rule admissible in $\mathbf{S3}'_{LP}$. By this I mean any rule that is an instance of the following general form.

$$(\mathrm{COND})\ \frac{\mathfrak{A} \triangleright \mathfrak{B}(\alpha) \qquad \mathfrak{A}(\beta) \triangleright \mathfrak{B}}{\mathfrak{A}(\alpha \supset \beta) \triangleright \mathfrak{B}}$$

---

[11] See Paoli (2002, §2.2). Multisets are like sets except that they keep track of how many times a given member appears.

[12] In the systems for LP and K3, two additional conditional rules each are derivable; I omit them here for reasons of space.

[13] Both Fjellstad (2016) and Wintein (2016) take it as a desideratum that each operational rule should refer to only one connective.

4.1 Systems with dual negation and shift rules

Both defects can be remedied by implementing the key idea underlying the systems of Fjellstad (2016) and Wintein (2016). This is to allow *dual* flip-flop behavior by altering the sequent structure. Fjellstad presents his system as a "dual two-sided sequent calculus," whereas Wintein presents his as a four-sided signed calculus. Here I follow Wintein, though I use notation that displays a "left" vs. "right" distinction implicit in his formulation, and replace his extra initial sequents with shift rules. (See the Appendix.)

To get systems $\mathbf{S4}_X$, let sequents take form $\Gamma; \Sigma \triangleright \Delta; \Theta$. As before, $\Gamma \triangleright \mathfrak{B}$ abbreviates $\Gamma; \emptyset \triangleright \mathfrak{B}$, but now $\mathfrak{A} \triangleright \Delta$ abbreviates $\mathfrak{A} \triangleright \Delta; \emptyset$. We modify $\mathbf{S3}_X$ to take advantage of left-right symmetry. $\mathbf{S4}_{FDE}$ consists of the following initial sequents and rules. Where a rule from §2.2 is included, $\mathfrak{A}(\alpha)$ on the left may be either $\Gamma, \alpha; \Sigma$ or $\Gamma; \Sigma, \alpha$ and $\mathfrak{B}(\alpha)$ on the right is $\Delta, \alpha; \Theta$ or $\Delta; \Theta, \alpha$.

1. *Initial sequents*

   For atomic α: (Id) $\alpha \triangleright \alpha$ (NegId**) $\emptyset; \alpha \triangleright \emptyset; \alpha$

2. *Operational rules*

   Negation rules

   $$(\neg L1) \ \frac{\Gamma; \Sigma \triangleright \Delta; \Theta, \alpha}{\Gamma, \neg\alpha; \Sigma \triangleright \Delta; \Theta} \qquad (\neg L2) \ \frac{\Gamma; \Sigma \triangleright \Delta, \alpha; \Theta}{\Gamma; \Sigma, \neg\alpha \triangleright \Delta; \Theta}$$

   $$(\neg R1) \ \frac{\Gamma; \Sigma, \alpha \triangleright \Delta; \Theta}{\Gamma; \Sigma \triangleright \Delta, \neg\alpha; \Theta} \qquad (\neg R2) \ \frac{\Gamma, \alpha; \Sigma \triangleright \Delta; \Theta}{\Gamma; \Sigma \triangleright \Delta; \Theta, \neg\alpha}$$

   Conjunction and disjunction rules

   $\wedge$L, $\wedge$R, $\vee$L, $\vee$R

3. *Structural rule*

   Weak

Wintein's systems for LP, K3, and classical logic are distinguished by additional initial sequents (as are Fjellstad's systems, once translated into the present format as indicated in the Appendix). Here, I instead use structural shift rules. To get $\mathbf{S4}_{LP}$, we add two *inward shift* rules.

LP, K3, and FDE as Substructural Logics

$$\text{(RightShiftL)} \frac{\Gamma, \alpha; \Sigma \triangleright \Delta; \Theta}{\Gamma; \Sigma, \alpha \triangleright \Delta; \Theta} \quad \text{(LeftShiftR)} \frac{\Gamma; \Sigma \triangleright \Delta; \Theta, \alpha}{\Gamma; \Sigma \triangleright \Delta, \alpha; \Theta}$$

To get $\mathbf{S4}_{K3}$, we add instead the dual pair of *outward shift* rules.

$$\text{(LeftShiftL)} \frac{\Gamma; \Sigma, \alpha \triangleright \Delta; \Theta}{\Gamma, \alpha; \Sigma \triangleright \Delta; \Theta} \quad \text{(RightShiftR)} \frac{\Gamma; \Sigma \triangleright \Delta, \alpha; \Theta}{\Gamma; \Sigma \triangleright \Delta; \Theta, \alpha}$$

Finally, to get $\mathbf{S4}_{CL}$, we include all four shift rules.

By allowing two pairs of flip-flop rules, $\mathbf{S4}_{FDE}$ derives the negated connective rules of $\mathbf{S3}_{FDE}$. Likewise, it derives these conditional rules:

$$(\supset\text{L1**}) \frac{\Gamma; \Sigma \triangleright \alpha, \Delta; \Theta \qquad \Gamma; \Sigma, \beta \triangleright \Delta; \Theta}{\Gamma; \Sigma, \alpha \supset \beta \triangleright \Delta; \Theta}$$

$$(\supset\text{L2**}) \frac{\Gamma; \Sigma \triangleright \Delta; \Theta, \alpha \qquad \Gamma, \beta; \Sigma \triangleright \Delta; \Theta}{\Gamma, \alpha \supset \beta; \Sigma \triangleright \Delta; \Theta}$$

And $\supset$L2**, which is an instance of the general form COND above, may be used to derive $\alpha \supset \beta; \alpha \triangleright \beta$.

$$\frac{\emptyset; \alpha \triangleright \beta; \alpha \qquad \beta; \alpha \triangleright \beta}{\alpha \supset \beta; \alpha \triangleright \beta} \supset\text{L2**}$$

The structural rules inadmissible in $\mathbf{S4}_X$ include, besides those we saw are inadmissible in $\mathbf{S3}_X$, dual rules involving a sequent's right-hand side.

$$\text{(ExchR)} \frac{\Gamma; \Sigma, \triangleright \Delta, \alpha; \Theta, \beta}{\Gamma; \Sigma \triangleright \Delta, \beta; \Theta, \alpha} \quad \text{(Cut*)} \frac{\Gamma; \Sigma \triangleright \Delta; \Theta, \alpha \qquad \Gamma, \alpha; \Sigma \triangleright \Delta}{\Gamma; \Sigma \triangleright \Delta}$$

$$\text{(ContR1)} \frac{\Gamma; \Sigma \triangleright \Delta, \alpha; \Theta; \alpha}{\Gamma; \Sigma \triangleright \Delta, \alpha; \Theta} \quad \text{(ContR2)} \frac{\Gamma; \Sigma \triangleright \Delta, \alpha; \Theta; \alpha}{\Gamma; \Sigma, \triangleright \Delta; \Theta, \alpha}$$

Examples parallel to those in Facts 1-3 show that ExchR is inadmissible in $\mathbf{S4}_{FDE}$, $\mathbf{S4}_{LP}$ and $\mathbf{S4}_{K3}$, while Cut* and ContR1 are inadmissible in $\mathbf{S4}_{K3}$, and ContR2 is inadmissible in $\mathbf{S4}_{LP}$.[14]

---

[14] For more on the status of cut rules, cf. Wintein (2016, pp. 526-8).

Lionel Shapiro

4.2 Many-valued semantics

Wintein gives a four-valued semantics for four-sided sequents. If we require that Θ be empty, and note that LP-valuations have range $\{t, f, b\}$, the result in the case of LP is essentially the semantics for three-sided sequents given in Shapiro (2016, note 16).[15]

Definition 3 *For any sets Γ, Σ, Δ, and Θ of sentences, $\Gamma; \Sigma \vDash_{4X} \Delta; \Theta$ iff there is no X-valuation v such that*
$v(\gamma) = t \text{ or } v(\gamma) = b \text{ for all } \gamma \in \Gamma, \text{ and}$
$v(\sigma) = t \text{ or } v(\sigma) = n \text{ for all } \sigma \in \Sigma \text{ and}$
$v(\delta) = f \text{ or } v(\delta) = n \text{ for all } \delta \in \Delta \text{ and}$
$v(\theta) = f \text{ or } v(\theta) = b \text{ for all } \theta \in \Theta$

The following soundness and completeness results obtain.

Proposition 4 $\Gamma; \Sigma \triangleright \Delta; \Theta$ *is derivable in* $\mathbf{S4}_X$ *iff* $\Gamma; \Sigma \vDash_{4X} \Delta; \Theta$.

Proof. To establish the soundness direction, it suffices to check that the initial sequents of $\mathbf{S4}_X$ correspond to 4X-consequences and that the rules preserve 4X-consequence. The completeness direction follows from the completeness of Wintein's systems (2016, Theorem 1), since his initial sequents and rules are all derivable in $\mathbf{S4}_X$.

Proposition 5 *(a)* $\Gamma; \emptyset \triangleright \Delta$ *is derivable in* $\mathbf{S3}_X$ *iff* $\Gamma; \emptyset \vDash_{4X} \Delta; \emptyset$ *and (b)* $\Gamma; \Sigma \triangleright \Delta$ *is derivable in* $\mathbf{S3}'_X$ *iff* $\Gamma; \Sigma \vDash_{4X} \Delta; \emptyset$.

Important note: it's *not* the case that $\Gamma; \Sigma \triangleright \Delta$ is derivable in $\mathbf{S3}_X$ iff $\Gamma; \Sigma \vDash_{4X} \Delta; \emptyset$. For example, $\emptyset; \alpha \vDash_{4LP} \alpha; \emptyset$, yet $\emptyset; \alpha \triangleright \alpha$ isn't derivable in $\mathbf{S3}_{LP}$. This shows that RightShift is inadmissible. And since $\emptyset \triangleright \alpha, \neg \alpha$ is derivable in $\mathbf{S3}_{LP}$, it also means that the inverse of ¬R isn't admissible, whereas by Lemma 1 of the Appendix it is admissible in $\mathbf{S3}'_{LP}$.

Proof. Again, soundness is easily checked, with $\Gamma; \Sigma \triangleright \Delta$ interpreted as the four-sided $\Gamma; \Sigma \triangleright \Delta; \emptyset$. Rather than show the completeness directions directly, or via Proposition 4, I give an argument that invokes the completeness of the two-sided $\mathbf{S2}_X$ (Proposition 1).

Completeness direction for (a): Suppose $\Gamma; \emptyset \vDash_{4X} \Delta; \emptyset$. Then by Definition 2 we have $\Gamma \vDash_X \Delta$. Proposition 1 now entails that $\Gamma \triangleright \Delta$ is derivable in $\mathbf{S2}_X$. But the initial sequents of $\mathbf{S2}_X$ are derivable in $\mathbf{S3}_X$, and the rules

---
[15]The only difference is that, as explained in note 9 above, the sequent structure employed in that paper has no sequent corresponding to $\emptyset; \Gamma \triangleright \Delta$.

of $S2_X$ are included in $S3_X$, when two-sided sequents $\Gamma \triangleright \Delta$ are interpreted as $\Gamma; \emptyset \triangleright \Delta$. Hence $\Gamma; \emptyset \triangleright \Delta$ is derivable in $S3_X$.

Completeness direction for (b): Suppose $\Gamma; \Sigma \vDash_{4X} \Delta; \emptyset$. Then it follows from Definitions 1 and 3 that $\Gamma; \emptyset \vDash_{4X} \Delta, \neg\Sigma; \emptyset$,[16] whence by Definition 2 also $\Gamma \vDash_X \Delta, \neg\Sigma$. Completeness of $S2_X$ now entails that $\Gamma \triangleright \Delta, \neg\Sigma$ is derivable in $S2_X$. But the initial sequents of $S2_X$ are again derivable in $S3'_X$, and the rules of $S2_X$ are again included, derivable or admissible in $S3'_X$. To show that $\neg\wedge R$ and $\neg\vee R$ are admissible, it suffices to show that the inverse of $\neg R$ is admissible. (See Lemma 1 of the Appendix.) Hence $\Gamma; \emptyset \triangleright \Delta, \neg\Sigma$ is derivable in $S3'_X$. By Lemma 1, so is $\Gamma; \Sigma \triangleright \Delta$.

Corollary 1 $\Gamma \triangleright \Delta$ *is is derivable in* $S4_X$ *iff* $\Gamma; \emptyset \vDash_{4X} \Delta; \emptyset$. *And the same holds for* $S3_X$ *and* $S3'_X$.

5 Conclusion

This paper has investigated sequent systems for LP, K3, and FDE that combine two features. First, they preserve versions of all classical rules. Second, the truth of a consequence claim $\Gamma \vDash \Delta$ corresponds to the derivability of a sequent with Γ alone on the left-hand side and Δ alone on the right-hand side. This allows derivations in the sub-classical systems to correspond closely to standard classical derivations. Moreover, we've seen that in some formulations, the systems for LP, K3, FDE, and classical logic differ only in the inclusion of structural rules that combine features of exchange and contraction. Including each such rule amounts to eliminating some distinction drawn by the nonstandard sequent structure. Including the full complement of structural rules accordingly results in a system for classical logic.

In discussion of logically revisionary approaches to paradox, those based on LP, K3, and FDE are often *contrasted* with "substructural approaches" (e.g. Beall & Ripley, 2011). Must we conclude that they should be reclassified as substructural approaches, given the senses we have seen in which these three logics qualify as substructural? If so, the substructural/structural distinction, when applied to approaches to paradox, would appear less interesting than it has been thought to be.[17]

---

[16] Here $\neg\Sigma$ is the set whose members are the negations of the members of Σ.
[17] See Shapiro (2016) for an extended discussion of this topic.

Appendix

A. Translating between systems

There are straightforward mappings between sequents derivable in the **S4** systems and those derivable in the systems **SK** of Wintein (2016) and $\mathbf{SC}_{RT}$ of Fjellstad (2016).

In **SK**, the **S4**-sequent $\Gamma; \Sigma \rhd \Delta; \Theta$ corresponds to the set of sets of signed sentences $\{\{t,b\} : \Gamma, \{t,n\} : \Sigma, \{f,n\} : \Delta, \{f,b\} : \Theta\}$. Wintein shows that for LP, K3, FDE, and classical logic, $\Gamma \vDash \Delta$ iff $\Gamma \rhd \Delta$ is derivable in the respective fragment of **SK**, where these are distinguished by their initial sequents. In the case of FDE, the initial sequents correspond to those of our **S4**. For LP, he adds the counterpart of $\emptyset; \alpha \rhd \alpha; \emptyset$, while for K3 he adds instead the counterpart of $\alpha; \emptyset \rhd \emptyset; \alpha$. For classical logic, he includes all of the above.

In $\mathbf{SC}_{RT}$, the **S4**-sequent $\Gamma; \Sigma \rhd \Delta; \Theta$ corresponds to different "dual two-sided sequents," depending on whether we are considering its derivability in $\mathbf{S4}_{LP}$ or in $\mathbf{S4}_{K3}$. Considered with regard to $\mathbf{S4}_{LP}$ it corresponds to $\Gamma \Rightarrow_R \Theta \parallel \Sigma \Rightarrow_T \Delta$. Considered with regard to $\mathbf{S4}_{K3}$ it corresponds to $\Sigma \Rightarrow_R \Delta \parallel \Gamma \Rightarrow_T \Theta$. Thus consequences in the two logics correspond to different derivable sequents: $\Gamma \vDash_{LP} \Delta$ iff $\Gamma \Rightarrow_R \emptyset \parallel \emptyset \Rightarrow_T \Delta$ is derivable in $\mathbf{SC}_{RT}$, whereas $\Gamma \vDash_{K3} \Delta$ iff $\emptyset \Rightarrow_R \Delta \parallel \Gamma \Rightarrow_T \emptyset$ is derivable. Finally, $\Gamma \vDash_{CL} \Delta$ iff $\emptyset \Rightarrow_R \emptyset \parallel \Gamma \Rightarrow_T \Delta$ is derivable.[18]

This allows Fjellstad to use the same three initial sequents for LP, K3, and classical logic (I ignore his building in of weakening and his first-order setting).

$$\emptyset \Rightarrow_R \emptyset \parallel \alpha \Rightarrow_T \alpha \qquad \alpha \Rightarrow_R \emptyset \parallel \emptyset \Rightarrow_T \alpha \qquad \emptyset \Rightarrow_R \alpha \parallel \alpha \Rightarrow_T \emptyset$$

The translation reveals that FDE-consequence (which Fjellstad doesn't discuss) can be recovered by removing one of these initial sequents: the first when consequence is read off derivable sequents in the manner that yields LP, or the second when it's read off in the manner that yields K3.

B. The rule ¬R in $\mathbf{S3}'_X$ is invertible

Lemma 1 *If* $\Gamma; \Sigma \rhd \Delta, \neg \alpha$ *is derivable in* $\mathbf{S3}'_X$*, then so is* $\Gamma; \Sigma, \alpha \rhd \Delta$.

---

[18] To be precise, Fjellstad's discussion concerns not a propositional language but a first-order language with a transparent truth predicate. Derivability of $\emptyset \Rightarrow_R \emptyset \parallel \Gamma \Rightarrow_T \Delta$ then corresponds to the nontransitive truth-theoretic consequence $\Gamma \vDash^{st}_+ \Delta$ of Ripley (2012), which coincides with classical consequence over the truth-free fragment.

Proof. Let $\vdash_{r,X} S$ mean sequent S is derivable in $\mathbf{S3}'_X$ with derivation height at most r. We show by induction that for all r, if $\vdash_{r,X} \Gamma; \Sigma \triangleright \Delta, \neg\alpha$ then $\vdash_{r,X} \Gamma; \Sigma, \alpha \triangleright \Delta$. The case $r = 0$ is easy, as no initial sequent has a negation on the right. Assume as inductive hypothesis that the conditional holds for all $n < r$, and suppose $\vdash_{r,X} \Gamma; \Sigma \triangleright \Delta, \neg\alpha$. We show $\vdash_{r,X} \Gamma; \Sigma, \alpha \triangleright \Delta$ by considering the cases.

Suppose the last step in deriving $\Gamma; \Sigma \triangleright \Delta, \neg\alpha$ doesn't have $\neg\alpha$ on the right as principal formula. We can use the inductive hypothesis on the rule's premise(s), and then apply the same rule again. The only special case to consider is where the last step uses LeftShift of $\mathbf{S3}'_{K3}$ to derive $\Gamma', \alpha; \Sigma' \triangleright \Delta, \neg\alpha$ from $\Gamma'; \Sigma', \alpha \triangleright \Delta, \neg\alpha$. Here, the inductive hypothesis gives us $\vdash_{r-1,K3} \Gamma'; \Sigma', \alpha \triangleright \Delta$, and a use of Weak shows $\vdash_{r,K3} \Gamma', \alpha; \Sigma', \alpha \triangleright \Delta$.

Suppose the last step is by Weak, with $\neg\alpha$ on the right a principal formula. If the step is the trivial one that derives $\Gamma; \Sigma \triangleright \Delta, \neg\alpha$ from itself, the inductive hypothesis suffices. In the non-trivial case, use an instance of Weak that differs only in introducing α to the right of the semicolon.

Finally, suppose the step is by $\neg R$ with $\neg\alpha$ as principal formula. If it comes from $\Gamma; \Sigma, \alpha \triangleright \Delta$, we're done. The other possibility is that it's from $\Gamma; \Sigma, \alpha \triangleright \Delta, \neg\alpha$, in which case the inductive hypothesis suffices.

References

Anderson, A. R., & Belnap, N. D. (1963). First degree entailments. *Mathematische Annalen, 149*, 302–319.

Anderson, A. R., & Belnap, N. D. (1975). *Entailment: The Logic of Relevance and Necessity, Vol. 1*. Princeton University Press.

Asenjo, F. G. (1966). A calculus of antinomies. *Notre Dame Journal of Formal Logic, 16*, 103–105.

Avron, A. (1991). Natural 3-valued logics: Characterization and proof theory. *Journal of Symbolic Logic, 56*, 276–294.

Baaz, M., Fermüller, C., & Zach, R. (1993). Dual systems of sequents and tableaux for many-valued logics. *Bulletin of the EATCS, 51*, 192–197.

Beall, J. (2011). Multiple-conclusion LP and default classicality. *Review of Symbolic Logic, 4*, 326–336.

Beall, J. (2016). There is no logical negation: True, false, both and neither. *Australasian Journal of Logic*, to appear.

Beall, J., & Ripley, D. (2011). Non-classical theories of truth. In M. Glanzberg (Ed.), *The Oxford Handbook of Truth*. Oxford University Press, to appear.

Došen, K. (1993). A historical introduction to substructural logics. In P. Schroeder-Heister & K. Došen (Eds.), *Substructural Logics* (pp. 1–30). Oxford University Press.

Fjellstad, A. (2016). Non-classical elegance for sequent calculus enthusiasts. *Studia Logica*, to appear.

Hjortland, O. T. (2013). Logical pluralism, meaning-variance, and verbal disputes. *Australasian Journal of Philosophy*, *91*, 355–373.

Kleene, S. C. (1952). *Introduction to Metamathematics*. North-Holland.

Paoli, F. (2002). *Substructural Logics: A Primer*. Kluwer.

Priest, G. (1979). The logic of paradox. *Journal of Philosophical Logic*, *8*, 219–241.

Priest, G. (2008). *An Introduction to Non-Classical Logic: From If to Is*. Cambridge University Press.

Ripley, D. (2012). Conservatively extending classical logic with transparent truth. *Review of Symbolic Logic*, *5*, 354–378.

Shapiro, L. (2016). The very idea of a substructural approach to paradox. *Synthese*, to appear.

Wintein, S. (2016). On all Strong Kleene generalizations of classical logic. *Studia Logica*, *104*, 503–545.

Lionel Shapiro
University of Connecticut
USA
E-mail: lionel.shapiro@uconn.edu

Truth via Satisfaction?

NICHOLAS J.J. SMITH[1]

Abstract: One of Tarski's stated aims was to give an explication of the classical conception of truth—truth as 'saying it how it is'. Many subsequent commentators have felt that he achieved this aim. Tarski's core idea of defining truth via satisfaction has now found its way into standard logic textbooks. This paper looks at such textbook definitions of truth in a model for standard first-order languages and argues that they fail from the point of view of explication of the classical notion of truth. The paper furthermore argues that a subtly different definition—also to be found in classic textbooks but much less prevalent than the kind of definition that proceeds via satisfaction—succeeds from this point of view.

Keywords: Truth, Satisfaction, Tarski, Model

1 Introduction

In presenting his now famous definition of truth, one of Tarski's aims was to give an explication of the ordinary notion of truth:

> What will be offered can be treated in principle as a suggestion for a definite way of using the term "true", but the offering will be accompanied by the belief that it is in agreement with the prevailing usage of this term in everyday language. Our understanding of the notion of truth seems to agree essentially with various explanations of this notion that have been given in philosophical literature. What may be the earliest explanation can be found in Aristotle's *Metaphysics*: "To say of what is that it is not, or of what is not that it is, is false, while to

---

[1] Thanks to John P. Burgess, Max Cresswell, Kit Fine, Rob Goldblatt, Tristan Haze, Ed Mares and Simon Varey for discussions and correspondence. Earlier versions of this paper were presented at the Australasian Association for Logic Annual Conference at the University of Sydney on 3 July 2015, at a Pukeko meeting at Victoria University of Wellington on 10 October 2015 and at Logica 2016 in Hejnice on 24 June 2016; thanks to the audiences for their comments. For the germ of one of the core ideas of this paper see Smith (2012, 503–4, n.23 and n.25); cf. also Smith (2016, 756, n.17).

say of what is that it is, or of what is not that it is not, is true."
... We shall attempt to obtain here a more precise explanation of
the classical conception of truth, one that could supersede the
Aristotelian formulation while preserving its basic intentions.
(Tarski, 1969, 63–4)[2]

Many have felt that Tarski achieved this aim:

> The two most famous and—in the view of many—most important examples of conceptual analysis in twentieth-century logic were Alfred Tarski's definition of *truth* and Alan Turing's definition of *computability*. In both cases a prior, extensively used, informal or intuitive concept was replaced by one defined in precise mathematical terms. ... in the view of many... Tarski's definition of truth is one of the most important cases of conceptual analysis in twentieth-century logic. (Feferman, 2008, 72, 90)

In this paper I shall argue that Tarski's approach to defining truth does not succeed *from the point of view of explication or conceptual analysis of the ordinary or classical notion of truth*. I thereby go further than Field in his well-known criticism of Tarski's definition. According to Field, the received view is that Tarski reduced truth to non-semantic notions. Field argues, on the contrary, that Tarski reduced truth to other semantic notions:

> My contrary claim will be that Tarski succeeded in reducing the notion of truth *to certain other semantic notions*; but that he did not in any way explicate these other notions, so that his results ought to make the word 'true' acceptable only to someone who already regarded these other semantic notions as acceptable. ... Tarski merely reduced truth to other semantic notions. (Field, 1972, 347, 348)

I shall argue that Tarski does not reduce *truth* to anything: he does not explicate or reduce truth (in the ordinary or classical sense) at all.

Tarski's core idea of defining truth via the notion of satisfaction has now found its way into the logic textbooks. I shall focus on standard textbooks—rather than Tarski's original papers—because they give us a cleaner version

---

[2]Cf. Tarski (1956, 153).

Truth via Satisfaction?

of Tarski's core idea and arguably have greater contemporary relevance.[3] My central concern is not the historical study of Tarski's contributions but the analysis of truth. My central claims are first that a successful analysis of the ordinary notion of truth cannot be achieved by proceeding via satisfaction in the Tarskian way and second that a subtly different definition of truth—also to be found in classic textbooks but much less prevalent than the kind of definition that proceeds via satisfaction—*does* yield an explication of the classical notion of truth.

2 Two ways of defining truth in a model

We consider a standard first order language with names (individual constants), and predicates of each arity.[4] A model of the language comprises a domain (a nonempty set), and an assignment of a referent (an object in the domain) to each name and an extension (a set of n-tuples of members of the domain) to each n-place predicate.

Let us now set out two ways of defining truth in a model: one that proceeds via satisfaction in the Tarskian way; and one that defines truth directly rather than via first defining satisfaction.[5]

First some preliminary definitions:[6]

- A value assignment $\mathfrak{v}$ on a model $\mathfrak{M}$ is a function from the set V of all variables in the language into the domain of $\mathfrak{M}$.

- $\mathfrak{M}^{\mathfrak{v}}$ is a model $\mathfrak{M}$ together with a value assignment $\mathfrak{v}$ on $\mathfrak{M}$

- A *term* is a name or variable

- Where $\underline{t}$ is a term, $[\underline{t}]_{\mathfrak{M}^{\mathfrak{v}}}$ is:

    - the referent of $\underline{t}$ on $\mathfrak{M}$, in case $\underline{t}$ is a name
    - the value assigned to $\underline{t}$ by $\mathfrak{v}$, in case $\underline{t}$ is a variable

---

[3]It is irrelevant to my argument whether the authors of these textbooks do or should care about conceptual analysis. Some people certainly care about the analysis of truth—and my point is that they will not find what they seek in the approach taken in these logic texts.

[4]Nothing apart from simplicity of presentation turns on the omission of function symbols.

[5]For the first definition, see e.g. Enderton (2001). For the second definition, see Jeffrey (1967), Boolos and Jeffrey (1989) and Boolos, Burgess, and Jeffrey (2007).

[6]Following Smith (2012, §8.4.2), underlining is used for metavariables; so $\underline{x}$ is any variable, $\underline{a}$ is any name, etc.

- Where $\underline{P}$ is a predicate, $[\underline{P}]_{\mathfrak{M}}$ is the extension of $\underline{P}$ on the model $\mathfrak{M}$

Now the definition of satisfaction:[7]

- $\underline{P^n t_1} \ldots \underline{t_n}$ is satisfied relative to $\mathfrak{M}^{\mathfrak{v}}$ iff $\langle [\underline{t_1}]_{\mathfrak{M}^{\mathfrak{v}}}, \ldots, [\underline{t_n}]_{\mathfrak{M}^{\mathfrak{v}}} \rangle \in [\underline{P^n}]_{\mathfrak{M}}$
- $\neg \alpha$ is satisfied relative to $\mathfrak{M}^{\mathfrak{v}}$ iff α is unsatisfied relative to $\mathfrak{M}^{\mathfrak{v}}$
- $(\alpha \wedge \beta)$ is satisfied relative to $\mathfrak{M}^{\mathfrak{v}}$ iff α and β are both satisfied relative to $\mathfrak{M}^{\mathfrak{v}}$
- $(\alpha \vee \beta)$ is satisfied relative to $\mathfrak{M}^{\mathfrak{v}}$ iff one (or both) of α and β is satisfied relative to $\mathfrak{M}^{\mathfrak{v}}$
- $(\alpha \rightarrow \beta)$ is satisfied relative to $\mathfrak{M}^{\mathfrak{v}}$ iff α is unsatisfied relative to $\mathfrak{M}^{\mathfrak{v}}$ or β is satisfied relative to $\mathfrak{M}^{\mathfrak{v}}$ (or both)
- $(\alpha \leftrightarrow \beta)$ is satisfied relative to $\mathfrak{M}^{\mathfrak{v}}$ iff α and β are both satisfied, or both unsatisfied, relative to $\mathfrak{M}^{\mathfrak{v}}$
- $\forall \underline{x} \alpha$ is satisfied relative to $\mathfrak{M}^{\mathfrak{v}}$ iff α is satisfied relative to $\mathfrak{M}^{\mathfrak{v}'}$ for every value assignment $\mathfrak{v}'$ on $\mathfrak{M}$ which differs from $\mathfrak{v}$ at most in what it assigns to $\underline{x}$
- $\exists \underline{x} \alpha$ is satisfied relative to $\mathfrak{M}^{\mathfrak{v}}$ iff α is satisfied relative to $\mathfrak{M}^{\mathfrak{v}'}$ for some value assignment $\mathfrak{v}'$ on $\mathfrak{M}$ which differs from $\mathfrak{v}$ at most in what it assigns to $\underline{x}$

And now the first definition of truth:[8]

- A wff is true (henceforth 's-true') on a model $\mathfrak{M}$ iff it is satisfied relative to $\mathfrak{M}^{\mathfrak{v}}$ for every value assignment $\mathfrak{v}$ on $\mathfrak{M}$.
- A wff is false (henceforth 's-false') on a model $\mathfrak{M}$ iff it is unsatisfied relative to $\mathfrak{M}^{\mathfrak{v}}$ for every value assignment $\mathfrak{v}$ on $\mathfrak{M}$.

Next we set out a second definition of truth—one that defines truth directly rather than via satisfaction. First some preliminary definitions:

- Where $\underline{a}$ is a name, $[\underline{a}]_{\mathfrak{M}}$ is the referent of $\underline{a}$ on the model $\mathfrak{M}$

---

[7] If a wff is *not* satisfied relative to $\mathfrak{M}^{\mathfrak{v}}$, we say that it is *unsatisfied* relative to $\mathfrak{M}^{\mathfrak{v}}$.

[8] We use the terms 's-true' and 's-false' (rather than plain 'true' and 'false') for the properties here defined, in order to avoid ambiguity later when we look at a different definition of truth.

Truth via Satisfaction?

- Where $\alpha(\underline{x})$ is a wff that has no free variables other than $\underline{x}$, $\alpha(\underline{a}/\underline{x})$ is the wff that results from $\alpha(\underline{x})$ by replacing all free occurrences of $\underline{x}$ by the name $\underline{a}$

- Where $\mathfrak{M}$ is a model and $\underline{a}$ is a name that is not assigned a referent in $\mathfrak{M}$, $\mathfrak{M}_o^{\underline{a}}$ is the model that is just like $\mathfrak{M}$ except that in it the name $\underline{a}$ is assigned the referent o

Now the definition of truth:[9]

- $\underline{P^n}\underline{a_1}\ldots\underline{a_n}$ is true on $\mathfrak{M}$ iff $\langle[\underline{a_1}]_{\mathfrak{M}}, \ldots, [\underline{a_n}]_{\mathfrak{M}}\rangle \in [\underline{P^n}]_{\mathfrak{M}}$

- $\neg\alpha$ is true on $\mathfrak{M}$ iff α is false on $\mathfrak{M}$

- $(\alpha \wedge \beta)$ is true on $\mathfrak{M}$ iff α and β are both true on $\mathfrak{M}$

- $(\alpha \vee \beta)$ is true on $\mathfrak{M}$ iff one (or both) of α and β is true on $\mathfrak{M}$

- $(\alpha \rightarrow \beta)$ is true on $\mathfrak{M}$ iff α is false on $\mathfrak{M}$ or β is true on $\mathfrak{M}$ (or both)

- $(\alpha \leftrightarrow \beta)$ is true on $\mathfrak{M}$ iff α and β are both true on $\mathfrak{M}$ or both false on $\mathfrak{M}$

- $\forall\underline{x}\alpha(\underline{x})$ is true on $\mathfrak{M}$ iff for every object o in the domain of $\mathfrak{M}$, $\alpha(\underline{a}/\underline{x})$ is true on $\mathfrak{M}_o^{\underline{a}}$, where $\underline{a}$ is some name that is not assigned a referent in $\mathfrak{M}$

- $\exists\underline{x}\alpha(\underline{x})$ is true on $\mathfrak{M}$ iff there is at least one object o in the domain of $\mathfrak{M}$ such that $\alpha(\underline{a}/\underline{x})$ is true on $\mathfrak{M}_o^{\underline{a}}$, where $\underline{a}$ is some name that is not assigned a referent in $\mathfrak{M}$

Note some key points about these two definitions of truth:

- Satisfaction is defined relative to a model and a variable assignment. S-truth and truth are defined relative to a model.[10]

- Satisfaction and s-truth are defined for all wffs. Truth is defined only for closed wffs.[11]

- Truth and s-truth have the same extension amongst closed wffs.

---

[9] If a closed wff is *not* true on $\mathfrak{M}$, we say that it is *false* on $\mathfrak{M}$.

[10] S-truth is satisfaction by the model and *all* variable assignments thereon. It is by quantifying over the variable assignments in this way that we get rid of them—kick away the ladder—and get a notion of s-truth defined relative to a model only.

[11] (i) A closed wff is one that contains no free occurrence of any variable; an open wff is one

3 The classical conception of truth

We are interested in whether the definitions of truth in the previous section can serve as explications of the ordinary or classical notion of truth. It is time to say a little more about this notion.[12]

The core idea is that truth is *saying it how it is*. A claim is true if things are the way it claims them to be; it is false if things are not as it claims them to be.

This idea goes back at least as far as Plato and Aristotle:

> SOCRATES: But how about truth, then? You would acknowledge that there is in words a true and a false?
> HERMOGENES: Certainly.
> SOCRATES: And there are true and false propositions?
> HERMOGENES: To be sure.
> SOCRATES: And a true proposition says that which is, and a false proposition says that which is not?
> HERMOGENES: Yes, what other answer is possible? (Plato, c.360 BC)

> ...we define what the true and the false are. To say of what is that it is not, or of what is not that it is, is false, while to say of what is that it is, and of what is not that it is not, is true (Aristotle, c.350 BC, Book IV (Γ) §7)

This is just a rough guiding idea about truth. The rough idea has been used to motivate more precise, detailed theories of truth—some of which (such as certain versions of the correspondence theory of truth) are quite contentious. My interest in this paper is not in such detailed theories: it is in the basic guiding idea of truth as saying it how it is.

---

that is not closed. (ii) S-truth is well-defined for all wffs but some open wffs might be neither s-true nor s-false, relative to a given model. (iii) Some proponents of the definition of truth via satisfaction restrict the *term* 'truth' to closed wffs—but this does not change the fact that the property of satisfaction relative to all variable assignments is one that open and closed wffs can possess. We return to this point in §5. (iv) In the definition of truth (i.e. the second definition—not the definition of s-truth), the clause for atomic wffs covers only wffs containing names (not variables) and the clauses for quantified wffs cover only wffs where no variable other than the one in the quantifier occurs free in the remainder of the formula.

[12] I sometimes call the notion of truth under discussion the 'ordinary' notion of truth and sometimes call it the 'classical' notion—but note that our interest in this paper is in the analysis of *this notion* (whatever one wants to call it): it is not in the question whether this notion is the 'folk notion' of truth.

Truth via Satisfaction?

Note the following comment from Tarski:

> we are not interested here in 'formal' languages and sciences in one special sense of the word 'formal', namely sciences to the signs and expressions of which no material sense is attached. For such sciences the problem here discussed has no relevance, it is not even meaningful. (Tarski, 1956, 166)

A formula in a purely formal, uninterpreted language does not make a claim—it does not set up a condition that the world may meet or fail to meet. Hence the question of truth—in the sense in which we are interested—does not arise for such formulas. The question of truth arises only for *contentful* sentences: only when a sentence says something does the question arise whether things are the way it says they are.[13]

4 Explicating truth I

I submit that the second definition of truth in §2 explicates the ordinary notion of truth. In particular, it spells out more precisely what 'saying it how it is' amounts to—and it incorporates the key insights that (i) what it is for a sentence to *say it how it is* varies depending on the *form* of the sentence and (ii) ultimately reduces to the case of *atomic sentences*. The definition contains one case for atomic sentences and then one case per logical operator—hence point (i)—and it is a recursive definition where the truth or falsity of any sentence is ultimately grounded in the truth or falsity of certain atomic sentences—hence point (ii). Let's look at these points in a little more detail. Each clause in the definition can be seen as telling us what it is for a certain kind of sentence to *say it how it is*. The first clause tells us that for Pa to say it how it is is for the thing picked out by a to be in the set of things (that have the property) picked out by P.[14] The second clause tells us that for a negation $\neg \alpha$ to say it how it is is for α to say it how it isn't. The third clause tells us that for a conjunction to say it how it is is for both conjuncts to say it how it is; the fourth clause tells us that for a disjunction to say it how it is

---

[13]Cf. Hodges (1985–6, 147–8): "The issue is simply this. If a sentence contains symbols without a fixed interpretation, then the sentence is meaningless and doesn't express a determinate thought. But then we can't properly call it true or false." This is Hodges's gloss on Frege (1971, 98): "A proposition that holds only under certain circumstances is not a real proposition."

[14]For Rab to say it how it is is for the pair of things picked out by a and b (in that order) to be in the set of pairs of things (that stand in the relation) picked out by R; and so on.

is for at least one disjunct to say it how it is; the fifth and sixth clauses tell us similar things, mutatis mutandis, for conditionals and biconditionals. The seventh clause tells us that for a universally quantified sentence to say it how it is is for every *particularisation* of it—one for each thing in the domain—to say it how it is. The core idea here is that 'everything is ϕ' says it how it is iff *this* is ϕ, and *this* is ϕ, and so on—through all the things there are. The way the definition spells out this idea is as follows. Consider a name that nothing currently has—say (for the sake of example) 'Rumpelstiltskin'. Then for 'Everyone in the room was born in Tasmania' to say it how it is is for 'Rumpelstiltskin was born in Tasmania' to say it how it is—no matter who in the room we name 'Rumpelstiltskin'. Finally the eighth clause tells us that for an existentially quantified sentence to say it how it is is for at least one particularisation of it to say it how it is.

Note how the truth/falsity of every sentence is explained in terms of the truth/falsity of sentence(s) with fewer logical operators. Hence we eventually get down to atomic sentences—and for them, the truth condition evidently encapsulates the idea of 'saying it how it is'. Thus the definition as a whole provides a genuine explication of the classical notion of truth. The definition is mathematically precise and—in the ways just discussed—it provides genuine insight into the idea of 'saying it how it is'.

Interestingly, point (i) above—which I regard as a key insight of the explication—is the basis of one of Prior's *criticisms* of Tarski:

> When the presupposed definition of 'satisfaction' is examined, however, it will be found that this definition of truth has a further defect. Satisfaction can only be defined in the following roundabout way (I again give the thing roughly): 'x is included in y' is satisfied by the pair of classes a, b if and only if a is included in b; 'not-y' is satisfied by any group of classes which does not satisfy y; 'x or y' is satisfied by any group of classes which either satisfies x or satisfies y; and a function preceded by the universal quantifier is satisfied, etc. Such a piecemeal definition of satisfaction means a similarly piecemeal definition of truth, when it is all spelt out; and the more complex the language considered the more pieces there will be. I know there are plenty of quite un-Tarski-like people who will be entirely happy about this—people who contend that even in 'everyday or colloquial' language the word 'true' has different meanings when applied to sentences of different sorts, so that it can have

a single meaning only in the sense of a disjunction of these. My own understanding of ordinary language is quite otherwise; there are no doubt dozens of different ways of deciding whether a given sentence is true, but what it means for a sentence to be true is pretty much the same throughout, and pretty much what was suggested at the beginning of this discussion. (Prior, 1957, 408–9)

I take it that Prior is referring at the end of this quotation to his earlier comment "We generally say that a sentence is true if it says that something is so, and it is so" [406]. In response to this criticism, however, I'd say that 'true' *does* have the same meaning throughout the second definition of truth in §2: it means (just as Prior says) *saying it how it is*. However, what exactly 'saying it how it is' amounts to differs depending on the form of the sentence—and one way in which the formal definition is an advance over and explicates the guiding idea is precisely that it reveals this.

5 Explicating truth II

I shall now argue that the first definition of truth in §2 does not explicate the classical notion of truth. The key point to observe is that open wffs can be s-true (relative only to a model—not relative to a model *and* a variable assignment). For example, in a model in which the extension of P is the entire domain, the wff Px is s-true; and in any model whatsoever, the wff $\forall x Px \rightarrow Py$ is s-true. However open wffs do not *say anything* (relative only to a model)—they do not *make any claim*—and so they cannot (on the classical conception of truth) be true or false. (I do not mean that the open wff all by itself does not say anything. I mean that even given a model it does not say anything. For example, even given a domain and an extension for P, Px and $\forall x Px \rightarrow Py$ do not say anything, do not make any claim.) The *reason* they do not say anything is that some symbols in them—the free occurrences of variables—'have no material sense attached to them'. So the problem here turns on the very point that Tarski himself noted in the passage quoted in §3: when no material sense is attached to the expressions in a formula—in this case, the free occurrences of variables (remember that we are talking about open wffs relative to models, not relative to models and variable assignments)—the question of truth "has no relevance, it is not even meaningful".[15]

In short, there are formulas that have the property s-truth but that cannot be regarded as true (in the classical sense). So s-truth does not provide an explication of the classical notion of truth.

Some textbook authors withhold the term 'true' from open wffs: they say only that a closed wff is true (on a model) iff it is satisfied relative to all variable assignments (on that model). But this move does not help with the present problem. Even if we restrict the *term* 'true' to closed wffs, the *property* that we call 'truth' when closed wffs have it (viz., satisfaction relative to all variable assignments) can't yield an explication of truth (in the ordinary sense) because that property is one that open wffs can also have—and they cannot be true in the ordinary sense (because they make no claim). Declining to *call* an open wff 'true' when it has this property does not take away from the fact that one has defined a property that open wffs can possess, and then called a closed wff 'true' when it has *that property*. My point is that because open wffs can also have this property, possessing this property cannot make something (not even if it is a closed wff) *true* in the ordinary sense.

6 Value Assignments vs Sequences of Objects

Where a wff is true or false relative to one thing—a model—a wff is satisfied or unsatisfied relative to *two* things: a model $\mathfrak{M}$ *and* something else ('X'). One option for X is, as we have seen in §2, a value assignment on $\mathfrak{M}$. A second option for X is a denumerable (i.e. countably infinite) sequence of objects from the domain of $\mathfrak{M}$.[16]

Given a countable infinity of variables and an enumeration of them, the two approaches are interchangeable. To get a sequence of objects, given a value assignment, we transfer the enumeration of the variables to objects in the domain, via the value assignment function. To get a value assignment, given a sequence of objects, we assign the nth object in the sequence to the nth variable.

---

[15] Cf. Hodges (2004, 99). My reminders throughout this paragraph that we are talking about open wffs relative to models—not models and variable assignments—should not be taken to suggest that relative to a model and a variable assignment, any open wff *does* say something. I discuss this issue in §7.

[16] See e.g. Mendelson (1987) and Hunter (1971). Note that the domain need not be infinite: the same object may appear at more than one place in the sequence.

Truth via Satisfaction?

The sequence approach goes back to Tarski himself. Tarski writes at the outset "In this construction I shall not make use of any semantical concept if I am not able previously to reduce it to other concepts" (Tarski, 1956, 152–3). Milne (1999, 151–2) regards an assignment of values to variables as a semantic notion and says that the approach via sequences is a "stroke of genius" that enables this semantic notion to be avoided. I'm not sure (a) that the idea of an assignment of values to variables is a semantic notion—nor (b) that, if it were, the approach via sequences would really avoid semantic notions. On (a): cf. the discussion in §7 below of the point that a model plus a value assignment does *not* make any wff say something or make a claim. On (b): Imagine an old-fashioned school dance, where the headmaster wants to avoid doing anything as vulgar as pairing up boys and girls—so he assigns each boy a number, and each girl a number, and then the students match numbers. If there really were something wrong with pairing up boys and girls, then could this coy approach via numbering seriously be regarded as having avoided wrongdoing?

Of course, to settle issues (a) and (b), we'd need a full discussion of what makes a notion 'semantic'. However we do not need to settle these issues here. The key point for purposes of this paper is that moving from value assignments to sequences does not help with the problem raised in §5. As far as the analysis of truth is concerned, the sequence version just adds a further step of indirection to the variable-assignment version. This certainly doesn't make the definition any better from the point of view of explication of the classical notion of truth; if anything it makes it worse.

7   Notational variants?

I imagine someone might object as follows:

> This is a storm in a teacup. The two approaches to defining truth presented in §2 are simply notational variants of one another. An open wff relative to a model *and* a value assignment can be seen this way: the free occurrences of variables have effectively been made into names and then assigned referents (by the value assignment).[17] So this is exactly like the second approach to defining truth—where we have new names (that

---

[17] Cf. Field (1972, 349): "The idea is going to be to treat the variables, or at least the free variables, as sort of "temporary names" for the objects assigned to them."

are not assigned a referent on the original model) and extended models that assign them referents.

There are two points to make in response to this objection.

First, even if it were true that satisfaction relative to a model plus value assignment were simply a notational variant on truth as defined in the second way in §2, this would not affect my argument: for even if there were some *other* property involved in the first definition of truth (e.g. the property of satisfaction relative to a model and a variable assignment) that *did* explicate the ordinary notion of truth, that would not affect my point that the property of *s-truth* (satisfaction relative to *all* variable assignments) does *not* explicate the ordinary notion of truth.

Second, we cannot, in general, regard assigning a value to a variable as the same thing as viewing the variable as a name and assigning it a referent. This gloss works fine for free occurrences of variables—but not for bound occurrences. Yet note that a value assignment assigns values to *variables*—not to *occurrences* of variables. It is fine to say that Px says that Bill is P, relative to a value assignment that assigns Bill to x. But what about $\forall x Rxy$, relative to a value assignment that assigns Bill to x and Ben to y? It does *not* say 'Everything bears R to Ben'. That would be to ignore the assignment of Bill to x. It actually says something like 'Every Bill is such that Bill bears R to Ben'—which makes no sense at all.[18] The hope (of the imagined objector) was that a model *plus* a value assignment makes *any* wff say something or make a claim—and that satisfaction (relative to a model and a variable assignment) is just like truth as defined in the second way in §2 and hence does explicate truth in the classical sense. However this thought does not pan out. There is a dilemma here (for the imagined objector). Without a variable assignment (i.e. relative only to a model), open wffs have no content—they do not make claims. But *with* a variable assignment (regarded as turning variables into names and assigning them referents), wffs with bound occurrences of variables become meaningless.

---

[18]Cf. Shoenfield (1967, 13): "Most of our previous remarks about variables are false when applied to the x in $\forall x(x = 0)$. This formula has only one meaning, while $x = 0$ has many meanings. We get a particular meaning of $x = 0$ by substituting 2 for x; but if we substitute 2 for x in $\forall x(x = 0)$, we get the meaningless expression $\forall 2(2 = 0)$."

8 Conclusion

I have considered two definitions of truth: one that proceeds via satisfaction and one that does not. (I have furthermore considered two versions of the first definition: one that appeals to variable assignments and one that appeals to sequences.) I have argued that the first definition fails from the point of view of explicating the classical notion of truth and that the second definition succeeds from this point of view.

Note that these conclusions are quite specific. For example, I have not argued that the second definition is the only possible definition of truth that can serve as an explication of the classical notion. So what about other definitions of truth in the mainstream textbooks—and in the broader literature? A discussion of all these approaches would take far more space than I have available here. It would also raise a host of new issues that have not featured in the discussion in this paper. For example, consider the approach taken by Robbin (1969),[19] who assumes that the set of names in the language is the very same set as the domain of the model—or the approach taken by Robinson (1951, 1963, 1966), who assumes a set of names of arbitrary transfinite cardinal number and a one-one correspondence between the domain of the model and the set of names. Neither of their definitions of truth faces the problem I have raised for the first definition presented in §2 above. However one might still think that they are less than ideal as explications of the ordinary notion of truth, because the ordinary notion applies to languages in which the names used to talk about objects and the objects talked about are distinct, and to languages that do not contain names for every object. However these are arguments that would have to be made: they bring in new considerations and do not follow automatically from the arguments of this paper.

References

Aristotle. (c.350 BC). Metaphysics. (In Barnes (1984), vol.2.)
Barnes, J. (Ed.). (1984). *The Complete Works of Aristotle: The Revised Oxford Translation*. Princeton NJ: Princeton University Press.
Boolos, G. S., Burgess, J. P., & Jeffrey, R. C. (2007). *Computability and Logic* (Fifth ed.). Cambridge: Cambridge University Press.

---

[19]Thanks to Rob Goldblatt for pointing me to this book.

Boolos, G. S., & Jeffrey, R. C. (1989). *Computability and Logic* (Third ed.). Cambridge: Cambridge University Press.

Enderton, H. B. (2001). *A Mathematical Introduction to Logic* (Second ed.). San Diego: Harcourt/Academic Press.

Feferman, S. (2008). Tarski's conceptual analysis of semantical notions. In D. Patterson (Ed.), *New Essays on Tarski and Philosophy* (pp. 72–93). Oxford: Oxford University Press.

Field, H. (1972). Tarski's theory of truth. *Journal of Philosophy*, 69(13), 347–375.

Frege, G. (1971). *On the Foundations of Geometry and Formal Theories of Arithmetic*. New Haven: Yale University Press. (Translated by Eike-Henner W. Kluge.)

Hamilton, E., & Cairns, H. (Eds.). (1963). *The Collected Dialogues of Plato*. New York: Pantheon Books.

Hodges, W. (1985-6). Truth in a structure. *Proceedings of the Aristotelian Society*, 86, 131–51.

Hodges, W. (2004). What languages have Tarski truth definitions? *Annals of Pure and Applied Logic*, 126, 93–113.

Hunter, G. (1971). *Metalogic: An Introduction to the Metatheory of Standard First Order Logic*. Berkeley: University of California Press.

Jeffrey, R. C. (1967). *Formal Logic: Its Scope and Limits*. New York: McGraw-Hill Book Company.

Mendelson, E. (1987). *Introduction to Mathematical Logic* (Third ed.). Monterey CA: Wadsworth & Brooks/Cole.

Milne, P. (1999). Tarski, truth and model theory. *Proceedings of the Aristotelian Society*, 99, 141–67.

Plato. (c.360 BC). Cratylus. (In Hamilton and Cairns (1963).)

Prior, A. (1957). Critical notice of *Logic, Semantics, Metamathematics* by Alfred Tarski. *Mind*, 66(263), 401–10.

Robbin, J. W. (1969). *Mathematical Logic: A First Course*. New York: W.A. Benjamin.

Robinson, A. (1951). *On the Metamathematics of Algebra*. Amsterdam: North-Holland Publishing Company.

Robinson, A. (1963). *Introduction to Model Theory and to the Metamathematics of Algebra*. Amsterdam: North-Holland Publishing Company.

Robinson, A. (1966). *Non-Standard Analysis*. Amsterdam: North-Holland Publishing Company.

Shoenfield, J. R. (1967). *Mathematical Logic*. Reading MA: Addison-Wesley.

Smith, N. J. (2012). *Logic: The Laws of Truth*. Princeton: Princeton University Press.

Smith, N. J. (2016). Truthier than thou: Truth, supertruth and probability of truth. *Noûs, 50*(4), 740–58.

Tarski, A. (1956). The concept of truth in formalized languages. In *Logic, Semantics, Metamathematics: Papers from 1923 to 1938* (pp. 152–278). Oxford: Clarendon Press. (Translated by J.H. Woodger.)

Tarski, A. (1969). Truth and proof. *Scientific American, 220*(6), 63–77.

Nicholas J.J. Smith
Department of Philosophy, The University of Sydney
Australia
E-mail: `nicholas.smith@sydney.edu.au`

Free Choice Permission in STIT

FREDERIK VAN DE PUTTE[1]

Abstract: We argue for a new approach to free choice permission in the context of a-temporal **STIT** logic. According to our analysis, an agent has a free choice permission w.r.t. two propositions φ and ψ iff (a) the agent is permitted to see to $\varphi \wedge \neg\psi$ and (b) the agent is permitted to see to $\psi \wedge \neg\varphi$. The primitive notion of permission we use is the dual of one of Horty's operators for "ought to do" from (Horty, 2001). We argue that the approach improves on existing proposals in various ways.

Keywords: Free choice permission, Deontic logic, STIT logic, Multi-agency

1 Introduction

Jane goes to the fruit buffet and asks the waiter which pieces of fruit she can take. The waiter replies: "you may take an apple or a pear". We take this statement to be equivalent to "Jane is free to choose between taking an apple and a pear" or to "Jane has a free choice between taking an apple and taking a pear". Let us represent such statements formally as $\mathsf{F}_j(apple, pear)$, where $apple, pear$ are states of affairs (that Jane takes an apple, resp. a pear) and j is shorthand for Jane. In this case, we say that the agent (i.c. Jane) has a free choice permission (henceforth FCP) w.r.t. the two options $apple, pear$. Likewise, an agent may have a FCP w.r.t. three, four, or more options.

Our aim in this paper is to propose a new semantics for the operator F_j, one that arguably improves on existing accounts. We focus on the binary case for the sake of simplicity; however, all our observations and our own proposal generalize readily to a finite number $n \geq 2$ of choices.

Mind that we take FCP here to concern a *normative* claim, i.e. it is not simply a descriptive claim about what the agent can choose (as in "you can

---

[1]Research for this paper was funded by the Flemish Research Foundation (FWO-Vlaanderen). The basic ideas for this paper were obtained during a research stay in Bayreuth (July 2016), which was co-funded by the PIOTR project (RO 4548/4-1). Many thanks to Olivier Roy, Igor Sedlar, and Nathan Wood for stimulating discussions, as well as more concrete suggestions.

choose to kill your wife or not, but there is no reason whatsoever to kill her"), but it is a claim about what the agent is *permitted* to choose. In the remainder, we moreover presuppose that we are working with a *single* normative system that grounds claims concerning FCP.

Another very important point is that FCP, as we think of it, concerns a given agent and a concrete deliberative context for that agent, i.e. a situation in which that agent has to make a choice among a number of options that are "real". By the latter we mean that choosing one such option entails that the option is effectively realized. So our problem is: given that the agent has to make a choice, when can we say that the agent is permitted to choose between φ and ψ?

Perhaps it makes sense to claim that one (everyone, someone, ...) is free to order a pizza or spaghetti at piazza San Marco in Venice, but these are not two options that present themselves to you at this very moment (unless of course you happen to be reading this paper at San Marco). So in this sense *you* do not have the FCP *here and now* to order a pizza or spaghetti at San Marco.[2] We explicitly leave such forms of "free choice permission" out of the picture in this paper.

To fully appreciate this point, consider the following example from Sedlár (2016): suppose that according to the laws of your country, you are permitted to travel abroad and you are permitted to vote. These two actions are clearly distinct, meaning that you can do one without the other. Still, it seems to make little sense to say that you have a free choice between traveling abroad and voting. The two are simply not related, and hence we do not think in terms of making a choice between either.

Suppose however that you are in a specific situation in which traveling abroad is one of your options and voting is another. For instance, you may have to decide here and now between going to a conference, thus missing out on the elections, and staying home so that you can vote; there happens to be no way you can combine both. Assume moreover that not voting is permitted in your country, and likewise, that you are not obliged to go to the conference either. There is no other feasible option that would be way

---

[2]One might argue that this claim really communicates that *once* you are at San Marco, you have the free choice between pizza and spaghetti, and (perhaps) moreover that you are permitted to go to San Marco in the first place. To model such claims, one could extend our current analysis, e.g. by adding a possibility operator which ranges over alternative deliberative contexts and (perhaps) also a similarly wide-ranging permission operator. But even if one disagrees, it still makes sense to focus on the concrete here-and-now FCP as we do, before trying to tackle other (probably more complicated) notions.

better than each of these options. Then, on the approach we advocate here, it *does* make sense to say you are free to choose between voting and traveling abroad.

In other words, whether we can speak of a genuine choice between two options is a matter of context. In our account, the options of agents are made explicit within the models, whence we can explicate necessary and sufficient conditions for a FCP. This means in turn that FCP does not only depend on the norms that apply in the context, but also to the available choices of the agent — we return to this point below.

2 Some general observations

To get our analysis off the ground, let us make a number of observations about FCP and the way it relates to permission in natural language. Some of these are not original, but we collect them here for ease of reference.[3]

(I) $F_\alpha(\varphi, \psi)$ does not entail that every way of making φ, resp. ψ true, is permitted.[4] In other words, $F_\alpha(\varphi, \psi)$ does not entail that either φ or ψ is "sufficient for permissibility".[5]

To understand this claim, suppose that whoever takes a pear or an apple should also take a napkin (since the apples and the pears are very juicy). Jane may well take a pear or an apple without taking a napkin. That mere fact by no means entails that she is no longer free to choose between taking an apple and taking a pear. In fact, "you are free to choose between an apple and a pear, but do take a napkin in case you take either" sounds perfectly consistent. It would be strange to say that in such cases, you are really only free to choose between taking an apple and a napkin, or taking a pear and a napkin.

---

[3] Although we call them "observations" and although they are clearly inspired by natural language, these points should not be taken as purely descriptive facts of the matter. Rather, they are salient properties of what we consider a useful and natural concept of free choice permission, one that allows us to report on the permissibility of an agent's choices in a given situation.

[4] A similar remark is made by Giordani and Canavotto (2016, p. 89): "we are ordinarily allowed to choose between alternative actions even if there are ways of performing such actions that lead to a violation of the law".

[5] See (Van De Putte, 2016) for a formal investigation of this notion of sufficiency.

(II) A free choice permission w.r.t. φ and ψ – relative to a given body of norms – entails that both choosing φ and choosing ψ are permitted here and now, relative to the same body of norms.

"You are free to choose between taking an apple or a pear, but actually all apples must remain untouched" sounds self-contradictory – at least, if we interpret both parts of the sentence as expressing information about one and the same normative system. When someone makes such a claim, we will automatically infer that the speaker wants to distinguish between two normative systems (e.g. the law regarding apples and pears on the one hand, and the rules of the house regarding apples and pears on the other).

(III) That φ and ψ are permitted does not entail that there is a FCP.

There are at least two ways this can be argued against:

(iii.a) φ may entail ψ. It may be permitted to take a piece of fruit, and it may also be permitted to take an apple, but it doesn't make sense to infer from this that you are free to choose between taking a piece of fruit and taking an apple.

(iii.b) It may be obligatory that φ whenever ψ. For instance, it may be obligatory that you take an apple whenever you take a pear. In that case, even if both are permissible, one cannot infer that you are free to choose between taking an apple and taking a pear.

One may argue that if φ and ψ are mutually exclusive (and i.c. when $\psi = \neg\varphi$), then the permissibility of both entails that there is a FCP. Here, our stance is more subtle: we will argue (in Sections 4 and 5) that this depends essentially on the notion of permissibility one is using.

(IV) FCP is always relative to an (a group of) agent(s): it is always the permission *of an agent* to choose between a number of things.

It makes no sense to say there is a free choice between X and Y, unless one refers (implicitly) to one or more agents that are permitted to make this choice. Also, FCPs of one agent α need not coincide with FCPs of another agent β. Jack may be free to choose whether his back door is left open at night, but his neighbour Daniel clearly does not have a FCP w.r.t. this same state of affairs. This distinguishes FCP from the concept of permission that is at stake in sentences like "it is permitted that your car is parked here", where it does not matter who brings it about that the car is parked.

(V) $F_\alpha(\varphi, \psi)$ entails that α can in fact choose between φ and ψ, that both are "live options" for α.

"You are free to choose between winning or losing the lottery, but there is no way you can choose among either" seems to make little sense – again, when it is taken to report a permission that applies here and now. In this sense, FCP is a mixed notion, since it presupposes not only normative claims, but also claims about the possible choices of agents.

3 Existing accounts of FCP

In the deontic logic literature, FCP is usually considered problematic in view of our Observation (**II**). That is, when FCP is formalized by $P(\varphi \vee \psi)$, one cannot accomodate this observation within Standard Deontic Logic (**SDL**).[6] Adding a corresponding axiom schema would trivialize the logic, whence one is bound to look for alternative accounts of permission or disjunction in order to handle such inferences.

Mind however that permissions of disjunctions do not behave uniformly in natural language. Often they do behave like FCPs; but often they don't:

(a) If we negate a sentence of the form "it is permitted that φ or ψ", then that permission behaves like a disjunction of permissions after all. That is, "it is not permitted that you take an apple or a pear" usually means "it is not permitted that you take an apple *and* it is not permitted that you take a pear". This is not at all the same as saying that "you are not free to choose between having an apple or a pear" – perhaps someone else is to decide, or perhaps you can only take an apple.[7]

(b) Embedded permissions over a disjunction are also sometimes interpreted as disjunctions of permissions. That is, "You may take an apple or a pear, but ask the waiter which of both" usually communicates uncertainty about the norms at hand, rather than a FCP.

Note also that, even if **SDL** does not validate the inference from $P(\varphi \vee \psi)$ to $P\varphi \wedge P\psi$, it does validate other seemingly plausible principles. For

---

[6]**SDL** is just the normal modal logic **KD**, with $\Box$ read as "it is obligatory that" and $\Diamond$ as "it is permitted that".

[7]Thanks to Malte Willer for drawing our attention to this problem (during a Q&A session at the DEON2016 conference).

instance, from "it is permitted to take an apple and a pear", a competent language speaker will infer that "it is permitted to take an apple". Rejecting all such inferences, simply in order to be able to validate the inference from $P(\varphi \vee \psi)$ to $P\varphi \wedge P\psi$, can hardly be seen as a proper solution to the problem of free choice permission.

This is not in itself an argument against reducing FCP to a combination of *some* concept of permission and *some* concept of disjunction. It does show that one cannot simply defend such an account by referring to the behavior of permission in natural language. Quite to the contrary: having various formal operators of permission around seems inevitable, if one is to explain the various ways utterances such as "may", "can", "it is permitted that" and the like behave and interact with "not", "and", and "or" in natural language.

But there is more. Even when intended solely as a formal account of FCP, the existing proposals are insufficient. To argue for this in full would require that we go over every such proposal, spell out the relevant formalities, and argue why it gives counterintuitive properties for F. For reasons of space we cannot do this here; we will however point out in brief terms what our main worries are.

First, one general weakness of existing accounts is that they do not make it explicit which agent has a given FCP. Perhaps these approaches can be enriched so that one obtains an agent-relative concept of FCP, which accomodates observations **(IV)** and **(V)**; but this remains to be done. Mind that, even if one abstracts from the agent in question – as this is often done in deontic logic –, one should still be able to distinguish between states of affairs that the agent can see to, and states of affairs that are beyond its abilities; at least if one agrees with observation **(V)**.

Second, it is shown in (Hansson, 2013) that any intensional account of FCP in terms of unary permission and disjunction will give extremely counterintuitive results, due to the validity of replacement of classical equivalents (RE). By (RE), we can e.g. infer from the fact that it is permitted to either take an apple or not, $P(a \vee \neg a)$, that it is also permitted to either take a napkin or not, $P(n \vee \neg n)$. But the free choice interpretation of the former statement clearly need not imply the free choice interpretation of the latter.

Hansson's observation still leaves room for alternative accounts. One of them is to translate FCP straight into a *conjunction* of permissions: $F(\varphi, \psi) =_{df} P\varphi \wedge P\psi$ — let us call this the *conjunctive account* of FCP. This is in line with Makinson (1984), who claims that the so-called paradox of FCP only arises due to a mistranslation of natural language into the deontic formalism.

However, in view of our Observation **(III)**, this simple version of the conjunctive account does not allow us to fully characterize FCP. In Section 4 we will consider more refined conjunctive accounts. There, it will become clear that also these variants fall short of capturing FCP, when based on the standard concepts of permission taken from the literature. In Section 5 we will however show that, when P is itself an agent-dependent notion of permission, one variant of the conjunctive account does work.

Another alternative – which is suggested by Hansson himself – is to treat FCP as a primitive, binary operator. We will not have much to say about this approach here; it may well be promising as a "minimalistic" account of FCP, omitting explicit operators for agency and ability.[8] Still, as we show in the remainder of this paper, for the specific notion of FCP in a deliberative context it *is* possible to reduce FCP to a suitable combination of unary permission and classical connectives. Such an account moreover has the advantage that it clearly links the various well-known notions of permission to FCP, thereby providing an explanation of (some of) its behavior.

Finally, one may question the very idea that a logic for FCP should be intensional and hence closed under (RE). That is, one may advocate a hyper-intensional account of permission, following Anglberger, Faroldi, and Korbmacher (2016) and Fine (2016). For instance, Fine works with a distinct set of "ideal" actions, where actions are to be understood as states (ordered according to part-whole). $P\varphi$ is then true iff every action that is in compliance with φ is contained in some ideal action. Since the set of actions in compliance with $\varphi \vee \psi$ equals the union of the set of actions in compliance with φ and the set of actions in compliance with ψ, we immediately get $P(\varphi \vee \psi) \equiv (P\varphi \wedge P\psi)$. This account also validates the intuitive principle that $P(\varphi \wedge \psi) \vdash P\varphi$.

However, the problem remains here that $P(\varphi \vee \psi)$ and $P\varphi \wedge P\psi$ are equivalent, which runs counter to our Observation **(III)** if we take $P(\varphi \vee \psi)$ as the definiens of $F(\varphi, \psi)$.[9] It is also not clear how "actions" are to be individuated in this framework, when applied to the simple cases that we

---

[8]The only work in this direction that we are aware of is the unpublished manuscript (Sedlár, 2016). Here, Hansson's suggestion is worked out into a full formal system with operators for permission that have an arbitrary arity. One of the problems of this formal system is that it does not account for our Observation **(V)**. We decided to leave a full discussion of this for another occasion.

[9]One may counter this objection by defining $F(\varphi, \psi)$ as $P((\varphi \wedge \neg\psi) \vee (\psi \wedge \neg\varphi))$, but in the context of the hyperintensional logics from (Fine, 2016), negation seems too weak to fully express mutual exclusiveness of the involved states of affairs.

present in Section 4. Even if this does not amount to a wholesale rejection of the hyperintensional account of FCP, it does show that some more work needs to be done in order to obtain a full explication of FCP.

4 Permissions and choice in STIT logic

In the remainder, we will focus on a very simple notion of a deliberative context. It is our conviction that we should at least try to get our interpretation of the language right for these simple models; once we are there, we can start looking at more complex structures.

The models we obtain are simple versions of the BT+AC-frames known from (Belnap, Perloff, Xu, & Bartha, 2001) and (Horty, 2001), stripped from temporal aspects, and enriched with a deontic component. They can also be seen as multi-agent, one-shot games. The deontic component consists in specifying which outcomes of the game are permissible (normatively ok, acceptable, legal) and which are not. Intuitively speaking, one may think of those outcomes as satisfying all norms that apply in the context at hand.[10]

Language Let $\mathcal{S} = \{p, q, \ldots\}$ be a set of sentential variables and $Agt = \{\alpha_1, \ldots, \alpha_n\}$ be a set of n distinct agents. Our formal language is defined by the following BNF, where $p \in \mathcal{S}$ and $\alpha \in Agt$:

$$\mathcal{W} := p \mid \neg \langle \mathcal{W} \rangle \mid \langle \mathcal{W} \rangle \vee \langle \mathcal{W} \rangle \mid \Diamond \langle \mathcal{W} \rangle \mid \mathsf{P} \langle \mathcal{W} \rangle \mid \mathsf{P}^s \langle \mathcal{W} \rangle \mid [\alpha] \langle \mathcal{W} \rangle$$

The connectives $\wedge, \supset, \equiv$ are defined in the standard way. $\Diamond$ is an existential modality (read as "it is possible that"); P represents weak permission and P^s represents "strong permission" (also known as "deontic sufficiency"). $[\alpha]\varphi$ expresses that "α's choice guarantees that φ is the case".[11]

Frames and Models We assume a set W of possible worlds, and a non-empty set $P \subseteq W$ of permissible worlds. W represents the *modal base*, i.e. the set of all possible outcomes, regardless of what the agents choose. To model choices, we moreover need for each agent α a partition $\mathcal{C}_\alpha$ of W into *choice cells* $X \subseteq W$. Frames are thus triples of the type $\langle W, P, \{\mathcal{C}_\alpha\}_{\alpha \in Agt} \rangle$.

---

[10] Accordingly, the operators for weak and strong permission defined below should be seen as weak, resp. strong counterparts of "must" or "obligatory", rather than of "ought". See (McNamara, 1996) for a discussion of these various modalities and their behavior in natural language.

[11] $[\alpha]$ is also referred to as the *Chellas stit* in the literature on STIT logic. See e.g. (Horty, 2001) for a discussion of this and other operators for agency in STIT.

Free Choice Permission in **STIT**

Where $w \in W$ and $\alpha \in Agt$, we let $\mathcal{C}_\alpha(w)$ denote the unique $X \in \mathcal{C}_\alpha$ such that $w \in X$, i.e. the choice that is actually taken by α at w.

We assume the condition of *independence of agents*: if $X_1 \in \mathcal{C}_{\alpha_1}$ and ... and $X_n \in \mathcal{C}_{\alpha_n}$ then $X_1 \cap \ldots \cap X_n \neq \emptyset$. This condition ensures that no given agent α_i can prevent another agent α_j from making any of its choices.

We say that a *model* is a quadruple $M = \langle W, P, \{\mathcal{C}_\alpha\}_{\alpha \in Agt}, V \rangle$, where $\langle W, P, \{\mathcal{C}_\alpha\}_{\alpha \in Agt} \rangle$ is a frame and $V : \mathcal{S} \to \wp(W)$ is a valuation function. The semantic clauses for the modal operators are as follows (where $w \in W$):

(SC1) $M, w \models \Diamond \varphi$ iff there is a $w' \in W$ such that $M, w' \models \varphi$
(SC2) $M, w \models [\alpha]\varphi$ iff $M, w' \models \varphi$ for all $w' \in \mathcal{C}_\alpha(w)$
(SC3) $M, w \models \mathsf{P}\varphi$ iff there is a $w' \in P$ such that $M, w' \models \varphi$
(SC4) $M, w \models \mathsf{P}^s\varphi$ iff, for all $w' \in W$ such that $M, w' \models \varphi$, $w' \in P$

That α is able to enforce φ can be expressed by $\Diamond[\alpha]\varphi$ — see (Horty, 2001, Section 2.3) for an elaborate discussion of this approach to the logic of ability. We can express and abbreviate the claim that "α has a choice between φ and ψ" as follows:

$$\mathsf{C}_\alpha(\varphi, \psi) =_{df} \Diamond[\alpha](\varphi \wedge \neg\psi) \wedge \Diamond[\alpha](\psi \wedge \neg\varphi)$$

Defining FCP Given these operators, various ways to define F suggest themselves, as shown in Table 1.

(a)	$\mathsf{P}\varphi \wedge \mathsf{P}\psi$	(f)	$\mathsf{P}^s(\mathsf{C}_\alpha(\varphi, \psi))$
(b)	$\mathsf{P}(\varphi \wedge \neg\psi) \wedge \mathsf{P}(\psi \wedge \neg\varphi)$	(g)	$\mathsf{P}^s([\alpha](\varphi \wedge \neg\psi) \vee ([\alpha](\psi \wedge \neg\varphi))$
(c)	$\mathsf{C}_\alpha(\varphi, \psi) \wedge \mathsf{P}(\varphi \wedge \neg\psi) \wedge \mathsf{P}(\psi \wedge \neg\varphi)$	(h)	$\mathsf{P}^s((\varphi \wedge \neg\psi) \vee (\psi \wedge \neg\varphi))$
(d)	$\mathsf{P}[\alpha](\varphi \wedge \neg\psi) \wedge \mathsf{P}[\alpha](\psi \wedge \neg\varphi)$	(i)	$\mathsf{P}^s(\varphi \vee \psi)$
(e)	$\mathsf{P}(\mathsf{C}_\alpha(\varphi, \psi))$	(j)	$\mathsf{P}^s(\varphi \vee \psi) \wedge \mathsf{C}_\alpha(\varphi, \psi)$

Table 1: Some possible ways to model FCP

However, neither of these formalizations are adequate w.r.t. each of our observations **(I)**-**(V)**, as we argue in the remainder of this section.

Ad (a)-(d). Note first that the definitions in (a) to (d) are of increasing strength. Here, we will show that even (d) does not suffice to speak of free choice.

Frederik Van De Putte

Example 1 α and β both have two choices: either take an apple, or don't take it. There are two apples in the fruit basket, so they can in fact both take one. It is permitted that one apple is taken; however, it is not permitted that both apples are taken. (Perhaps some third party should still have the option of taking an apple, upon arrival.) Note that $\mathsf{P}[\alpha]apple_\alpha$ and $\mathsf{P}[\alpha]\neg apple_\alpha$ are both true (here, $apple_\alpha$ stands for "α gets an apple"). Indeed, in case β does not take the apple and α does take the apple, we end up in a permitted state. However, it seems incorrect to say that α has a free choice in this scenario, normatively speaking. Indeed, the only choice α can make that ensures that we end up in an acceptable world, is not to take the apple. One may even argue that in this case, if α really wants to take an apple, then α should discuss this with β so that they can coordinate their choices.[12]

The example shows that weak permission, in combination with agency and the notion of choice as we formalized them here, cannot adequately account for our intuitions regarding FCP. In fact, they point to a more general problem with any such reduction: that it is permitted *for an agent* α to see to it that such-and-such is the case, cannot be reduced to what the agent does or chooses in any of the permissible states. This observation mirrors Horty's observations from Horty (2001, Section 3.4.2); we will return to it in the next section. Mind that this feature of FCP is strongly linked to our observation **(IV)**.

Ad (e). Likewise, one cannot just model FCP in terms of the weak permissibility of choices. Obviously, this cannot be done in the present framework, since $\mathsf{C}_\alpha(\varphi, \psi)$ and $\mathsf{P}(\mathsf{C}_\alpha(\varphi, \psi))$ are equivalent in it.[13] But even in purely informal terms, such an analysis does not seem to make sense.

To see why, suppose α is permitted to carry a loaded gun when walking on the street. Suppose moreover that α runs into an innocent person β. Let us assume that at this point, one of α's options is to kill β (suppose, for the sake of argument, that α is perfectly aware that he can do this, that it is just a matter of taking the gun, aiming it, and pulling the trigger). Clearly, α does not have a FCP to kill β or not; killing β is simply impermissible. However,

---

[12]Alternatively, one might feel that *conditional on* β's not taking an apple, α does have a FCP, or that in the worlds where β actually takes no apple, α has this FCP. We will not go into this argument in detail, but briefly point out in Section 5.2 how such conditional or world-relative FCPs can be modeled according to our approach.

[13]This is an immediate consequence of the fact that the abilities of α are the same at every point in a model.

Free Choice Permission in **STIT**

that does not all of a sudden make it impermissible for α to carry a loaded gun, and hence to *have the choice* to kill β or not.

Ad (f)-(j). Note that among these definitions, (g) is the weakest.[14] We now show that even (g) is not necessary in order to have a FCP. The example can be seen as a precise illustration of our earlier observation that FCP does not imply sufficiency for permissibility (cf. Observation **(I)**).

Example 2 *α has a choice between four states: take nothing, take an apple and a napkin, take an apple and no napkin, take a napkin and no apple. We assume moreover that there is exactly one state corresponding to each of these choices. The only states that are permissible are the ones where α takes nothing, or where α takes an apple and a napkin. It seems intuitive to claim in this example that α is free to choose between taking an apple or not. Likewise, "you are free to choose between taking an apple and a pear, but in case you take a pear, you have to take a napkin as well" seems perfectly consistent.*

5 Free choice permission

5.1 Permitted to see to

The above examples and discussion suggest an alternative account of FCP that is directly based on deontic STIT logic (Horty, 2001). In this paper we present a simplified, a-temporal version of it. We first define a deontic preference relation on arbitrary sets of states $X, X' \subseteq W$. That is, let $X \preceq X'$ iff for all $w \in X, w' \in X'$, w is permissible or w' is impermissible.[15]

Where α is given, we now define a more fine-grained preference relation between between choices X, X' of α. To do so, we need to introduce the auxiliary definition of all choices of a group of agents $A \subseteq Agt$. That is, where A consists of exactly n distinct agents $\alpha_1, \ldots, \alpha_n$, we let $\mathcal{C}_A =_{df} \{X_1 \cap \ldots \cap X_n \mid X_1 \in \mathcal{C}_{\alpha_1}, \ldots, X_n \in \mathcal{C}_{\alpha_n}\}$. Note that, in view of the *independence of agents*-condition, $\mathcal{C}_A$ is a partition of W. As before, we let $\mathcal{C}_A(w)$ denote the (unique) member X of $\mathcal{C}_A$ such that $w \in X$.

We define:

---

[14] To see why, note that in general, if φ is stronger than ψ, then $\mathsf{P}^s \psi$ implies $\mathsf{P}^s \varphi$.

[15] If we use a more refined semantics, e.g. one where each world is assigned a value within the interval $[0, 1]$, then obviously also the preference relation $\preceq$ can be further refined. See (Horty, 2001) for how this would work.

$X \sqsubseteq X'$ iff for every $Y \in \mathcal{C}_{Agt-\{\alpha\}}, X \cap Y \preceq X' \cap Y$

In other words, option X is at least as good as option X' iff, whatever the other agents do, the world we end up with when X is chosen will always be at least as good as the world we end up with when X' is chosen.

Let $\sqsubset$ be the strict counterpart of $\sqsubseteq$, defined in the usual way. One may now define a set of permissible choices of α as follows:

$X \in \mathcal{C}_\alpha$ is *permissible* iff there is no $X' \in \mathcal{C}_\alpha$ such that $X' \sqsubset X$

Note that if $\mathcal{C}_\alpha$ is finite, then there will be always at least one permissible $X \in \mathcal{C}_\alpha$ (relative to a given frame). In general there might be infinite descending chains of "ever better actions". In such cases one can refine the definition of permissible choices as follows: X is permissible iff there is no other choice X' that is itself $\sqsubset$-minimal within $\mathcal{C}_\alpha$, and for which $X \sqsubset X'$.[16]

We now introduce a new operator P_α for "α is permitted to see to it that", with the following semantic clause:

(SC5) $M, w \models \mathsf{P}_\alpha \varphi$ iff there is a permissible choice X of α such that $X \subseteq |\varphi|^M$.

Thus, $\mathsf{P}_\alpha \varphi$ means that there is a choice X of α which implies that φ is guaranteed, and there is no choice Y of α that is strictly better than X.

5.2 Relation to Horty's analysis of ought to do

Using Horty's ought-operator O_α for "dominance act utilitarianism" (Chapter 4 of (Horty, 2001)), one can define our P_α as $\neg \mathsf{O}_\alpha \neg [\alpha]$. That is, $\mathsf{P}_\alpha \varphi$ is true iff there is some "minimal" (in our terms, strongly permissible) choice that guarantees φ, iff it is not the case that every optimal choice of α is such that it is not the case that α sees to it that φ.

Following Horty, one may also define variants of these operators. We will not go into detail here but just explain the main idea for two of them. First, one can consider conditional variants. In our Example 1, a conditional operator allows one to express such things as "given that α does not take an apple, β is permitted to take an apple." This can be done by defining the preference relation over β's choices in terms of their intersection with the

---
[16] See e.g. (Van De Putte & Straßer, 2014) for an elaborate discussion of such constructions that deal with non-smooth preference relations.

Free Choice Permission in STIT

truth set of the antecedent of the conditional, i.e. that α does not take an apple. See (Horty, 2001, Chapter 5, Section 1) for the formal details.

Second, one can define operators for permission based on Horty's analysis of "orthodox act utilitariansm" (Horty, 2001, Chapter 5, Section 4). The main difference here is that what the agent is permitted to do depends on the specific point of evaluation $w \in W$. That is, we compare two actions X, X' of α by looking at the intersection of both with the set $\mathcal{C}_{Agt-\{\alpha\}}(w)$. So for instance, looking again at Example 1, in the worlds where β takes no apple, α is permitted to take an apple.

5.3 Free choice permission

So far we have only considered a unary operator that allows us to express what an agent α is permitted to do. Once there, we apply a by now familiar trick in order to express (binary) free choice permission:

$$\mathsf{F}_\alpha(\varphi, \psi) =_{df} \mathsf{P}_\alpha(\varphi \wedge \neg\psi) \wedge \mathsf{P}_\alpha(\psi \wedge \neg\varphi)$$

In words, you have the free choice between φ and ψ (in a particular deliberative situation) if and only if you are permitted to see to it that $\varphi \wedge \neg\psi$ and you are permitted to see to it that $\psi \wedge \neg\varphi$.

This implies that you are able to guarantee each of φ and ψ, and that they are distinct. It does however not imply that φ and ψ are disjoint, or that their conjunction is not permitted.

It can be easily checked that this definition accomodates each of our observations **(I)-(V)**:

Ad (I) Immediate in view of the existential quantification in the semantic clause for P_α. There may well be another *impermissible* choice X of α which also guarantees either φ or ψ.

Ad (II) Suppose that $\mathsf{F}_\alpha(\varphi, \psi)$ holds in a model; we prove that $\mathsf{P}\varphi$. By the definition of F_α, $\mathsf{P}_\alpha(\varphi \wedge \neg\psi)$. Hence by the semantic clause for $\mathsf{P}\alpha$, there is a permissible action X of α such that $X \subseteq \|\varphi\|^M$. We know that $X \cap P \neq \emptyset$ – otherwise X would not be permissible[17] – and hence $\|\varphi\|^M \cap P \neq \emptyset$. But this means exactly that $\mathsf{P}\varphi$ holds in the model.

Ad (III) In the model that corresponds to Example 1, $\mathsf{P}(apple_\alpha)$ and $\mathsf{P}(\neg apple_\alpha)$ are both true, but $\mathsf{F}_\alpha(apple_\alpha, \neg apple_\alpha)$ fails, since $\mathsf{P}_\alpha(apple_\alpha)$ fails.

---

[17] For this step, it is crucial that $P \neq \emptyset$.

Ad (IV) Immediate in view of the way F_α is defined. Note that we can easily accomodate cases where $F_\alpha(\varphi, \psi)$ holds, but $F_\beta(\varphi, \psi)$ fails.[18]

Ad (V) Immediate in view of the definition of $F_\alpha(\varphi, \psi)$ and the semantic clauses of P_α, $\Diamond$, and $[\alpha]$.

So although free choice permission implies weak permission, the latter does not allow us to define F, in line with our examples from Section 4. Interestingly, we have the following relation between FCP and strong permission:

$$P^s(\varphi \wedge \neg\psi), P^s(\psi \wedge \neg\varphi), C_\alpha(\varphi, \psi) \vdash F_\alpha(\varphi, \psi) \qquad (1)$$

The converse implication fails, as it should in view of our earlier discussion. So although FCP does not reduce to sufficiency for permissibility, the former can still be derived from the latter whenever the agent has the corresponding abilities.

6 Conclusion and outlook

In the preceding, we argued for a new approach to free choice permission, in which the agents and their choices are represented explicitly within the models and formal language. We showed that this approach matches each of our basic observations concerning FCP, in contrast to existing formal accounts.

A lot of open issues remain; we mention just three here. First and foremost, it remains to be seen if and how the resulting logic (including the operators for weak and strong permission, P and P^s) can be axiomatized.[19] Second, we mentioned some variants of Horty's deontic STIT logic, whose application to FCP is in need of further consideration. Third and last, we deliberately restricted the focus to a very specific and simple notion of FCP in this paper; one next step is to ask whether we can also get a grip on more "loose" or complex types of free choice permission which do not refer to one specific deliberative context.

---

[18] In fact, whenever $\alpha \neq \beta$, $F_\alpha(\varphi, \psi) \supset \neg F_\beta(\varphi, \psi)$ is a theorem of our logic. This follows from the fact that $P_\alpha \varphi$ entails $\Diamond[\alpha]\varphi$ and the independence of agents-condition, cf. Section 4.

[19] As we noted above, the fragment of our logic without P and P^s coincides with the a-temporal fragment of (one of) the logic(s) of *ought to do* in (Horty, 2001). An axiomatization for these logics can be found in (Murakami, 2005).

References

Anglberger, A. J., Faroldi, F., & Korbmacher, J. (2016). An exact truthmaker semantics for permission and obligation. In O. Roy, A. Tamminga, & M. Willer (Eds.), *Deontic Logic and Normative Systems* (pp. 16–31). College Publications.

Belnap, N., Perloff, M., Xu, M., & Bartha, P. (2001). *Facing the Future: Agents and Choice in Our Indeterminist World*. Oxford University Press.

Fine, K. (2016). Compliance and Command I and II. *Unpublished manuscript*. (https://nyu.academia.edu/KitFine)

Giordani, A., & Canavotto, I. (2016). Basic action deontic logic. In O. Roy, A. Tamminga, & M. Willer (Eds.), *Deontic Logic and Normative Systems* (pp. 80–92). College Publications.

Hansson, S. O. (2013). The varieties of permission. In D. Gabbay, J. Horty, X. Parent, R. van de Meyden, & L. van der Torre (Eds.), *Handbook of Deontic Logic and Normative Systems* (Vol. 1, pp. 195–240). College Publications.

Horty, J. F. (2001). *Agency and Deontic Logic*. Oxford University Press.

Makinson, D. (1984). Stenius' Approach to Disjunctive Permission. *Theoria*, 50(2-3), 138–147.

McNamara, P. (1996). Must I do what I ought? (or will the least I can do do?). In M. A. Brown & J. Carmo (Eds.), *Deontic Logic, Agency and Normative Systems* (p. 154-173). Springer London.

Murakami, Y. (2005). Utilitarian deontic logic. In *Advances in Modal Logic* (Vol. 5, p. 211-230).

Sedlár, I. (2016). Generalized permissions and free choice. *Unpublished manuscript*.

Van De Putte, F. (2016). That will do: Logics of deontic necessity and sufficiency. *Erkenntnis*. (in print, published online)

Van De Putte, F., & Straßer, C. (2014). Preferential semantics using non-smooth preference relations. *Journal of Philosophical Logic*, 43(5), 903–942.

Frederik Van De Putte
Ghent University
Belgium
E-mail: frederik.vandeputte@ugent.be

A Hierarchy of Logical Constants

ALEXANDRA ZINKE

Abstract: The paper provides a new argument against the classical invariance criterion for logical terms: if all terms with a permutation invariant extension qualify as logical, then for any arbitrary true contingent sentence K of the meta-language, there would be a logically true object-language sentence 'φ' such that K follows from the sentence 'φ is true'. Thus, many logically true sentences would be *a posteriori*. To prevent this fatal consequence, we propose to alter the invariance criterion: not only the term's extension, but also its semantic clause must satisfy certain invariance conditions. The paper ends with the observation that the new criterion makes explicit the dependency of the classification of terms into logical and non-logical ones at the different levels of the Tarskian hierarchy of languages.

Keywords: Logical Constants, Permutation Invariance, Interpretational Definition of Logical Truth, Tarski

The interpretational definition of logical truth, tracing back to Tarski (1936), says that a sentence is logically true if and only if it is true under all interpretations of the non-logical terms. As has often been observed, the definition presupposes a delineation of the logical from the non-logical terms. Depending on where we draw the line, different sentences will turn out to be logically true. This is best illustrated with respect to two extreme cases: If no term is considered to be a logical constant, then no sentence whatsoever will qualify as logically true. If we take all constants to be logical constants, all truths will qualify as logical truths.

The criteria usually proposed to demarcate the logical from the non-logical constants divide into proof-theoretic and semantic criteria. I will not enter into the discussion of proof-theoretic versus semantic attempts to define the logical constants here, but exclusively focus on the most prominent, and most promising, semantic demarcation criterion, namely the criterion of permutation invariance.

In a 1966 lecture, Tarski presents the core idea of this approach:

> Consider the class of all one-one transformations of the space, the universe of discourse, or "world" onto itself. What will

be the science which deals with the notions invariant under the widest class of transformations? I suggest that they are the logical notions, that we call a notion "logical" if it is invariant under all possible one-one transformations of the world onto itself. (Tarski, 1966/1986, p. 149)

Logical notions should be topic-neutral and hence must not distinguish between different objects. The criterion of permutation invariance seems not only to allow for a precise mathematical formulation, but to be philosophically well motivated.

Notably, Tarski applied the invariance criterion not primarily on the level of linguistic entities, but used it to define logical *notions*, where a notion is supposed to be some set-theoretic entity. However, the characterization of a logical notion provides the basis of the classical invariance definition of a logical constant:

(C) A constant c is a *logical constant* iff its extension is a logical notion.[1,2]

According to (C), all the paradigm cases of logical constants, i.e. the truth-functional sentence connectives and the standard quantifiers, qualify as logical constants. On the other hand, no individual constant and only very special predicate and relation constants qualify as logical terms.[3]

It is widely agreed that the criterion of permutation invariance provides necessary conditions for a term to be a logical constant: if the extension of a term is not permutation invariant, then it is not a logical constant.[4] However, hopes that the criterion also provides sufficient conditions for a term to be a logical constant have been frustrated by Timothy McCarthy's prominent counter example, the N-*operator*.

---

[1] As terms usually have different extensions in different domains, we should relativize the notion of permutation invariance to domains. The definition of a logical constant then reads as follows: a constant c is a *logical constant* iff for all domains $\mathfrak{D}$, c's extension on $\mathfrak{D}$ is a logical notion with respect to $\mathfrak{D}$. However, as the reference to a domain is irrelevant for the following discussion, we will here stick to the simpler definition (C).

[2] For different versions, technical refinements, and applications of the invariance criterion, see e.g. van Benthem (1989); Bonnay (2008); Feferman (1999); McCarthy (1987); Sher (1991).

[3] For example, only two predicates, namely the universal and the empty predicate, and exactly four binary relations, i.e. the universal relation, the empty relation, the identity relation, and the diversity relation qualify as logical constants according to (C). We get similar results for relations of higher arity. For details, see Tarski (1966/1986, pp. 150–151).

[4] See, however, Dutilh Novaes (2014) who argues that there are logical constants that do not satisfy the criterion of permutation invariance.

A Hierarchy of Logical Constants

1 Counter example to the criterion of permutation invariance

In his "The Idea of a Logical Constant" McCarthy introduces the one-place sentential operator 'N' characterized as follows:

$\mathfrak{M} \models N\varphi$ iff ((K and $\mathfrak{M} \not\models \varphi$) or (not-$K$ and $\mathfrak{M} \models \varphi$))
(where K is a true sentence of the meta-language).[5]

As K is true, '$N\varphi$' is true in a model $\mathfrak{M}$ just in case 'φ' is not true in $\mathfrak{M}$. Thus, the N-operator expresses the same operation as the classical negation operator. As the operation of negation is permutation invariant, so is the operation expressed by the N-operator. However, if we treat 'N' as a logical constant (and, of course, also '$\leftrightarrow$' and '$\neg$'), the sentence '$N\varphi \leftrightarrow \neg\varphi$' is logically true according to the interpretational definition: the only term that is open for reinterpretation is 'φ', and '$N\varphi \leftrightarrow \neg\varphi$' remains true under all interpretations of 'φ'. But, so McCarthy's (1981) argument goes, '$N\varphi \leftrightarrow \neg\varphi$' shouldn't be declared logically true, as, for contingent K, the sentence '$N\varphi \leftrightarrow \neg\varphi$' is not necessary:

> [I]n any counterfactual situation [...] in which 'K' is false, the sentence '$N\varphi \leftrightarrow \neg\varphi$', though a logical truth, is false. However, it seems a reasonable constraint on any theory of validity for a language L that a sentence of L that it construes as a logical truth is not possibly false in L. As the choice of 'N' as a logical constant does violence to this constraint, I suggest that we are entitled to conclude that 'N' is not, after all, a logical constant [...]. (p. 515)

Let me illustrate McCarthy's argument with respect to one particular instance of the N-operator, the N_s-operator, which results from substituting the meta-language sentence 'snow is white' for K:

$\mathfrak{M} \models N_s\varphi$ iff ((snow is white and $\mathfrak{M} \not\models \varphi$) or (snow is not white and $\mathfrak{M} \models \varphi$))

As snow is white, '$N_s\varphi \leftrightarrow \neg\varphi$' is true. If '$N_s$' is a logical constant, '$N_s\varphi \leftrightarrow \neg\varphi$' is trivially true under all interpretations of the non-logical

---
[5]This is a disambiguated version of the definition we find in McCarthy (1981, p. 514). For similar examples, see e.g. Gómez-Torrente (2002); Hansson (1997); McGee (1996).

constants and thus logically true. However, in a world in which snow is not white, the sentence '$N_s\varphi \leftrightarrow \neg\varphi$' is false. Thus, so McCarthy concludes, we should not accept 'N_s' as a logical constant. To foster the transparency of the discussion of McCarthy's argument, let me introduce a much simpler operator which serves the same purpose. Here it is:

$\mathfrak{M} \models \bullet\varphi$ iff K
(where K is a true sentence of the meta-language).

This only provides a schema for an operator. Different substitutions for K yield different operators. (Although there are many instances of '$\bullet$', I will nevertheless also speak of *the* operator '$\bullet$'.) One such operator is '$\bullet_s$', defined as follows:

$\mathfrak{M} \models \bullet_s\varphi$ iff snow is white.

As K may only have true instances by definition, '$\bullet$' has the same extension as the tautology operator: its extension is the function that maps every set of variable assignments on the set of all variable assignments. According to definition (C), the operator '$\bullet$' is, just like the tautology operator, a logical constant. But if '$\bullet$' is a logical constant, the sentence '$\bullet\varphi$' is true under all interpretations of the non-logical terms: the only term open for reinterpretation is 'φ', but '$\bullet\varphi$' is true regardless of the truth value of 'φ'. The sentence '$\bullet\varphi$' would thus be logically true according to the interpretational definition. However, if K is contingent, '$\bullet\varphi$' could have been false. As logical truths must be necessary, '$\bullet$' seems to constitute a counter example to (C).

This style of argument has appeared convincing to many. However, in a recent paper, Gil Sagi claims that it rests on a conflation of the modal status of a sentence and the modal status of the proposition expressed by a sentence.[6] She concedes that the sentence '$\bullet\varphi$' may be false, but also stresses that this does not infringe the necessity of the proposition expressed by '$\bullet\varphi$'. The principle that a logically true sentence cannot be false is aiming at the modal status of what the sentence says, not at the modal status of the sentence itself. As meaning is a matter of contingency, any sentence, also a logically true one, can be false. (The sentence 'ψ or not ψ' is false in a world where, say, 'or' stands for conjunction. But this does, of course, not disqualify the sentence 'ψ or not ψ' from the logically true ones.)

According to Sagi, the following holds. The proposition expressed by a logical truth must not possibly be false, i.e. a logical truth cannot be made

---
[6] See Sagi (2015).

A Hierarchy of Logical Constants

false by a change in the worldly facts. However, a logically true sentence need not be true if we consider different interpretations of the terms occurring in it. Indeed, a logical truth can be made false by changing the meaning of the terms. The reason that the sentence '$\bullet\varphi$' violates the necessity principle is that, if K were false, '$\bullet$', and therefore the sentence '$\bullet\varphi$', would have a different meaning. However, the semantic contingency of '$\bullet\varphi$' does not entail that the proposition actually expressed by '$\bullet\varphi$' is only contingent. Sagi concludes that there is as yet no reason to deny '$\bullet\varphi$' the status of a logical truth and hence no reason to abandon the invariance criterion.

For illustration, let us review Sagi's critique with respect to our sample instance of '$\bullet$', i.e. '$\bullet_s$'. McCarthy says (or rather: would say) that '$\bullet_s\varphi$', though logically true if '$\bullet_s$' is accepted as a logical constant, is false in a world where snow is not white. Sagi agrees, but contests that this violates the principle of the necessity of logical truths properly understood. If snow were not white, the operator '$\bullet_s$' would no longer stand for the tautology operator, but for the contradiction operator. The sentence can be made false, but this does not render the proposition actually expressed by the sentence contingent. Irrespectively of the contingency of the color of snow, given that snow is white, what '$\bullet_s\varphi$' says must be the case.

I agree with Sagi's critique. McCarthy's argument indeed rests on a confusion of the modal status of a sentence with the modal status of the proposition it describes. However, I think that there is *another* argument showing that '$\bullet$' can be rehabilitated as a counterexample to (C). If we accept '$\bullet$' as a logical constant, then it holds that for any arbitrary contingent sentence K of the meta-language, there is an instance of the operator such that ' '$\bullet\varphi$' is true' would imply K, although the sentence '$\bullet\varphi$' is logically true. Thus, we could deduce an arbitrary contingent truth from some logical truth's being true. To say it differently: some logical truths could only be known given knowledge of some contingent truth and thus fail to be *a priori*. As this is untenable, we must not accept '$\bullet$' as a logical constant.

Assume for *reductio* that '$\bullet$' is a logical term. Then '$\bullet\varphi$' is logically true. From '$\bullet\varphi$' being true, and the definition of '$\bullet$', you can easily infer K:
(1) '$\bullet\varphi$' is true
(2) $\mathfrak{M}^* \models \bullet\varphi$ \qquad (from (1), where '$\mathfrak{M}^*$' stands for the intended model)[7]
(3) For all $\mathfrak{M}$: ($\mathfrak{M} \models \bullet\varphi$ iff K) \qquad\qquad (from definition of '$\bullet$')
∴ K

---
[7] The intended model is the model in which all and only the actually true sentences are true.

Recall that the only constraint on K is that it be a true sentence of the meta-language. The argument then shows that you can infer an arbitrary truth from a logical truth's being true (and the definition of the logical terms). This is of course untenable.

Again, consider the argument with regard to our favorite instance of '$\bullet$', i.e. '$\bullet_s$'. It can be inferred from '$\bullet_s\varphi$' being true that snow is white. Thus, if '$\bullet_s$' is treated as a logical constant, as the invariance criterion requires, and consequently '$\bullet_s\varphi$' is accepted as a logical truth, a contingent truth can be inferred from the truth of some specific logical truth: we know that '$\bullet_s\varphi$' is true only if we know that snow is white. '$\bullet_s\varphi$' is then *a posteriori*. As this is absurd if '$\bullet_s\varphi$' is a logical truth, '$\bullet_s$' must not be accepted as a logical constant and the invariance criterion must be abandoned.[8]

The operator '$\bullet$' is a counterexample to the invariance criterion, and so is McCarthy's own example, the N-operator. If the N-operator is accepted as a logical constant, then '$N\varphi \leftrightarrow \neg\varphi$' is a logical truth. From '$N\varphi \leftrightarrow \neg\varphi$' being true, and the definition of 'N', one can infer K, where K is a contingent sentence of the meta-language.[9]

2 Permutation invariance reinterpreted

We have to amend the definition of a logical constant such that it does not render '$\bullet$' or McCarthy's N-operator logical constants. Definition (C) only takes into account whether or not a constant has a permutation invariant extension. It thus automatically treats the tautology operator and '$\bullet$' alike. Already McCarthy suggests a way out by proposing that the logical status of a term is not dependent on its extension only, but also on how the term is introduced:

> The general point suggested by these observations is that the logical status of an expression is not settled by the functions it introduces, independently of how those functions are specified. (McCarthy, 1981, p. 516)

---

[8] To avoid any possible misunderstanding, let me stress that *some* contingent truths are indeed plausibly inferred from some logically true sentence's being true. The truth of a logical truth may imply some contingent *semantic* facts. For example, from ' "$\varphi \lor \neg\varphi$" is true' and the definition of '$\neg$' (as well as the meaning of 'is true' and the role of quotation marks), one can infer the contingent fact that '$\lor$' does not stand for classical conjunction. This does of course not make '$\lor$' a non-logical term.

[9] The essence of this argument against the invariance criterion may well be found already in McCarthy (1981, p. 516). If so, I have presented only an old argument in a new guise.

A Hierarchy of Logical Constants

McCarthy does not elaborate on this idea, so let us try to do so.[10]

An operator or connective of a formal language L^n is specified in a meta-language L^{n+1}. L^{n+1} contains (i) translations of the expressions of L^n and (ii) expressions that are names for the expressions of L^n as well as terms needed to talk about the syntactic or semantic properties of L^n. We call the first fragment of the meta-language the *improper meta-language*, and the second one the *proper meta-language*. An operator is introduced by its *semantic clause*, i.e. the expression stating its truth conditions. For example, the truth conditions of the connective '$\wedge$' are usually spelled out as follows: $\mathfrak{M} \models \varphi \wedge \psi$ iff ($\mathfrak{M} \models \varphi$ and $\mathfrak{M} \models \psi$) (where '$\varphi$' and '$\psi$' are arbitrary formulas of the object-language). Thus, the semantic clause of the connective '$\wedge$' is '$\mathfrak{M} \models \varphi$ and $\mathfrak{M} \models \psi$'. The truth conditions of '•' are given as follows: $\mathfrak{M} \models \bullet\varphi$ iff K. The semantic clause of '•' is thus just K.

We define a meta-language sentence as *invariant under all admissible interpretations* iff it has the same truth value under all interpretations of the non-logical terms of the meta-language. If one subscribes to the interpretational definitions of logical truth and logical falsity with respect to sentences of the meta-language, a sentence is invariant under all admissible interpretations iff it is a logical truth or a logical falsehood of the meta-language. By means of the notion of invariance under interpretations, we can modify definition (C) of logical constants as follows:

(C*) A constant of the language L^n is logical iff

(i) its extension is a logical notion, and

(ii) its semantic clause is invariant under all interpretations of the non-logical terms of the improper meta-language L^{n+1}.

Let us see which terms are classified as logical constants by the modified definition (C*). Obviously, (C*) agrees with the original definition (C) in the classification of individual terms and predicates. These constants do not have a semantic clause in the relevant sense and thus the second condition is trivially fulfilled. The connectives and quantifiers provide more interesting cases.

Before we discuss the paradigm cases of logical constants, let's check whether (C*) can cope with '•'. As we know, all instances of '•' have the same permutation invariant extension: as only true sentences may be substituted for K, all instances of '•' have the same extension as the tautology

---

[10] For an alternative recent development of this idea, see Woods (forthcoming).

operator. The first condition on logical constants is thus fulfilled. Is the semantic clause of '•' interpretation invariant, such that '•' also fulfils the second criterion? Obviously, the result will vary for different instances of '•'. All instances of '•' that are generated by an interpretation invariant K will be declared logical constants, all others will be categorized as non-logical. We already observed that a meta-language sentence K is interpretation invariant just in case it is logically true or logically false. As we assumed K to be true, it follows that exactly those instances of '•' are logical constants according to (C*) that have as their semantic clause a logical truth of the meta-language. All other instances of '•', e.g. '•$_s$', will be excluded from the logical constants: the meta-language sentence 'snow is white' is true, but there is an interpretation that makes it false.

The modified criterion for logical constants can thus handle the 'ordinary' instances of '•'. However, some instances of '•' still qualify as logical constants, namely those with an interpretation invariant K. Is this problematic? I do not think so. While arguing against treating '•' as a logical term, we assumed that the meta-language sentence K should not be inferable from the fact that some logical truth of the object-language is true. This principle is, of course, not plausible if K itself is logically true. A logical truth is inferable from anything. It thus seems unproblematic to accept as logical constants those instances of '•' that are generated by substituting K with a logical truth of the meta-language.

With regard to the various instances of '•', (C*) yields the intended results. Note, however, that I introduced this operator as a bare-bone version of McCarthy's original example. Let me therefore also discuss McCarthy's more complicated example, the N-operator. We have already seen that the semantic value of the N-operator is permutation invariant; on all domains it has the same semantic value as the negation operator. Thus, the operator is a logical constant iff '(K and $\mathfrak{M} \not\models \varphi$) or (not-$K$ and $\mathfrak{M} \models \varphi$)' is invariant under all admissible interpretations of the improper meta-language, i.e. under all interpretations of K.[11] Note first that those instances of the N-operator that result from an interpretation invariant K can be accepted as logical constants without problems. If K is an interpretation invariant truth, it may well be implied by anything. So, let us focus on those instances of the N-operator that are generated by an interpretation variant sentence of the meta-language. To evaluate whether the semantic clause '(K and $\mathfrak{M} \not\models \varphi$)

---

[11] Here I implicitly assume that 'and', 'or' and 'not' are logical constants of the meta-language and may not be reinterpreted. See below for a discussion.

A Hierarchy of Logical Constants

or (not-K and $\mathfrak{M} \models \varphi$)' is then invariant under interpretations, we must distinguish two cases: the case in which it holds that $\mathfrak{M} \not\models \varphi$ and the case in which it holds that $\mathfrak{M} \models \varphi$.[12] If it holds that $\mathfrak{M} \not\models \varphi$, then the whole semantic clause is true. However, as there is an interpretation which makes K, and thereby the whole semantic clause, false, the semantic clause is not interpretation invariant. If, on the other hand, it is the case that $\mathfrak{M} \models \varphi$, the semantic clause is false. Then, however, there is an admissible interpretation of the meta-language under which K is false, thereby rendering the whole semantic clause true. Thus, again, the semantic clause is interpretation variant. (C*) excludes McCarthy's N-operator from the logical constants.

So far, the modified definition of logical constants yields very intuitive results. It rules out the problematic instances of '•' and of the N-operator. We still have to see, however, whether the paradigm cases of logical constants, i.e. the truth-functional connectives and the standard quantifiers, satisfy the modified condition. Here is, exemplarily, the satisfaction condition for conjunction: $\mathfrak{M} \models \varphi \wedge \psi$ iff $\mathfrak{M} \models \varphi$ and $\mathfrak{M} \models \psi$ (where 'φ' and 'ψ' are arbitrary formulas of the object-language). The conjunction operation is, of course, permutation invariant; consequently, the first condition of the modified definition is satisfied. Whether the second condition is fulfilled depends on whether '$\mathfrak{M} \models \varphi$ and $\mathfrak{M} \models \psi$' is interpretation invariant. Note, however, that the 'and', i.e. an expression which belongs to the meta-language, has to be understood as a pure truth-functional connective here. It must not be understood as having any additional, e.g. temporal or causal, meaning. Otherwise, the clause would not yield the truth conditions of our purely truth-functional object-language connective '$\wedge$'. To indicate this, we symbolize it with '$\curlywedge$'. The question to be answered is thus whether '$\mathfrak{M} \models \varphi \curlywedge \mathfrak{M} \models \psi$' is interpretation invariant.

'$\mathfrak{M} \models \varphi$' and '$\mathfrak{M} \models \psi$' are expressions of the proper meta-language and thus not open for reinterpretation. The only critical expression is '$\curlywedge$'. If we treat it as a logical constant of the meta-language, it is not open for reinterpretation and the expression '$\mathfrak{M} \models \varphi \curlywedge \mathfrak{M} \models \psi$' is trivially interpretation invariant. If we do not treat it as a logical constant, it is open for reinterpretation. We can then reinterpret '$\curlywedge$' as, e.g. expressing the operation of disjunction. In a model $\mathfrak{M}$ with '$\mathfrak{M} \models \varphi$' and '$\mathfrak{M} \models \psi$', the sentence '$\mathfrak{M} \models \varphi \curlywedge \mathfrak{M} \models \psi$' is then false according to the intended interpreta-

---

[12]Importantly, we do not consider cases in which the expressions '$\mathfrak{M} \not\models \varphi$' or '$\mathfrak{M} \models \varphi$' are reinterpreted. As these expressions belong to the proper meta-language, they are held fixed. However, we must check whether the semantic clause is permutation invariant for both the case in which '$\mathfrak{M} \not\models \varphi$' holds and the case in which '$\mathfrak{M} \models \varphi$' holds.

tion, i.e. according to the interpretation that interprets '⅄' as conjunction, and true in an interpretation that interprets '⅄' as disjunction. Thus, the semantic clause of the object-language connective of conjunction would not be interpretation invariant and, therefore '∧' would not qualify as a logical constant. To sum up: whether or not the object-language connective '∧' is a logical constant depends on whether or not the meta-language connective '⅄' is a logical constant.

Let me illustrate the situation with a further paradigm case of a logical constant, the universal quantifier, which is defined as follows: $\mathfrak{M} \models \forall x \varphi$ iff $\prod a \, \mathfrak{M}a/x \models \varphi$ (where '$\prod$' is the universal quantifier of the meta-language). The universal quantifier, of course, satisfies the first condition for logical constants: its extension is permutation invariant. Whether it also satisfies the second condition depends on whether the expression '$\prod a \, \mathfrak{M}a/x \models \varphi$' is interpretation invariant. '$\mathfrak{M}a/x \models \varphi$' belongs to the proper meta-language and is thus not open for reinterpretation. (As 'a' is bound by the universal quantifier of the meta-language, its interpretation is irrelevant for the truth value of '$\prod a \, \mathfrak{M}a/x \models \varphi$' anyhow.) The crucial expression is '$\prod$', i.e. the universal quantifier of the meta-language. Only if it is a logical constant of the meta-language, the semantic clause of the universal quantifier of the object-language is interpretation invariant. Thus, whether the universal quantifier of the object-language is a logical constant depends on whether the universal quantifier of the meta-language is a logical constant: either both are, or neither is.[13]

To determine whether the object-language connectives and quantifiers are logical constants, we first have to determine whether the relevant connectives and quantifiers of the meta-language are logical. So, do the meta-language connectives and quantifiers themselves satisfy the two conditions specified by (C*)? They obviously have a permutation invariant extension and thus fulfill the first criterion on logical constants. Do they also fulfill the second condition, i.e. are *their* respective semantic clauses interpreta-

---

[13] This is not quite right. There are of course various ways to spell out the truth-conditions of the connectives and quantifiers. For example, we need not use the meta-language term 'and' in the semantic clause of the conjunction connective. We could just as well state the truth-conditions of '∧' as follows: $\mathfrak{M} \models \varphi \wedge \psi$ iff (not (not $\mathfrak{M} \models \varphi$ or not $\mathfrak{M} \models \psi$)). If we do so, the logical status of '∧' does not depend on the logical status of the meta-language conjunction connective, but on the logical status of the meta-language terms for disjunction and negation. I do not want to establish a one-one correspondence between the classification of terms in the object-language and the meta-language. Rather, I think that the logical status of the object-language expressions depends on the logical status of those expressions of the meta-language used in their respective semantic clauses.

A Hierarchy of Logical Constants

tion invariant? – The answer obviously depends on the again higher-order logical terms. The semantic clause of, say, the conjunction term of the meta-language L^1 is given in the meta-meta-language L^2. Whether the semantic clause of the conjunction term of the meta-language is interpretation invariant depends on whether the conjunction term of the meta-meta-language is a logical constant. Obviously, we are entering into a regress here. To determine whether the conjunction term of language L^n is a logical constant, we already have to presuppose a demarcation of the logical from the non-logical terms of the language L^{n+1}.

As a first reaction, one might try to prevent this regress by resorting to an alternative conception of the meta-language. It seems that the regress only arises if we assume that the language L^{n+1} contains only a translation of the object-language, not the object-language itself. But we can also construct the hierarchy of languages such that the language L^{n+1} literally contains the language L^n. Under this second conception, there are not two conjunction terms, '$\wedge$' and '$\curlywedge$', of which the first belongs to the language L^0 and the second to the language L^1, but those terms are identified: there is just one conjunction term which is contained in all languages of the hierarchy. In this picture, the postulated regress does not get off the ground. However, this alternative model concerning the relation between object-language and meta-language does not yield any relief because we are then faced with a circularity problem. Whether or not '$\wedge$' is a logical constant now depends on whether or not '$\curlywedge$' is a logical constant, where '$\wedge$' and '$\curlywedge$' are identical. The definition of a logical constant would no longer be well-founded.

There seems to be no easy way out, but we needn't give way to despair. Perhaps we are not confronted with a vicious definitional regress, but rather with an interesting feature of the hierarchy of languages. Some logical properties of the language L^n depend on those of L^{n+1}. This observation can also be turned upside down: if we classify certain constants of the language of level L^n as logical constants, we thereby also subscribe to a corresponding bisection of terms of the language of level L^{n+1}. If we accept, say, the conjunction term of L^0 as a logical constant, and specify its truth conditions in the common direct way, i.e. by using the conjunction term of L^1, we must accept the conjunction term of L^1 as a logical constant. Otherwise, the semantic clause of '$\wedge$' would not be interpretation invariant, and '$\wedge$' couldn't be considered a logical constant. Thus, the demarcation of the logical terms in the language of level L^{n+1} also depends on the demarcation of the logical terms of L^n. The demarcations are interdependent. We get, so to speak, the whole hierarchy of logical constants at once. The bisection of terms into logical and non-logical ones transcends the level of the language.

References

van Benthem, J. (1989). Logical constants across varying types. *Notre Dame Journal of Formal Logic, 30*, 315–342.
Bonnay, D. (2008). Logicality and invariance. *Bulletin of Symbolic Logic, 14*, 29–68.
Dutilh Novaes, C. (2014). The undergeneration of permutation invariance as a criterion for logicality. *Erkenntnis, 79*, 81–97.
Feferman, S. (1999). Logic, logics, logicism. *Notre Dame Journal of Formal Logic, 40*, 31–55.
Gómez-Torrente, M. (2002). The problem of logical constants. *Bulletin of Symbolic Logic, 8*, 1–37.
Hansson, W. H. (1997). The concept of logical consequence. *Philosophical Review, 56*, 365–409.
McCarthy, T. (1981). The idea of a logical constant. *Journal of Philosophy, 78*, 499–523.
McCarthy, T. (1987). Modality, invariance, and logical truth. *Journal of Philosophical Logic, 16*, 423–443.
McGee, V. (1996). Logical operations. *Journal of Philosophical Logic, 25*, 567–580.
Sagi, G. (2015). The modal and epistemic arguments against the invariance criterion for logical terms. *Journal of Philosophical Logic, 112*, 159–167.
Sher, G. (1991). *The Bounds of Logic*. MIT Press.
Tarski, A. (1936). On the concept of logical consequence. Avaible in Logic, Semantics, Metamathematics: Papers from 1923 to 1938; 1969: 409–421.
Tarski, A. (1966/1986). What are logical notions? *History and Philosophy of Logic, 7*, 143–154.
Woods, J. (forthcoming). Characterizing invariance. *Ergo*.

Alexandra Zinke
University of Konstanz, Department of Philosophy
Germany
E-mail: alexandra.zinke@uni-konstanz.de